Ethical Issues in Behavioral Research

To Marie-Ange

Ethical Issues in Behavioral Research

A *Survey*

Allan J. Kimmel

First published 1996
2468 1097531

Blackwell Publishers Inc
238 Main Street
Cambridge, Massachusetts 02142,
USA

Blackwell Publishers Ltd
108 Cowley Road
Oxford OX4 1JF
UK

Library of Congress Cataloging-in-Publication Data
Kimmel, Allan J.
   Ethical issues in behavioral research / Allan J. Kimmel.
      p   cm.
   Includes bibliographical references and indexes.
   ISBN 1–55786–394–6 (alk. paper). – ISBN 1–55786–395–4 (pbk. alk. paper)
      1 Psychology – Research – Moral and ethical aspects.   2 Human experimentation in psychology – Moral and ethical aspects   3 Animal experimentation – Moral and ethical aspects.   I. Title.
BF76.4.K55 1996
174'.915 – dc20   95–37163
                                                                              CIP

British Library Cataloguing in Publication Data
A CIP catalogue record for this book is available from the British Library.

Typeset in 10½ on 12 pt Sabon by Photoprint, Torquay, Devon.
Printed in Great Britain by T. J. Press Ltd., Padstow, Cornwall.

This book is printed on acid-free paper

And it is best to err, if err one must
As you have done, upon the side of trust.

Tartuffe, by Molière (English translation by Richard Wilbur).
New York: Harcourt, Brace and World, 1963

Contents

Boxes

Tables

Foreword

To someone like myself, who received his graduate training in psychology in 1947–51, the idea of a text on ethical issues in behavioral research is still wondrously strange. Prior to the 1960s, there was no conception of research ethics as a formal component of the training and practice of behavioral scientists. Not that researchers were oblivious to ethical concerns: clearly, they did not feel free to cause or risk physical harm to research participants, or to disclose confidential information. Beyond that, there were various proactive efforts in those years to raise the moral quality of research – such as the development of action research, which fostered the active participation of individuals and communities in studies that addressed their own needs and problems. But the idea that systematic attention to investigators' moral obligations to research participants and to the larger society is an integral component of the research process had not yet taken hold.

As a result, certain deceptive, invasive, manipulative, subtly coercive, and otherwise ethically questionable procedures – which are now matters of active concern in the field, as readers of this book will discover – were routinely used in those days, because of their methodological advantages, with little attention to their implications for the rights of research participants or the integrity of societal values. Here and there, questions were raised about these practices, but such questions tended to be viewed as intrusions on the scientific enterprise rather than legitimate parts of it.

This state of affairs began to change in the 1960s for various reasons, discussed by Allan Kimmel. First, several cases involving serious abuse of research participants came to light, which led to official inquiries, public debate, national commissions, and eventually government regulations. The most notorious cases were in the field of biomedical research, but there were also several controversial pieces of behavioral research (described in some detail in the present volume) that focused public and professional attention on the ethics of research. Second, the logic of the various human rights movements that sprang up in the 1960s was extended to the rights of human subjects – of the individuals, groups, communities, and societies that provided the data for research. (The issue of animal rights – as discussed in chapter 8 – did not enter into the debate until the next decade.) Framing the debate in human rights terms helped to direct particular attention to the exploitation of powerless, disadvantaged, and captive populations, whose informed consent to research procedures could more easily be bypassed. Third, the reexamination of established methodological assumptions, in which many social scientists engaged in the 1960s, provided a hospitable context for raising ethical issues. In social psychology, for example, the growing recognition that the experimental situation involves a real human relationship, to which subjects bring their own expectations and purposes, provided an opening for questioning both the effectiveness and the ethical acceptability of deception in our experiments. The relationship between ethical and methodological issues is one of the important recurring themes in this book.

In the ensuing years, the role of ethical considerations in the work of behavioral scientists has become markedly different from what it was in my graduate student days. The change is reflected in the detailed guidelines for research ethics that have been developed by various professional associations around the world, as documented by Allan Kimmel; in the growing literature and curricular offerings on research ethics; and, above all, in the government regulations that mandate ethical review of research proposals and in the institutional review boards at universities and other research centers – at least in the United States – that enforce these regulations. The regulations and review boards require researchers to think about the ethical implications of their procedures and to develop appropriate mechanisms for obtaining informed consent, for assuring the welfare of research participants, for protecting the confidentiality of data, for providing adequate feedback to the individuals and communities who make the research possible.

Cynics might account for the observed changes by the fact that adherence to the regulations is a condition for obtaining government funding for research. There is no doubt some truth in this proposition: the fear of losing research funds has a way of concentrating the minds of researchers and research administrators. Institutionalization – the formal requirement to spell out how ethical concerns will be handled and to obtain committee approval – also contributes to maintaining at least minimal ethical standards. But I am convinced that, beyond the regulations and the routinized formalities, there has been a genuine change in consciousness about ethical issues within the behavioral research community.

The evidence presented in this book is not entirely reassuring about the depth of the changes that have occurred. For example, deception is still widely used in psychological research; subject pools at university departments do not always ensure free choice and commensurate benefits to research participants; in different areas of applied research participants' privacy continues to be compromised. Where the rules circumscribe a particular practice but do not absolutely prohibit it – as in the use of deception – there is often a tendency to push the limits. The minimal safeguards stipulated by the regulations have a way of becoming maximal when translated into practice. It is still too often the case that ethical considerations are treated as afterthoughts or as obstacles to be gotten out of the way so that the researcher's "real" work can proceed. In short, the ethical dimension has not been fully internalized by the research community.

Yet there have been enormous changes in the atmosphere, of which the present book is an outstanding manifestation. Ethical issues are now a legitimate part of the research enterprise. They are discussed within the profession, included in the curriculum, and empirically investigated. Government regulations, institutional mechanisms, and professional guidelines are all contributing to raising consciousness about these issues. The next frontier is conceptualizing ethical concerns as an integral component of the research process itself, on a par with theoretical and methodological concerns – based on the proposition that *what* we learn through our research is intimately linked to *how* we learn it. It is in this context that I value this book's thoughtful, balanced, and scholarly exposition, as a model not only for advanced students, but also for practicing behavioral scientists.

Herbert C. Kelman
Harvard University

Preface

When I originally conceived the idea of writing this volume on research ethics, I envisioned a final product quite different from the one presently in the reader's hands. My original intent was to write a brief (less than a hundred pages), introductory level paperback for undergraduate psychology students to be used as a supplement in research methods courses. Fortunately, my original contacts at Blackwell had something else in mind, a more thorough, in-depth volume applicable to senior honors students and graduate students who themselves will be undertaking individual research projects, not only in psychology but in related behavioral and social science disciplines as well. I say "fortunately" because in retrospect I realize that it would not have been possible to complete the briefer overview that I originally had in mind and at the same time done justice to the many complex ethical and methodological issues involved in conducting worthwhile research. I now feel that a brief introduction to the issues would have misled readers into believing that the ethical issues are overly simplistic and the ethical dilemmas more readily resolvable than actually is the case.

Although I was successfully dissuaded from writing a short introductory volume on ethical issues for psychology undergraduates, I am confident that beginning level students in psychology, sociology, and related fields can benefit from this volume's extensive coverage of ethical issues and methodological approaches in the conduct of behavioral research. However, the primary audience that I envisioned while working on this book is one consisting of advanced undergraduates and graduate students in the behavioral sciences. This is a book that was written from the perspective of pragmatic and common-sense considerations, emphasizing actual (and hypothetical) research cases, in the hope that it may serve as a useful resource during all phases of the research process. With this in mind, the book should also be of benefit to many research professionals.

Let me briefly explain why I believe this volume represents an important contribution to the field. Currently, there does not appear to be an up-to-date text on research ethics. While there are several texts dealing with ethical issues in psychology and other helping professions, these volumes are either dated or they provide more generic surveys of issues that emerge in biomedical, therapeutic, consulting, and testing contexts (in addition to research issues). In the past fifteen years, ethical guidelines have been revised, new codes have appeared, additional ethical problems have been identified in applied and non-experimental areas, and unique approaches to resolving ethical dilemmas have been proposed. As an indication of the voluminous work that has been added to the literature, this book contains nearly 250 references that have appeared since 1982.

Despite the growing level of concern about ethical issues in the behavioral sciences during the past two decades, there does not appear to have been a corresponding increase in emphasis on research ethics at the college or university level. A 1984 survey conducted by the Committee for the Protection of Human Participants in Research (CPHPR) revealed that there were few courses offered by psychology departments on research ethics and there is little evidence suggesting that this situation has changed during the past decade.

The serious exploration of ethical issues for students of psychology tends to occur as a topical area within the research methods and experimental psychology courses. Thus, the primary sources of information about research ethics for psychology students are their textbooks in such courses and their course instructors. However, recent content analyses of psychology textbooks have revealed that the coverage of ethical issues in psychological research tends to be incomplete and highly selective. All too often, I am afraid, the "ethical issues" topic on course syllabi gets relegated to the "if time permits" category. Following an analysis of psychology textbooks published after 1979, one investigator concluded "that it is unlikely that students learn much about research ethics from these texts and will rarely be exposed to arguments opposing deceptive research" (Korn, 1984, p. 146). In this light, it is hoped that the present volume will provide a more complete and balanced coverage of research ethics than that which students have been receiving. The availability of this book, at least in part, might ameliorate the apparent lack in formal ethical training of behavioral scientists by stimulating university instructors to devote more time in their courses to research ethics and to begin to offer courses on the topic.

Towards satisfying these goals, this volume includes numerous case studies of controversial studies from a variety of disciplines, offers suggestions for resolving ethical conflicts and developing research alternatives, and compares professional ethical principles, governmental research guidelines, and self regulation. Informed consent, the weighing of costs and benefits, deception and debriefing, privacy and confidentiality, investigator/participant role relationships, the humane treatment of animals in laboratory research, and professional responsibilities in the accurate reporting of research findings are among the topics given central attention in this book.

When one considers the wide-ranging considerations involved in carrying out an investigation with human or animal subjects, it may appear that the research process is akin to a "high wire act," in which there are certain potential benefits and risks inherent in the activity that must be weighed. Like the inexperienced tightrope walker who freezes in the middle of a performance, unable to continue forward and too scared to turn back, researchers also may find themselves confronting unanticipated dilemmas and dealing with them in ineffective ways once an investigation is underway. Like other balancing acts, the research process typically is equipped with "safety nets" for the protection of researchers and their participants, in the form of ethical review committees, informed consent forms, and the like in order to avoid serious undesirable consequences. In the spirit of the compromises that frequently must be reached in conducting an investigation, an attempt is made throughout this book to strike a balance between the rights and welfare of the participants of behavioral research and the rights of researchers to seek a fuller understanding of behavior.

Finally, I have selected the Molière quote as an epigraph for the book for several reasons, but mostly because of its emphasis on trust. Trust is a word that frequently appears in discussions of ethical issues in research, with good reason. Trust lies at the heart of virtually every decision that must be made by the researcher, and all human participants in the research process depend on the trust of others at all levels. Research subjects trust the researcher to treat them with dignity and respect, to protect their well-being, and to safeguard them from potential dangers or risks of harm. Researchers trust their subjects to maintain honesty in their responding, to respect the seriousness of the research enterprise, and to maintain their promises not to reveal certain aspects of a study to future participants. Society lends its trust to researchers to pursue worthwhile research questions which stand to benefit humanity, to protect participants from research abuses, and to maintain honesty and objectivity throughout the research process. In turn, researchers trust society to apply research findings cautiously and in an ethically responsible manner. The scientific community expects individual researchers to carry out their research in ways deemed appropriate by other members of the profession, to report their findings honestly and accurately, and to place the interests of others ahead of personal ambition in the conduct of research. Researchers trust their peers who serve on ethics committees and journal review boards to evaluate their research objectively and fairly. And so on. A breach of trust in any of these relationships will inevitably give rise to ethical problems.

This book is intended as a guide for researchers in anticipating ethical problems before they emerge, coping with them once they have become apparent, and perhaps avoiding them altogether. There are no guarantees that one will choose the most reasonable course of action that protects the greatest number of interests, or that ethical dilemmas will be resolved in the fairest and most reasonable ways. My main hope in having written this book is that it will contribute at least in some small way to shaping the values of individual researchers so that their judgments are made, in the words of Molière, "upon the side of trust."

Allan J. Kimmel
Paris, France

Acknowledgments

There are many people to whom I am deeply indebted for their assistance, inspiration, and support, and without whom this volume never would have come to fruition. I am sincerely grateful that Susan Milmoe, a past Executive Editor at Blackwell, recognized the need for a comprehensive text on research ethics and granted me the opportunity to write it. I thank Nathalie Manners and Hilary Scannell for their diligent and capable editorial work and for guiding the project through to publication. This volume also greatly benefited from the feedback provided by three anonymous reviewers who commented on earlier drafts of the manuscript.

I owe a great deal of thanks to Peter Hogan, Robert Keefer, and Margot Kempers, three colleagues who provided invaluable assistance in gathering reference material which I did not have ready access to in France. More importantly, their friendship and support during the writing of this book were sincerely appreciated. I also thank Elyette Roux for her insights into marketing research issues and Fred Scholz for assisting in the gathering of additional background material. I am sincerely indebted to Marie-Ange Gudefin for her help in translating the French, Swiss, and Spanish ethical codes, and for her enduring confidence, enthusiasm, and emotional support.

Among others to whom I am indebted are Leslie Hogan and her colleagues at the Clark University library for their assistance in furnishing critical referencing information. Thanks also to Robert Foley and his staff at the Fitchburg State College library for conducting a literature search during the early stages of my preparation for writing. A number of persons kept me appraised of ongoing developments in the shaping of various professional ethical codes, particularly Stacey Cunningham (American Psychological Association), Alan Kraut (American Psychological Society), Barbara Melber (American Sociological Association), Heinz Schuler (German Association of Professional Psychologists), Carole Sinclair (Canadian Psychological Association), Michèle Carlier (CNRS, France), and Alain Létuvé (French National Syndicate of Psychologists).

Ralph Rosnow and Herbert Kelman are two persons who stand out as having had a substantial influence on the shaping of many of the ideas presented in this book and for having sparked my interest in research issues. It is becoming something of a habit for me to thank Ralph in print for his intellectual stimulation and friendship. As for Herb Kelman, I am honored that the person who has significantly contributed to the ethical sensitivities in contemporary psychology agreed to be associated with this volume by contributing a foreword.

The author and publishers are grateful for permission to use the following copyright material: American Marketing Association, Code of Ethics, 1962, revised 1972, Chicago, IL; American Sociological Association, Code of Ethics, 1989, Washington, DC; The British Psychological Society, Code of Conduct, Ethical Principles and Guidelines, 1995, Leicester, UK; The British Psychological Society, Guidelines for the Use of Animals in Research, 1995, Leicester UK; The Canadian Psychological Association, Canadian Code of Ethics for Psychologists, revised edition, 1991, Old Chelsea, Quebec. On p. 206 is reproduced "Public Announcement" from *Obedience to Authority* by Stanley Milgram. Copyright © 1974 by Stanley Milgram. Reprinted by permission of HarperCollins Publishers, Inc., and Tavistock Publications.

1

Introduction:
Why Research Ethics?

Americans recoiled when the Nazis conducted brutal experiments on humans. But as the world was learning of those horrors, US scientists injected plutonium into 18 people without their informed consent to see how the element that fuels atomic bombs reacts in the body. The identities of these human guinea pigs were hidden for almost 50 years. Until now.

Eileen Welsome, *The Plutonium Experiment*

Albert Stevens. Elmer Allen. Eda Schultz Charlton. John Mousso. These are the names of but a few of the persons whose lives were tragically altered as a result of their participation in secret radiation experiments carried out by the United States Energy Department during the 1940s and 1950s. In contrast to earlier reports suggesting that the government-sponsored research involved fewer than 50 chronically ill patients, it now is estimated that thousands of relatively healthy human subjects were studied in several hundred experiments (Hilts, 1994b; Kong, 1994). Among the participants, who received no apparent medical benefits and in some cases suffered harmful consequences as a result of the experimental treatments, were pregnant women, infants, terminally ill patients, poor people, and other vulnerable groups (see Table 1.1).

Details regarding the number and breadth of the radiation experiments began to emerge in 1994 following a six-month inquiry by a government committee created to investigate Energy Department files and other relevant documents. Although there had been earlier

Table 1.1 Human radiation experiments

Studies	Example	Purpose
Fifteen studies involving about 130 blacks	Tulane University and New Orleans' Charity Hospital researchers created and cut off blisters on subjects' forearms, then injected radioactive mercury	To study action of mercurial diuretic, used for patients with congestive heart failure, but study subjects did not have that health problem. Officials say patients volunteered
Eight studies involving about 140 infants and children	Washington University medical researchers injected children, most of them healthy newborns or premature infants with radioactive sodium	To study "sodium space," or electrolyte balance. One independent expert viewed the study as useful in understanding how to prevent infant death from diarrhea, but added it probably would not be approved today
Seven studies involving about 880 women, most of them pregnant	Columbia University researchers gave women tracer doses of radioactive sodium	To study sodium balance during pregnancy. A spokeswoman said the university has had a policy of getting consent from research participants since the 1950s
Eleven studies involving about 100 terminally ill patients	Beth Israel Hospital researchers injected tracer doses of radioactive sodium into terminally ill patients	To study a blood plasma substitute widely used during World War II. Level of consent obtained from patients and their families unknown
Three studies involving about 530 psychiatric patients	Researchers from University of California at Berkeley gave tracer doses of radioactive isotopes to Peruvian natives living at high altitude and to prisoners	To study blood metabolism in people living at high altitudes (in the case of the Peruvians) and to study body water (in the case of prisoner volunteers). Level of consent obtained from the Peruvians unknown

Breakdown of studies involving 1800 people from vulnerable populations who were given radioactive isotopes. The research offered no apparent medical benefit to the participants.
Source: Dolores Kong, "1800 tested in radiation experiments," *The Boston Globe*, 20 February 1994, pp. 1, 22. Reprinted courtesy of *The Boston Globe*.

indications that ethically dubious radiation studies had been conducted, serious attention to the research initially was stimulated by a series of articles which appeared in a New Mexico newspaper, the *Albuquerque Tribune*, focusing on one experiment in which 18 patients were injected with plutonium in order to track its course in the human body. The exposé was compiled by Eileen Welsome, a distinguished journalist who had begun investigating the experiment in 1987 after discovering a footnote about it in a declassified government report on animal experimentation at a military base. Welsome eventually was able to trace the identities and backgrounds of some of the research participants, and her articles provided a human face to what many now view as a scientific tragedy. In her words, "I was shocked by the experiment. Once I got to know some of the stories behind the patients and actually met their relatives, I became even more shocked" (Welsome, 1993). While there is still much to be learned about the circumstances surrounding the radiation experiments, the research has been condemned as a blatant example of ethical abuse that has tarnished the reputation of American science worldwide (e.g. Bell, 1995).

The radiation studies bear certain similarities to another government-sponsored research project that to many observers represents one of the most notorious cases of ethical misuse and abuse of human subjects in the United States. This is the Tuskegee syphilis study, the longest non-therapeutic experiment on humans in medical history (Jones, 1993; Thomas and Quinn, 1991). The Tuskegee study was begun in 1932 by the US Public Health Service to investigate the long-term effects of untreated syphilis in 399 semi-literate black men in Macon County, Alabama. Although these men were diagnosed as having syphilis, treatment for the disease purposely was withheld so that the researchers could observe its "natural" course. Also included in the study were 201 black men without syphilis who served as control subjects.

As in many of the radiation experiments, the Tuskegee participants were not correctly informed about the nature of the non-therapeutic experiment and crucial information was withheld from them. The syphilitic subjects never were told about the true nature of their disease; instead, they were informed that they were being treated for "bad blood," a term allegedly understood by the researchers to be a local euphemism for syphilis, but which the local blacks associated with unrelated ailments. In order to encourage continued participation in the study, the researchers promised the subjects "special free treatments" (which included painful, non-therapeutic spinal taps) on a periodic basis. In this way, the researchers were able to monitor the participants over time.

It was not long before it became clear that the infected subjects were suffering from more medical complications than the controls. By the mid-1940s, mortality rates were found to be twice as high for the untreated syphilitic subjects. The researchers estimated that between 28 and 100 deaths as of 1969 were due to syphilis-related conditions. Nevertheless, treatment was not provided until after the experiment was exposed in a 1972 *New York Times* article.

Because many details of the radiation and syphilis experiments were kept secret, it is perhaps understandable that the studies were allowed to continue for as long as they did without arousing public outrage about the lack of existing mechanisms for protecting the research participants. Contrary to the belief that there was little debate about the appropriateness of the radiation experimentation during the 1950s, government documents have revealed that serious ethical discussions had indeed taken place at high levels of the military and scientific establishment (Hilts, 1994b). However, despite some protests from officials within the Atomic Energy Commission, the experiments continued, some of the most dangerous involving the total-body irradiation of cancer patients. Further, a decision was reached to withhold all documentation referring to the experiments that could result in legal suits or adversely affect public opinion. As for the syphilis research, the investigators apparently believed that they were serving the best interest of humanity by carrying out important research on a deadly disease (Jones, 1993). Of course, it is difficult to imagine such research even being contemplated today without a recognition of the potential dangers they pose to unsuspecting research participants.

Within the behavioral sciences there are no known cases of research abuse in which comparable risks have seriously endangered the lives of human subjects. However, for anyone with at least a smattering of training in psychology, sociology, and related disciplines, the phrase "research ethics" conjures up some equally dramatic images. Many students of psychology known something about Stanley Milgram's studies of behavioral obedience, in which subjects were placed in the highly uncomfortable role of teacher, responsible for the delivery of severe "shocks" to a confederate learner. Others may have learned about the Stanford prison study, conducted by Philip Zimbardo and his colleagues, a study in which male college students participated in a simulated prison scenario playing the roles of prisoners or guards; the situation quickly developed into a psychological and emotional nightmare for several of the participants. Many of us have seen troubling pictures of research animals in textbooks and in the mass media depicting apparently painful surgical implants in monkeys, mice, and other animals presumably sacrificed at the conclusion of the studies. In the realm of sociological research, many Americans are aware of some of the details of John Griffin's disguised participant observation, an informal investigation in which he himself served as the sole subject. Griffin, a white man, altered his appearance and lived for several months as a "black" in the southern United States. His experiences, as recounted in his famous book, *Black Like Me,* offer a chilling insight into the treatment of American blacks during the 1950s (Griffin, 1961).

Whatever additional concerns these situations raise for us, it is no doubt clear that in each case several ethical questions emerge. Were appropriate safeguards taken by the researchers to protect the well-being of their subjects? Should these studies have been conducted in the first place? What are the potential risks or harms involved in conducting such research, and what are the anticipated benefits or gains? Are such investigations justified by their benefits, regardless of the risks involved? Do researchers have a right to inflict pain on innocent research animals? Are there laws or guidelines that prohibit such research? These kinds of questions help us understand the issues involved in research ethics.

Research Ethics Defined

At the start, these questions seem to have something to do with the acceptability of one's conduct or, more specifically, appropriateness of one's approach in pursuing scientific truths. Indeed, "ethics" and the related term "morality" have similarly developed from terms that pertain to accepted (i.e. customary or usual) practice (Reese and Fremouw, 1984). The word "ethics" is derived from the Greek *ethos,* meaning a person's character or disposition, whereas the related term "morality" is derived from the Latin *moralis,* meaning custom, manners, or character. Philosophers typically describe moral judgments as involving matters of right or wrong, ought or ought not, a good action or a bad one. Whenever the question "Should I conduct this study?" is raised in the context of research, a moral issue is at stake.

Moral philosophers and others are apt to make subtle distinctions between the similar terms "moral" and "ethical," and to use these terms in different ways. For our purposes, these subtle distinctions may not be very useful. Perhaps the most basic way to differentiate between the terms is as follows. In contrast to moral concerns, which question whether specific acts are consistent with accepted notions of right or wrong, the term "ethical" often is used to connote rules of behavior or conformity to a code or set of principles (Frankena, 1973; Reynolds, 1979). As an illustration, consider a case in which a psychologist deceived her research subjects but in so doing did not violate her profession's codified rules of proper behavior, as detailed in psychology's accepted ethical guidelines. In subsequent chapters we will see that some professional ethics codes do indeed condone the use of deceit under certain specified circumstances. Thus, although the psychologist may not have crossed the bounds of ethical propriety as defined by her profession's code of conduct, we may still feel that her behavior was not right in a moral sense by arguing that deceiving others can never be justified. This case does not necessarily imply a distinction between the terms, but rather a difference in the principles used to judge the psychologist's behavior (the profession's standards and those of a critic of deception). The psychologist's behavior was viewed as proper according to one set of principles (those by which professional psychologists are guided), but not according to another (a broader and more general set of moral principles). Thus, the terms "ethical" and "moral" may be used interchangeably to refer to rules of proper conduct, although we may prefer to distinguish between them in a context where codified principles are relevant.

Most people would agree that the terms ethical and moral are related to values, and may choose to describe the terms as referring to behaviors about which members of society hold certain values (Reese and Fremouw, 1984). When defining ethics from the perspective of values, a major emphasis is placed on ethical decisions (Diener and Crandall, 1978). This approach to ethics suggests that the ethical or moral researcher makes judgments about research practices on the basis of his or her own values, and not by blindly following codified ethical guidelines. However, the ethical decision maker is one who is educated about ethical guidelines, and is able to carefully examine the various alternatives and take responsibility for his or her choices. In essence, choices for conduct in each research situation are related to the decision maker's values, and these values must be weighed carefully when important decisions are to be made.

When moral problems reflect an uncertainty about how to balance competing values it is proper to speak of the situation as an ethical (or moral) dilemma (Smith, 1985). An ethical dilemma is apparent in research situations in which two or more desirable values present themselves in a seemingly mutually exclusive way, with each value suggesting a different course of action that cannot be maximized simultaneously. It will become apparent throughout this book that many of the ethical issues that arise in behavioral research result from conflicting sets of values involving the goals, processes, or outcomes of an investigation. For example, one very common dilemma in psychological research emerges when investigators attempt to observe behavior as it naturally occurs in various social settings (a methodological approach known as "naturalistic observation"). In such situations, researchers must weigh the scientific requirements of validity (i.e. in obtaining an objective and valid measure of behavior) against the ethical importance of informing participants about their intentions prior to the observation in order to protect subjects' privacy rights. This ethical dilemma involves an important ethical procedure known as "informed consent," which will be discussed in detail in subsequent chapters.

Historical Background: The Rise of Ethical Concerns in the Behavioral Sciences

As an expression of our values and a guide to achieving them, research ethics help to ensure that our scientific endeavors are compatible with our values and goals, through shared guidelines, principles, and unwritten laws. As recently as the 1960s, however, researchers in the behavioral sciences proceeded with few, if any, serious concerns about the ethical implications of their investigations. Discussions about ethical issues were rare at the time, although studies involving deceptive manipulations, invasions of privacy, and threats to confidentiality were commonplace and altogether consistent with traditional ways of conducting scientific research. In fact, discussing ethical issues was likely to be seen as an indication that one had not outgrown a pre-scientific nature, bringing into the scientific domain issues that did not belong there (see Kimmel, 1988b). In other words, to have ethical concerns about methodological procedures that were so widely accepted and utilized was taken as a sign that one was not yet ready to participate fully in the scientific enterprise. Today, students in the behavioral sciences are trained in a very different kind of environment than the one in which previous generations were trained. Ethical issues now are part and parcel of a broader framework of scientific issues that comprise the critical training of future researchers.

It was during the 1960s that behavioral scientists seriously began to attend to ethical issues pertinent to the conduct of their investigations. For example, in the discipline of social psychology, attention and debate focused on the issue of deception, largely because at that time most social psychological research on topics ranging from attitudes (e.g. Festinger and Carlsmith, 1959) to aggression (e.g. Berkowitz, 1962, 1969; Geen, 1968) and obedience (e.g. Milgram, 1963) involved laboratory experimentation based on the use of deception. Elsewhere, critics both within and outside the behavioral sciences began to grapple with issues arising in cross-cultural research, research in foreign areas, intelligence testing, and the like. These issues raised questions about the impact the research findings would have on the status of the groups from which data were derived. For the last quarter century, researchers have wrestled with these and other ethical issues that arise throughout the research process, from the initial stages of hypothesis generation to the final stages of reporting and application of the results of the study.

There are several possible reasons for the increase of attention to ethical issues in psychology and related disciplines, and many of these reasons will be discussed in detail in the following chapters. One that bears mentioning at the outset has to do with the increasing number of cases involving human subject abuse that came to public attention during the 1960s. For example, in one of the first cases to receive widespread publicity, researchers at the Jewish Chronic Disease Hospital in Brooklyn, New York, injected live cancer cells into 26 chronically ill, elderly patients who were receiving care from the hospital staff (Edsall, 1969; Schuler, 1982). The researchers hoped to demonstrate that the ill patients would react to the foreign cells by rejecting them rather quickly, as had been shown in experiments with healthy volunteer prisoners. This result indeed was obtained, but when word got out about the researchers' invasive and potentially life-threatening procedure people reacted with shock and dismay. Many agreed that the apparent fact that none of the patients developed cancer could not justify the methods used by the researchers to test their hypotheses. Although some of the patients were told that something was going to be done to them of an experimental nature, none was informed about the cancer cell injections, nor were any of them asked to give written consent for the procedure. Thus, the participants were unaware of the nature of the risks imposed by the study.

Although the interventions utilized by behavioral researchers tend to be far less extreme than in the case described here, it is a common practice for researchers to withhold crucial information so as not to bias their subjects' reactions to the experimental treatments. This sort of practice (often combined with research deceptions of an active sort) raises some complex ethical dilemmas in studies involving human subjects, and represents one of many conflicts between the methodological requirements of a study and societal standards that we will consider throughout this book.

Before leaving the New York cancer cell study, the reader might be curious to learn about the fate of the researchers who were responsible for the investigation. The two researchers who had received federal funds for conducting the study subsequently were found guilty of fraud, deceit, and unprofessional conduct, and were censured and placed on a year's probation by the Board of Regents of the State University of New York (Faden and Beauchamp, 1986). Ironically, a few years later, one of the researchers was elected president of the American Association for Cancer Research.

The cancer cell study was but one of many similar biomedical experiments that had been carried out without the informed consent of human participants (see Beecher, 1959, 1966). Several of these studies (in addition to the cancer cell research) are documented in psychiatrist Jay Katz's exhaustive anthology and casebook, *Experimentation With Human Beings*, which was published in 1972. Katz provides research examples from a number of fields, including sociology, psychology, medicine, and law, and traces the origins of political concern about the use of humans in research to the experiments conducted on prisoners in Nazi concentration camps during World War 2.

The publication of Katz's book coincided with public disclosures of the Tuskegee syphilis study, which was described at the outset of this chapter. People reacted in disbelief that the Tuskegee research could have been carried out over such an extended period, especially with governmental approval, and many critics openly questioned why a study of the long-term effects of syphilis was necessary in the first place. After all, it is widely known that throughout history many famous figures have met their demise from the illness, from the Chicago gangster Al Capone to the French writer Guy de Maupassant. Medical annals are filled with case reports describing the effects of untreated syphilis. However, it must be considered that progress in science proceeds largely through systematic, controlled observations; researchers cannot rely solely on anecdotal reports and isolated cases. By carefully studying a large number of infected individuals, the Tuskegee researchers had an opportunity to obtain more objective and precise measurements for gaining insight into the nature of a serious disease. We also must keep in mind that the researchers who carried out investigations like those described here likely did not intend to cause harm or pose dangers to their research subjects, and no doubt viewed their methods as justified by the expected gains from conducting the studies. This "ends justify the means" variant of moral reasoning is one that continues to underlie the ethical decision making of many contemporary behavioral researchers. In some cases, however, as undoubtedly occurred in the cancer cell, syphilis, and radiation studies, researchers can be blinded to the potential risks posed by a research methodology while at the same time overemphasizing the anticipated merits of their investigations.

In the following chapters, we will consider the various ethical guidelines that have been developed by governments and professional associations in recent years; but it suffices to say at this juncture that behavioral and biomedical researchers now are required by most ethical codes to carefully weigh the expected benefits against the potential risks before deciding to proceed with a study. In studies where more than minimal risks are evident, research subjects must give their informed agreement beforehand that they understand the risks and that they agree to participate in the investigation.

The studies described above could not be legitimately carried out today because the ethical climate in the United States and other industrial nations has changed. For example, research involving even minimal levels of radiation would now be required to go through a rigorous review process in order to determine the adequacy of the study design and to make sure that proper mechanisms for protecting subjects' rights were in place (Kong, 1994). As suggested above, it was public awareness of earlier investigations like the New York cancer cell study and the Tuskegee syphilis study that in part led to stricter ethical standards for research. More specifically, such studies raised the awareness of public agency officials about the actual and potential risks to human participants in clinical investigations. They also highlighted the uncertain legal position of government agencies and functioned as a significant precedent in the eventual development of federal research guidelines (Faden and Beauchamp, 1986; Frankel, 1975). For example, the Tuskegee experiment indirectly led to the formation of a US governmental commission responsible for formulating some of the initial federal guidelines for behavioral and biomedical research (see chapter 2).

While these cases primarily occurred in the context of biomedical research, controversial cases also came to light in other disciplines. For example, in political science, a government-sponsored study known as Project Camelot gave rise to a wide-reaching ethical controversy (Levin, 1981; Reynolds, 1979). Project Camelot was a US$6 million project conducted under the auspices of the US Department of Defense by the Special Operations Research Office of the American University, Washington, DC.

Beginning in December 1964 (shortly after the United States had sent marines to the Dominican Republic), the project sponsors hoped to determine the conditions that gave rise to internal conflict in various Latin American countries, and the effects of actions taken by the indigenous governments to deal with those preconditions. In short, the goal of the research was to provide a systematic description of the events that preceded, occurred during, and followed either a peaceful or a violent change in government (Horowitz, 1967). The investigation was to involve various surveys and other field studies in Latin American countries, and a number of social scientists from various disciplines were recruited to serve as research consultants. However, the project was quickly condemned on ethical grounds by social scientists who viewed the project as a blatant attempt by the Department of Defense to intervene in the internal affairs of Latin American countries by sponsoring research designed to reduce the likelihood of revolution. The study was canceled in June 1965, prior to initiation of the actual fieldwork. One outcome of the controversy was a presidential communication instructing the US State Department to review all federally funded investigations involving research activities in other nations potentially affecting foreign policy.

Project Camelot is a particularly interesting case because it demonstrates the potential drawbacks of research conducted under the sponsorship of organizations such as the army, whose primary function is that of control. It also suggests how ethical issues can intensify as social scientists attempt to orient their research in the direction of broader human concerns while accepting support from specific social organizations (Kimmel, 1988a; Sjoberg, 1967).

Another landmark case of controversial social research is the Wichita Jury Study, begun in 1954 by a group of law professors from the University of Chicago. In an attempt to study the adequacy of the jury decision-making process, the researchers secretly used hidden microphones to record the jury deliberations of six separate civil cases. The study was carried out with the prior knowledge and approval of opposing counsel and the judges involved in the cases, but the jurors, defendants, and plaintiffs were not informed about any aspect of the research. It was agreed that the recordings would not be listened to until after the cases had been closed and all appeals exhausted, and that care would be taken to ensure the anonymity of all participants involved.

When the public became aware of the taped jury deliberations, a national uproar ensued (Faden and Beauchamp, 1986; Reynolds, 1979). The primary criticism centered on the methods used to collect the information for the study, given that secret jury deliberations are essential to the American judicial system and guaranteed by the Seventh Amendment to the US constitution. Serious questions also were raised about the confidentiality and privacy of those who were unaware of their participation in the research, and the possibility that general knowledge of the existence of the project might influence the decisions of other juries. The investigators defended their research by emphasizing the special care taken to minimize any effects on the jury members or the parties to the civil cases being considered. They also suggested that the project's successful completion could have resulted in recommendations for positive change in the judicial decision-making process or else could have boosted confidence in the current jury system. Interestingly, it was the controversy surrounding the Wichita study that ultimately resulted in legislation prohibiting intrusions upon the privacy of juries, including the tape recording of their deliberations (Faden and Beauchamp, 1986). (For further details concerning the Wichita Jury Study, see Katz, 1972 and Reynolds, 1979.)

Overview and Illustrative Examples

In each behavioral and social science discipline, it is possible to point to one or two controversial studies, like Project Camelot in the political realm and the Wichita Jury Study in the legal spectrum, that served to sensitize researchers within those fields and society in general about important ethical issues. In the remainder of this chapter, some additional studies are summarized in order to point out several of the ethical issues to be discussed in subsequent chapters.

Ethical issues can arise in laboratory or non-laboratory settings, in survey research, in the recruitment, selection and study of human or animal subjects, and in the reporting of one's research findings. The ethical issues arising in each of these research contexts comprise a primary focus of chapters 3 through 9. (Chapter 2 and the appendixes 1 and 2 describe the professional and governmental guidelines that regulate research in the behavioral sciences.) We conclude this initial chapter by presenting some of the more compelling reasons for studying research ethics. In essence, these reasons reflect the rationale for the writing of this book.

Milgram's obedience experiments

Chapters 3 and 4 present an overview and analysis of the ethical issues that emerge in the conduct of laboratory research involving human participants. The Milgram obedience project, mentioned at the beginning of this chapter, is often cited as the behavioral research most responsible for contributing to sensitivities to ethical problems in the conduct of experimental laboratory research (e.g. Rosnow, 1981; Schuler, 1982; Steininger et al., 1984). The central issue discussed in reference to Milgram's studies is that of the problem of deception of subjects. A brief summary of the research and an overview of the ethical debate that ensued appears below; we will refer back to the obedience experiments periodically in other chapters.

In a series of laboratory experiments conducted between 1960 and 1964, Stanley Milgram, a social psychologist at Yale University, led his volunteer subjects to believe that they were giving dangerous electric shocks to innocent victims. These subjects were recruited to participate in an experiment allegedly on the effects of punishment on learning, and were assigned the task of teaching a list of words to a learner (actually an experimental "confederate") for a test of memory. The "teachers" (actual subjects) were required to "punish" the learner's mistakes by delivering increasingly stronger electric shocks (up to 450 volts).

As was typically the case in social psychology laboratory experiments at that time, Milgram's procedure was rigged; in this case, the learner did not actually receive shocks, but made a number of preplanned mistakes and feigned pain upon receiving the "shocks." For example, in one variation of the procedure, although the learner was in another room and could not be seen by subjects, his agonizing screams could be heard through the laboratory walls. As is now widely known, Milgram's actual intent was to observe the extent to which subjects obeyed the authority of the experimenter, who ordered subjects to proceed with the procedure despite their protests and the confederate's apparent agony. Milgram drew an explicit analogy between his experiments and Nazi atrocities committed in obedience to the commands of malevolent authorities during World War 2.

Contrary to the expectations of several experts consulted by Milgram (1964) prior to conducting his research, a high degree of obedience was obtained in the experiments. In one typical study, twenty-six out of forty subjects (65 percent) continued to obey the orders of the experimenter to the end of the procedure, despite their apparent signs of discomfort and stress (Milgram, 1963). This outcome was unsettling in that it demonstrated the extent to which ordinary citizens might engage in brutal behavior at the direction of a person in a position of authority.

Milgram described his initial obedience experiments in 1961 in an unpublished research report. It was not until he published a series of articles about the project in 1963 that serious attention began to be paid to it, launching an unprecedented discussion in the profession of psychology (Schuler, 1982). (The lay public first learned of the research in 1974 with publication of Milgram's book, *Obedience to Authority.*) Milgram received high praise for his ingenious experiments, and he was awarded the American Association for the Advancement of Science's (AAAS) prize for research in 1964. Although his studies have compelled others to conduct replications and variations over the years, typically it is Milgram's project that is most frequently discussed in the psychological literature pertaining to obedience to authority. But it is in the context of discussions about ethical issues in laboratory research that Milgram's studies have attracted the most ink.

Psychologist Diana Baumrind (1964) was the first to criticize Milgram's research on ethical grounds, but several other colleagues soon followed. Milgram (e.g. 1964, 1974) steadfastly defended himself in print, attempting to answer all of the most important points of criticism. In brief, because we will return to these issues in chapter 3, the critics argued that there were not adequate measures taken to protect the participants from undue harm; that the entire program should have been terminated at the first indications of discomfort on the part of the participants; and that the participants would be alienated from future participation in psychological research because of the intensity of experience associated with laboratory procedures (Baumrind, 1964; Kelman, 1967).

Milgram (1964) countered these assertions by arguing that adequate measures were taken to protect the participants from harm; that subjects were allowed to withdraw from the study at any time; and that the deception was thoroughly explained to the participants at the end of the experiment. He further maintained that there was no reason to terminate the research because there were no indications of injurious effects apparent in the subjects. In fact, subjects indicated in follow-up questionnaires that the experiment was worthwhile and showed that they were not alienated from psychological research.

Milgram offered another compelling argument in defense of his obedience studies by suggesting that condemnation of his research procedure may have been fundamentally tied to the disturbing results he obtained. That is, if his results had turned out differently and subjects had refused to obey the experimenter-authority, then critics like Baumrind might not have perceived the deception as morally objectionable. In fact, there is some support for this contention, suggesting that perceptions of Milgram's research may be contingent more upon the subjects' responses than the experimental manipulations used (Bickman and Zarantonello, 1978). Whether or not one agrees with Milgram's points, it is clear that the use of deceptive and stressful methodologies in laboratory studies are obvious sources of ethical dilemmas. Such dilemmas force the ethical researcher to consider alternative non-laboratory research approaches for pursuing the answers to scientific questions of interest.

The "tearoom trade" study

A basic alternative to laboratory research is for the behavioral scientist to observe human behavior in more natural settings. Chapter 5 presents an overview of some of the various research methodologies frequently used in field (i.e. non-laboratory) settings, and the ethical issues that they entail. One of the advantages to conducting behavioral research in the field is that observations can be carried out on unsuspecting subjects, thus providing the researcher with a more natural reading of human reactions than might be obtained in the laboratory. Of course, the drawback to this sort of approach is that certain forms of disguised field research often threaten the research participant's right to privacy.

A controversial investigation conducted by sociologist Laud Humphreys serves as an illustration of the ethical problems that can emerge in disguised research in public settings. For his doctoral research at Washington University, Humphreys sought to describe the interactions among men who congregated in "tearooms," the name given to public restrooms where homosexuals illegally gathered to engage in sexual activity. He also intended to learn about the lifestyles and motives of the men whom he observed. To do so, Humphreys assumed the role of "watchqueen" who was allowed to observe the sexual acts without actually taking part in them, in exchange for serving as a lookout to warn of approaching strangers. But Humphreys was on the lookout for more than approaching strangers – he also noted the license plate numbers of the automobiles owned by the tearoom participants. By pretending to be a market researcher, he next traced the registration numbers of the automobiles in order to determine his subjects' home addresses. One year later, Humphreys altered his appearance and, claiming to be a health service interviewer, questioned his subjects at home. Once responses were obtained, Humphreys destroyed the names of the participants so that their identities could not be revealed.

The results of the study revealed that a majority of Humphreys's subjects apparently led normal heterosexual lives as accepted members of their communities and did not view themselves as either homosexual or bisexual. Another finding was that many of the participants' marriages were characterized by a high degree of tension. Despite these interesting results, held by many as helpful in dispelling some of the stereotypes and myths about the gay community, the approach taken by Humphreys to obtain these results aroused strong criticism. Critics argued that although he had taken precautions to protect his subjects, he nevertheless failed to protect his subjects' right to privacy and used questionable strategies for obtaining information about them. This negative reaction to Humphreys's research was so intense that it prompted an (unsuccessful) attempt to have his doctoral degree rescinded.

Several specific ethical questions are raised by disguised field research like the tearoom trade study (Sieber, 1982c). To what extent is deception justified by the importance of an investigation? How might one go about studying illegal behavior in scientifically valid and ethically justifiable ways? How can researchers protect their subjects' right to privacy in studies in which it is impossible to first obtain their voluntary consent? In Humphreys's study, the primary ethical dilemma involved the conflict between the interests of the scientist in obtaining knowledge and the interests of society in the protection of privacy and other rights.

Women and Love: *the 1987 Hite Reports*

Although ethical dilemmas may often seem more salient in research methodologies that involve the direct observation and measurement of subjects' behavior, problems also are encountered when indirect measurement tools, such as surveys, are used. In chapter 6 we consider the ethical issues that emerge in applied research, including studies in which behavioral scientists investigate human behavior through the use of interviews and written questionnaires. Concerns about privacy rights appear again as relevant issues in this context, although in different ways than in field research. To what extent can survey researchers probe the private, personal lives of their subjects for research purposes, even when subjects' voluntary consent is obtained for doing so? Related and equally important issues in applied research in general involve protection of subject anonymity and response confidentiality.

In chapter 7, ethical issues in the recruitment and selection of research participants are described. These are issues that are relevant to all forms of human subject research, and bear a strong relationship to some subtle issues that arise in the context of survey research. One example of behavioral science research that serves to illustrate some of the issues covered in chapters 6 and 7 can be found in the survey investigations conducted by Shere Hite (1976–1987). In order to illustrate Hite's research approach we shall focus here on one of her more recent publications, *Women and Love: A Cultural Revolution in Progress* (1987), the final volume of a trilogy of Hite Reports on the intimate lives of men and women. In this book, Hite argues that most American women are unhappy with their romantic, intimate relationships with men. Further, she contends that it is the men who largely are to blame for the dissatisfaction of their partners: men do not listen to their female partners, they refuse to discuss their feelings and to address tensions in their relationships, and they often unfairly criticize and ridicule their romantic partners. Hite's assertions are based on the responses of more than 4500 women who completed and returned her lengthy mail questionnaire.

It is not surprising that the contents of *Women and Love* quickly generated controversy, given the social climate of growing tensions between American men and women at the time of the book's publication. However, the loudest criticism of the Hite Report was voiced by leading survey research specialists and social scientists who decried Hite's research methodology. These research experts agreed that the report's methodology was seriously flawed, in that the findings were not based on a representative sample of American women.

The fact that Hite obtained such a large number of completed questionnaires may seem impressive at first, until one considers the large number of questionnaires that were *not* returned. To complete her research, Hite mailed 100,000 questionnaires to various feminist groups (such as the National Organization of Women, abortion rights groups, and university women's centers), and other women's groups (church groups, garden clubs, etc.). Thus, the 4500 completed surveys represent a very low return rate of 4.5 percent, far below the standard return rate of between 40 and 70 percent for social science research surveys. (These figures reflect the fact that social scientists typically send follow ups after the first mailing of questionnaires. Hite's return rate is closer to the average rate of 2.5 to 3.0 percent for commercial direct mail surveys.) Given the low rate of response to her survey, Hite's report merely reflects the relationship status and corresponding reactions of a small, non-representative sample of women who voluntarily chose to answer the 127 essay questions that comprised her questionnaire. Nevertheless, Hite has claimed that the responses she obtained are "typical" and that her sample "quite closely mirrors" the adult female population in the United States (Hochschild, 1987). As such, she asks us to accept that 70 percent of women married for five years or more are having extramarital affairs, despite other reports by social scientists that the rate is closer to 30 percent. She also wants us to believe her finding that 92 percent of all women who divorce actually initiate the divorce proceedings.

Even Hite's staunchest critics appear to agree that she obtained some remarkable insights into the emotional toll of unsatisfying relationships experienced by the women who chose to reveal their life stories to her. Where most social scientists have problems, however, is in her claim that the stories are "typical." Instead, the critics suggest that these subjects represent a group of women who probably had suffered a very high level of emotional pain and were strongly motivated to respond to the survey questions. But simply having a great many completed questionnaires does not make them typical if it is only those who have suffered who feel strongly enough to respond. Hite answered her critics by arguing that she purposely eschews random sampling in order to get more honest and personal responses. Her error is in then claiming that her responses are representative of a larger population.

On one level, the Hite Report gives rise to ethical concerns regarding privacy rights. Given the very personal and private nature of the written responses obtained from her subjects, an important consideration is whether Hite took precautionary steps to protect the anonymity of her subjects and to maintain the confidentiality of their responses. Additionally, did her subjects experience a recurrence of emotional and psychological pain as a result of completing the many essays on the questionnaire? On another level, Hite's research demonstrates how ethical issues are an inevitable consequence of drawing unwarranted conclusions from one's research data. For example, Robert Groves, a sociologist who specializes in survey research, went so far as to criticize Hite's methodology as "the equivalent of medical malpractice in our field" and called the report "such a public display of incompetence that all others who are involved [in survey research] in a serious way are associated by default" (Mehren, 1987). Others similarly questioned the possible negative impact Hite's research would have on all other survey research, especially given the extensive publicity that the Hite Report received. These sorts of concerns also arose following publication of the first two Hite Reports which dealt, respectively, with female and male sexuality. Clearly, procedures that produce misleading results are not beneficial, and thus are questionable on ethical grounds.

Taub's deafferented monkeys

The research examples described in this chapter thus far have all involved human subjects. In chapter 8, the ethical issues pertaining to behavioral research with animals are addressed. In recent years, the ethics of experimentation on non-human species has become a highly charged topic of controversy and debate. Members and supporters of animal rights groups have created an atmosphere in which behavioral scientists now are pressured to justify research methods that may expose animal subjects to pain and suffering in the name of potential human benefit.

Various defenders of behavioral research on animals have cited a wide range of studies that have made significant contributions to the welfare of animals and humans (e.g. Cole, 1992; Miller, 1985). Among this research is the controversial work of physiological psychologist Edward Taub and his associates (Taub et al., 1965). Taub's research represented a follow up of some valuable early experiments conducted by Sherrington (1906). Using anesthetized dogs and cats as his subjects, Sherrington's research demonstrated the importance of sensory information obtained from receptors in muscles, tendons, and joints in controlling reflexes necessary for balance, walking, and other motor activities. In one series of experiments, for example, Sherrington deprived his animal subjects of sensations from a limb by cutting the sensory nerves that served it (Mott and Sherrington, 1895). This procedure, known as deafferentation, resulted in a permanent loss of use of that limb after the animal had recovered from the surgery, despite the fact that the motor nerve pathways to the limb still were intact. Sherrington was awarded a Nobel Prize for his work, which provided useful information about reflexes necessary for the development of traditional techniques for the rehabilitation of persons suffering from neuromuscular disorders.

Several years after Sherrington's experiments, Taub and his colleagues challenged the untested assumption that animals could not be trained to regain use of a deafferented limb. Taub hypothesized that experimental animals were falling far short of their full potential for recovery after deafferentation. Using monkeys as his subjects, Taub demonstrated that much of the functioning of a forelimb that had been lost due to damage to sensory pathways indeed could be regained by restraining the use of the animal's good limb (Taub et al., 1965). Apparently, the restraints motivated the animal to attend more closely to any sensory information still available from the deafferented limb, and to relearn through trial and error to make maximum use of it.

Taub's experiments clearly demonstrated that special training could enable deafferented animals to recover skills thought to be permanently lost. These findings, consistent with Taub's hypothesis of learned disuse, have contributed to successful efforts to treat patients who have lost the use of their limbs as a result of stroke, brain injury, and other neurological damage. Patients now are better able to maximize their potential for recovery when conventional methods of physical therapy are supplemented with special training procedures. Recently, for example, as a direct result of Taub's work, doctors have successfully treated patients who were paralyzed on one side by restraining the movements of their undamaged arm (Miller, 1985). These patients, who appeared to have reached the limits of improvement through more conventional methods, were able to significantly improve their control over their damaged limbs following the restraining procedure.

While it is clear that the research reported here has led to important human benefits, the experiments nevertheless were vehemently criticized by those whose primary concern was the well-being of the animal subjects. Sherrington, for example, was accused of having performed painful surgery on animals without the use of anesthetics, in spite of the fact that his publications clearly stated that surgical anesthesia had accompanied the procedures (Keen, 1914). Faring much worse than his predecessor, however, was Taub, whose research became one of the first major targets of the growing anti-vivisection movement in the United States. Anti-vivisectionists are persons who oppose the practice of subjecting living animals to surgical procedures in order to advance scientific knowledge. When a group of animal rights activists, known as People for the Ethical Treatment of Animals, learned of Taub's research they filed a formal complaint charging Taub of animal abuse (Holden, 1986).

At issue were eight rhesus monkeys in Taub's laboratory at the Institute for Behavioral Research in Silver Spring, Maryland that had the sensory nerves removed from their forelimbs. Seven other monkeys in the laboratory served as controls in experiments on nerve regeneration. The charges were filed based on evidence of negligent conditions gathered by five scientists who toured Taub's laboratory in 1981. The evidence included slides taken in the laboratory that showed monkeys in undersized cages with mutilations believed to have been self-inflicted due to the loss of sensation in limbs caused by the experimental procedures. A court convicted Taub of animal cruelty, placed his monkeys in the custody of the National Institutes of Health (NIH), and revoked the unused portion of the more than US$200,000 in grant money he had received from the government to conduct his research. The conviction eventually was overturned by the Maryland Court of Appeals, which ruled that the state animal-cruelty law did not apply to federal research programs such as that carried out by Taub. His reputation still intact among his peers, Taub was awarded a Guggenheim Fellowship in 1983 to support additional research on the sensory mechanisms of monkeys, further angering animal rights activists who feared for the well-being of his research animals. Since the controversy surrounding Taub's experiments, more stringent governmental policies regarding animal research have been instituted. There are lingering fears, however, that more limiting legislation supported by animal protectionists will curtail future human benefits derived from animal research.

The case of Sir Cyril Burt

In the final chapter of this book, we focus on some integral professional issues that pertain to the ethical review of behavioral research and the communication of research findings (chapter 9).

Objective ethical review and the fair and accurate reporting of results are essential ingredients of ethical scientific conduct. An example of research in which these ingredients apparently were lacking can be found in the work of British psychologist Sir Cyril Burt on the inheritance of intelligence.

For some time, research on the heritability of intelligence was considered to be a particularly controversial area of study because of suggestions that the observed IQ differences between blacks and whites were genetically based. That is, if research demonstrates that IQ scores are consistently higher in one race as opposed to another because of heredity, this finding could be used to support assertions that the higher scoring group is genetically superior to the other. Burt, the founding father of British educational psychology, had conducted a number of twin studies that figured prominently in the public debate about racial differences in intelligence.

For more than twenty years, Burt generated overwhelming amounts of data in support of his views about the relative importance of genetic factors as opposed to environmental ones in determining intelligence. For example, his studies of identical twins who were separated very early in life revealed high correlation values, despite the fact that the twins were reared in very different environments. Burt's data were not questioned at the time, because he was a scientist of considerable influence in the sphere of mental measurement, and was widely respected for his research and statistical skills. Following his death in 1971, however, a different picture of the man began to emerge, and it is now believed that Burt fabricated much of his data, perpetuating one of the most infamous frauds in the history of science (Evans, 1976; Hearnshaw, 1979).

Suspicions about Burt's research were first brought to the public's attention in a 1974 book written by Princeton University psychologist Leon Kamin, *The Science and Politics of IQ*. Kamin pointed out numerous inaccuracies, inconsistencies, and methodological implausibilities which he had discovered in Burt's twins data. In one example, Burt had reported similar key correlation values (to three decimal places) for three different sample sizes (21 in 1955, over 30 in 1958, and 66 in 1966), a consistency that defies the laws of mathematical possibility. Kamin also discovered indications that Burt may have taken data from one study and pieced them into other studies so that the results would more conveniently fit his genetics hypotheses. Of course, Burt is not around to defend himself against these various charges of scientific misconduct, so we may never know whether the statistical anomalies were a result of carelessness or a premeditated distortion of research evidence (Rosnow, 1981).

About the same time as Kamin's book appeared, Oliver Gillie, a London medical correspondent who had been commissioned to write a book on the heritability of intelligence, reported being unable to locate two of Burt's collaborators, despite an intensive search (Evans, 1976). Burt had published articles during the 1950s in which the names of two co-authors, Margaret Howard and J. Conway, appeared. It is now widely believed that these collaborators may never have existed and that Burt invented them in order to increase the credibility of his research reports.

After these initial reports of possible cheating by Burt, other researchers described their own suspicions, suggesting that his data were often too good to be true – that is, they were too often a perfect fit for his theories (Evans, 1976; McAskie, 1978). Leslie Hearnshaw, whose authoritative biography of Burt appeared in 1979, added that Burt went beyond fabricating data and research assistants, and also authored suspicious letters to prominent scientific journals (including the *British Journal of Statistical Psychology* for which Burt served as editor) using pseudonyms to conceal his identity. These letters typically expressed praise for the work of Sir Cyril Burt and sometimes attacked his critics, thus providing Burt with a convenient device for calling attention to his own achievements and for further expressing his views. On occasion, Burt used these letters as an excuse to reply in print under his own name. According to Hearnshaw (1979), well over half of the 40 "contributors" of reviews, notes, and letters to the *British Journal of Statistical Psychology* during Burt's editorship are unidentifiable, but show enough consistency in style and content to suggest that Burt was their author.

Since the initial accusations of scientific fraud, Burt's research and reputation have been defended by scholars who have argued that the evidence against him is not as convincing as the many sensationalist reports surrounding the scandal have suggested. For example, Jensen (1974; 1978), a strong hereditarian, maintained that the statistical anomalies in Burt's published research were few and minor (see also Eysenck, 1981). More recently, persuasive cases have been made that many of the charges against Burt were unfounded, and that some of his initial detractors may have been bent on character assassination (Fletcher, 1991; Joynson, 1989). Despite the accumulated evidence on both sides of the debate, we probably will never know with complete certitude whether Burt deliberately falsified his data, although it appears that the charges against him have been excessive (Green, 1992).

Nevertheless, many observers maintain the view that Burt probably was so committed to his stance on the nature–nurture debate that he chose to use data in ways that would convince others of what he already believed to be true. Based on his thorough assessment of the man, Hearnshaw (1979) concluded that Burt's apparent deceit in large part may have been a function of certain defects in character, which ultimately served to tarnish the reputation of an otherwise brilliant scientist. Although extreme cases of scientific fakery appear to be rare in the behavioral sciences, when they occur their effects can be serious and far reaching. For instance, the British school tier system was based largely on Burt's findings regarding the heritability of intelligence (Diener and Crandall, 1978).

Conclusion: Why Study Research Ethics?

In this chapter some well-known examples of ethically questionable research in the behavioral sciences and related fields have been described. These cases represent only a small portion of the many other examples we will examine in this book. This is not to suggest that such studies are indicative of the majority of research in the behavioral and social sciences. In fact, much research in these fields is free of ethical problems and poses no threat to the participants studied or to society at large. However, it is hoped that an examination of the research controversies will sensitize the reader to some of the important considerations that are relevant to research ethics, including the difficulties that confront researchers in ethical decision making, and the consequences of their decisions.

A primary goal of this chapter was to convey to the reader a better appreciation for the study of research ethics. Too often it seems that students (and, it is feared, seasoned investigators) in the behavioral sciences approach the subject matter of research ethics as comprised of dry philosophical discourse, and view the many ethical issues and dilemmas as mere roadblocks to the exciting "stuff" of research – investigating interesting behavioral problems, designing complex studies, observing behavior, interviewing people, and the like. The examples we have considered in the previous pages, however, should have presented a quite different picture of what this book is all about. Recognizing and attempting to resolve the ethical dilemmas that threaten scientific progress in a field of inquiry can be as challenging as developing a complex research strategy and design for testing a hypothesis of interest (see Blanck et al., 1992). Research that is shrouded by ethical questions ultimately can lead to the destruction of reputations and careers, can threaten funding for research projects, can give rise to radical movements (such as the anti-vivisection campaign), and can stimulate ongoing, passionate debate. As Diener and Crandall (1978, p. 2) have argued, rather than associating research ethics "with sermons about morality and with endless philosophical debate," they more appropriately should be viewed as "a set of dynamic personal principles that appeal to vigorous and active scientists who face difficult real-world ethical decisions."

The study of research ethics helps to prevent research abuses and it assists us in understanding our responsibilities as ethical scientists. In addition, ethically educated investigators can become better able to anticipate ethical problems in their studies before they occur, in order to avoid them entirely or else to become more skillful in coping with them. Too often, it seems that individuals do not fully appreciate the complexities of ethical dilemmas until they are caught up in them. By then, it may be too late to cope effectively with the problems in a reasoned and objective manner, and less than satisfactory compromises are likely to result.

In the next chapter the ethical guidelines and standards that have evolved over the years to protect against the research abuses of the past are described. Such principles are intended to provide a workable framework within which researchers who seek guidance can analyze the issues relevant to the demands of their research, and to make decisions that do not compromise either scientific truth or human rights.

2

Ethical Principles in Behavioral Research: Professional and Governmental Guidelines

A code of ethics is always an expression of a profession's self-comprehension.

from the German *Professional Code of Ethics*, 1986

The following statement appears at the beginning (p. 5) of the American Psychological Association's (APA, 1982) *Ethical Principles in the Conduct of Research With Human Participants* (amended, June 2, 1989 (APA, 1990, p. 394)):

> The decision to undertake research rests upon a considered judgment by the individual psychologist about how best to contribute to psychological science and human welfare. Having made the decision to conduct research, the psychologist considers alternative directions in which research energies and resources might be invested. On the basis of this consideration, the psychologist carries out the investigation with respect and concern for the dignity and welfare of the people who participate and with cognizance of federal and state regulations and professional standards governing the conduct of research with human participants.

As the preamble to psychology's first full set of ethical principles for human subject research (now revised), this statement is informative in several respects. It illustrates the commitment of psychologists to self-regulate their behavior on behalf of the public and the profession, and it effectively captures the tone and spirit of ethical guidelines for research in related disciplines, many of which similarly have adopted research principles that were inspired by and modeled after those formulated by the APA. Moreover, the APA research principles, which are included in a more encompassing, generic ethical code for American psychologists have served as a model for codification in other countries (Schuler, 1982; and see appendix 2 of this book). The APA preamble is noteworthy for pointing out that psychologists have a professional responsibility to evaluate carefully and thoroughly the ethics of their investigations. Ethical decision making, according to the APA, involves a set of balancing considerations as to how best to contribute to science and human welfare. Such decision making does not occur in a vacuum, but in the context of the shared principles of a profession, government regulations, and the characteristic values of the general public.

This chapter presents a summary of the various ethical principles and governmental regulations that guide behavioral scientists in the conduct of their research. In the following pages, we consider the scope of their coverage, provide some background as to their development, and clarify their various functions. A single set of ethical rules of conduct is not available to behavioral scientists, who may work in a variety of disciplines and settings, each having its own working set of standards for human and animal research. However, there are enough similarities among the principles for researchers to get a clear idea of the essential ethical concerns when conducting behavioral investigations. In this chapter the APA guidelines receive particular emphasis because they are widely viewed as representing the most extensive and elaborate statement of ethics among those adopted by behavioral scientists. Our overview of ethical regulation begins with a brief history of professional codes of ethics and concludes with a consideration of the role that government has played in the research process during recent years.

The Development of Professional Codes of Ethics

Federal regulations and professional codes of ethics have replaced the so-called unwritten ethic that presumed individuals would act in fair and compassionate ways with regard to the rights of others. Among the earliest ethical principles for regulating human research are those which were developed for research and therapy in medical fields. These medical principles can be considered as the forebears to those eventually developed for research in the behavioral sciences.

The current ethical principles for medicine were developed after

World War 2, as is similarly the case for the biomedical and behavioral research fields. It is possible to find much earlier versions of medical codes, including the Hippocratic oath, which dates back to Hellenic times. Thomas Percival's book *Medical Ethics*, published in 1803, was a significant contribution to ethics during the nineteenth century. The major emphasis in Percival's book was placed on the practitioner's obligation to provide appropriate medical treatment to others. Perhaps because scant systematic experimentation was being carried out at the time, Percival devoted little attention to the ethical issues pertinent to the doctor–patient relationship, such as the need to fully inform treated individuals of procedures and risks. His book later served as the basis for the American Medical Association's (AMA) Code of Ethics in 1847, and the 1903 and 1912 revisions (Katz, 1970). The AMA's ethics code has been refined and elaborated over the years, with its most significant changes occurring in response to the moral outcry over atrocities committed during World War 2 by scientists in Nazi Germany.

The Nuremberg Code of Ethics

The contemporary origins of formal ethical standards for guidance and control in the behavioral sciences can be traced to the Nazi abuses of human subjects during World War 2. As is now widely known, horrific experiments were conducted on prisoners in Nazi concentration camps; after the war, these scientific cruelties were revealed and documented at the Nuremberg trials (*United States* v. *Karl Brandt* et al.). Among the studies carried out by Nazi physicians were experiments involving healthy prisoners' reactions to various diseases (such as malaria and typhoid fever), poisons, extreme temperatures, and simulated high altitudes. In one of these experiments, subjects were immersed in freezing water to determine how long it would take for them to die. Other studies were conducted to observe the course of infections and to test the effectiveness of treatments for different types of experimentally inflicted wounds. Another set of studies involved the measurement, execution, and defleshing of more than 100 prisoners for the purpose of completing a university skeleton collection (Katz, 1972). The Nuremberg trials ended in August 1947 with either death sentences or extensive prison terms imposed on the defendants, each of whom had been found guilty of committing war crimes and crimes against humanity.

In recent years much has been written about the abhorrent acts committed by the Nazi physicians in the name of research, and even this apparently clear-cut illustration of ethical abuse has not escaped debate and controversy. For example, while Mitscherlich and Mielke (1960) have documented how some of the studies actually had contributed to medical knowledge, others have questioned the use of knowledge obtained through such inhumane means. More recently, attempts have been made to explain how the Nazi physicians were able to cope with their involvement in carrying out tasks that must have conflicted with their moral and professional values, citing complex medical-ethical rationalizations for their participation and prevailing German authoritarian political and social values (Lifton, 1986; Ravitch et al., 1987). Ravitch et al., in part, have described how the researchers may have deluded themselves into believing that they were undertaking noble medical and scientific work, a point that echoed earlier suggestions that the physicians felt a sense of duty to contribute to the German struggle for existence (Katz, 1972; Mitscherlich and Mielke, 1960).

An important outcome of the Nuremberg trials was the formulation of a code of ethics to prevent future atrocities like those perpetrated under the guise of "medical science" by the Nazi researchers. The ten-point Nuremberg Code did not proscribe investigation as such, but instead attempted to outline permissible limits for experimentation (see Box 2.1). For example, the first principle of the code is particularly important for introducing the concept of "voluntary, informed consent," describing it as "absolutely essential" in research with humans. According to this principle, voluntary consent to participate in a study must be obtained from human subjects who are fully informed of the nature and risks of the experimental procedure. The other principles further delineate the criteria for an ethically acceptable investigation, including the necessity for the research to be carried out by competent, highly qualified researchers; that the subject has the freedom to discontinue participation in a study at any time; that precautions are taken to avoid unnecessary physical and psychological risks; that the subject is protected from more serious harmful effects and even remote hazards; that preliminary experimentation with animals is carried out to determine whether a procedure might prove too risky to use with human subjects; and that a study is discontinued when previously unforeseen negative consequences for subjects become apparent.

It is difficult to imagine that the Nuremberg Code would have prevented Nazi atrocities had it been in place earlier, and, as we have seen in the case of the American radiation experiments and the Tuskegee syphilis study, additional examples of scientific cruelty have surfaced since the Nuremberg trials. The existence of ethical codes are no guarantee that research abuses will be prevented, but they do serve certain useful functions. By codifying a set of value judgments in a formal document, an ethical code defines the grounds by which a research procedure can be evaluated as morally right or wrong

Box 2.1

The Nuremberg Code

The following ten principles of the 1946 Nuremberg Code were taken from Katz (1972, pp. 305–6) and appear in abridged form.

1 The voluntary consent of the human subject is absolutely essential.
2 The experiment should be such as to yield fruitful results for the good of society, unprocurable by other methods or means of study, and not random and unnecessary in nature.
3 The experiment should be so designed and based on the results of animal experimentation and a knowledge of the natural history of the disease or other problem under study that the anticipated results will justify the performance of the experiment.
4 The experiment should be so conducted as to avoid all unnecessary physical and mental suffering and injury.
5 No experiment should be conducted where there is an a priori reason to believe that death or disabling injury will occur; except, perhaps, in those experiments where the experimental physicians also serve as subjects.
6 The degree of risk to be taken should never exceed that determined by the humanitarian importance of the problem to be solved by the experiment.
7 Proper preparations should be made and adequate facilities provided to protect the experimental subject against even remote possibilities of injury, disability, or death.
8 The experiment should be conducted only by scientifically qualified persons.
9 During the course of the experiment the human subject should be at liberty to bring the experiment to an end if he has reached the physical or mental state where continuation of the experiment seems to him to be impossible.
10 During the course of the experiment the scientist in charge must be prepared to terminate the experiment at any stage, if he has probable cause to believe that a continuation of the experiment is likely to result in injury, disability, or death to the experimental subject.

(Rosnow, 1981). With its adoption, the Nuremberg Code offered at least some degree of protection to research participants and subsequently stimulated ethical codification in various research disciplines, including the behavioral sciences.

Ethical standards for psychology

We began this chapter with the introductory statement that, until the most recent revision in 1992, preceded psychology's ethical standards for human subject research. The APA's *Principles* were formulated by an ad hoc committee in 1973 as psychology's first complete set of ethical guidelines for human subject research, although a more generic code detailing ethical considerations in research, teaching, and professional practice had been in place since 1953. The development of ethical standards for research in psychology is interesting historically because of the process by which the standards evolved and the changing political and social climate they reflected (Reese and Fremouw, 1984). As such, the process of ethical codification in psychology provides an informative case study in the evolution of professional guidelines for research.

The beginnings of a code of ethics in psychology date back to 1938 when the Committee on Scientific and Professional Ethics was formed by the APA to consider the advisability of drafting an ethical code for the profession. It was concluded that a standing committee should be appointed to consider complaints of unethical conduct by psychologists, but that it would be premature to attempt to legislate a formal ethics code. Work eventually was begun in 1947 to develop a written set of ethical guidelines, apparently as a result of several converging forces. Sensitivities about ethical issues had been raised by the Nuremberg trials, and the APA standing committee had by that time received several problematic cases regarding the conduct of professional psychologists in both research and clinical contexts.

An additional influence that World War 2 had on ethical developments in psychology came about as a result of psychologists' extensive involvement in and contributions to the war effort. Psychologists had been called on to initiate a variety of testing programs during the war, including the intelligence testing of thousands of recruits for screening and placement purposes, and the situational stress testing of other military personnel for sensitive intelligence and espionage missions. When the war ended, there were accelerated demands for research applications in civilian contexts, such as personnel testing and training in business and industrial settings, which inevitably gave rise to questions about privacy, confidentiality, the misuse of test results, and the like. The call for a code of ethics thus seems to have come about, at least in part, as a result of increased emphasis on the problems of professional psychology. The emergence of applied psychology essentially shattered the previously held assumption that research in the discipline was invariably ethical and value free, and would eventually serve in the best interests of human welfare (Diener and Crandall, 1978).

APA's empirical approach
A unique approach was taken by the APA to accomplish the formidable undertaking of developing an ethics code. A committee of experts was formed, but the group eschewed any attempt to generate deductively a set of guidelines on its own, as had been done by elite groups in some other professions. Instead, the committee solicited the participation of all members of the professional association, who were asked to describe specific situations that actually had raised ethical problems for them. Thousands of responses were received by the committee, and an attempt was then made to categorize and analyze them in order to determine inductively the primary areas of ethical issues encountered by psychologists in their work. Based on the critical incidents received, a set of provisional ethical standards was drafted and submitted to all APA members for review and comment. Proposals for the ethical standards were published in 1950 and 1951, and numerous reactions appeared in two important psychology journals, *American Psychologist* and *Psychological Bulletin*. The standards were revised and modified throughout this review process, and finally were approved in 1953.

The formal code, *Ethical Standards of Psychologists*, has since undergone numerous revisions, most recently beginning in February 1986, resulting in an amendment of the principles in 1989 (APA, 1990) and substantial changes in 1992. Since the 1981 revision, the title of the document has been changed to *Ethical Principles of Psychologists*. Each revision is undertaken by the APA Committee on Scientific Ethics and Professional Conduct and reviewed by the entire APA membership prior to adoption.

The 1953 standards were generic in that they presented ethical principles in areas of consideration related to research, teaching, and professional practice. Very little in the initial code pertained to ethical issues in research *per se*; in fact, only about 4 out of a total 171 pages of text dealt directly with the ethical use of human subjects in research. The discussion in those few pages pertained to research planning, executing, and reporting; responsibility regarding interpretation of research results to the general public; and the researcher's relationship to subjects.

The section on subjects presented four principles that focused on such issues as exposing research participants to harmful effects; deception and the withholding of information from subjects; confidentiality of data; and fulfillment of obligations to participants. The most prohibitive principles were those requiring that the identity of research participants only could be revealed with their "explicit permission" (thus emphasizing the importance of confidentiality), and that psychologists were "obliged" to fulfill all responsibilities to participants in return for their research participation. The 1959 revision added the requirements of voluntary informed consent and removal of negative consequences in studies in which harmful after-effects were expected.

It is not surprising that the initial reaction to psychology's first formal ethical standards was for researchers to ignore them or else not to take them very seriously (Vinacke, 1954). Like other early professional codes of ethics, psychology's initial ethics code was comprised of guidelines that, in retrospect, were overly general and largely unenforceable, perhaps reflecting the uncertainty of the period as to the necessity for explicit research guidelines (Sieber, 1982c). However, by the late 1960s a changing scientific climate within psychology had begun to emerge, and with it came a great deal of concern about the adequacy of the current standards as they pertained to human subject research. Laboratory experimentation with human subjects had increased dramatically, along with the regular use of deception and other ethically questionable research procedures. A number of methodological concerns were raised about the implications of these laboratory procedures, especially deception (see chapters 3 and 4). An interest in ethical issues seemed to go hand in hand with rising concerns over issues related to the methodology of experimentation. However, it really was not until the debate over some highly controversial studies, like the Wichita Jury Study and Milgram's obedience research, that psychologists seriously began to wrestle with the issues of deception, privacy, and confidentiality in human subject research.

The rise in social conscience and the emphasis on the protection of individual rights ushered in during the late 1960s gave rise to additional ethical problems as psychologists began to focus their investigations on socially relevant problems, such as violence and crime, drug abuse, and minority issues. To a certain extent, the greater attention given to the rights of research subjects can be seen as part of the larger human rights movement in the United States during that time. Thus, it probably was a combination of factors that led to a reevaluation in 1971 of the ethics of research in psychology.

The 1973 research principles

Repeating the method used to develop the original ethical standards, an *ad hoc* Committee on Ethical Standards for Research, chaired by Stuart W. Cook, developed a separate set of ethical standards dealing solely with human subject research. Once again, the empirical process used to develop the principles was a time consuming and painstaking one, involving the participation of numerous psychologists who were asked to complete questionnaires, to provide examples of research situations that had raised ethical questions for them, and to respond to two draft versions of the principles. Much heated debate ensued, with some psychologists arguing that the principles threatened to limit scientific freedoms (e.g. Gergen, 1973a), and others maintaining that the principles were not strict enough to provide protection for research participants (Baumrind, 1971, 1972). Eventually, an ethics document consisting of ten research principles was adopted by the APA Council of Representatives in 1972 and was published in the January 1973 issue of *American Psychologist*. The research standards collectively appeared as principle 9 in the *Ethical Principles of Psychologists* and were detailed in an APA booklet with sample cases and extensive commentary (the document referred to above as *Ethical Principles in the Conduct of Research With Human Participants*; APA, 1973, 1982).

The ten research principles (see Box 2.2) pertained to the following seven areas of concern: (1) making the decision to conduct an investigation (principles A–C); (2) obtaining subjects' informed consent to participate (principles D and E); (3) assuring freedom from coercion to participate (principle F); (4) maintaining fairness in the research relationship (principle F); (5) protecting subjects from physical and mental harm (principle G); (6) observing responsibilities to participants once the research is completed (principles H and I); and (7) guaranteeing anonymity of research participants and confidentiality of data (principle J).

Although we will consider several of the methodological implications of psychology's research principles in subsequent chapters, a few noteworthy points can be made here about the profession's first

Box 2.2

Ethical principles for human subjects research

A   In planning a study, the investigator has the responsibility to make a careful evaluation of its ethical acceptability. To the extent that the weighing of scientific and human values suggests a compromise of any principle, the investigator incurs a correspondingly serious obligation to seek ethical advice and to observe stringent safeguards to protect the rights of human participants.

B   Considering whether a participant in a planned study will be a "subject at risk" or a "subject at minimal risk," according to recognized standards, is of primary ethical concern to the investigator.

C   The investigator always retains the responsibility for ensuring ethical practice in research. The investigator is also responsible for the ethical treatment of research participants by collaborators, assistants, students, and employees, all of whom, however, incur similar obligations.

D   Except in minimal-risk research, the investigator establishes a clear and fair agreement with research participants, prior to their participation, that clarifies the obligations and responsibilities of each. The investigator has the obligation to honor all promises and commitments included in that agreement. The investigator informs the participants of all aspects of the research that might reasonably be expected to influence willingness to participate and explains all other aspects of the research about which the participants inquire. Failure to make full disclosure prior to obtaining informed consent requires additional safeguards to protect the welfare and dignity of the research participants. Research with children or with participants who have impairments that would limit understanding and/or communication requires special safeguarding procedures.

E   Methodological requirements of a study may make the use of concealment or deception necessary. Before conducting such a study, the investigator has a special responsibility to (1) determine whether the use of such techniques is justified by the study's prospective scientific, educational, or applied value; (2) determine whether alternative procedures are available that do not use concealment or deception; and (3) ensure that the participants are provided with sufficient explanation as soon as possible.

F   The investigator respects the individual's freedom to decline to participate in or to withdraw from the research at any time. The obligation to protect this freedom requires careful thought and consideration when the investigator is in a position of authority or influence over the participant. Such positions of authority include, but are not limited to, situations in which research participation is required as part of employment or in which the participant is a student, client, or employee of the investigator.

G The investigator protects the participant from physical and mental discomfort, harm, and danger that may arise from research procedures. If risks of such consequences exist, the investigator informs the participant of that fact. Research procedures likely to cause serious or lasting harm to a participant are not used unless the failure to use these procedures might expose the participant to risk of greater harm or unless the research has great potential benefit and fully informed and voluntary consent is obtained from each participant. The participant should be informed of procedures for contacting the investigator within a reasonable time period following participation should stress, potential harm, or related questions or concerns arise.

H After the data are collected, the investigator provides the participant with information about the nature of the study and attempts to remove any misconceptions that may have arisen. Where scientific or humane values justify delaying or withholding this information, the investigator incurs a special responsibility to monitor the research and to ensure that there are no damaging consequences for the participant.

I Where research procedures result in undesirable consequences for the individual participant, the investigator has the responsibility to detect and remove or correct these consequences, including long-term effects.

J Information obtained about a research participant during the course of an investigation is confidential unless otherwise agreed upon in advance. When the possibility exists that others may obtain access to such information, this possibility together with the plans for protecting confidentiality, is explained to the participant as part of the procedure for obtaining informed consent.

substantive set of principles. First, the principles deal only with the use of humans in psychological research; a separate set of guidelines for animal research constituted principle 10 of *Ethical Principles of Psychologists* (APA, 1990) and are described in detail in chapter 8. One unique aspect of the 1973 document (and subsequent revisions) is that it uses the term "participant" throughout, instead of the more common "subject" to refer to the individuals who are studied in psychological research. This choice of terminology was a conscious one on the part of the *ad hoc* committee, which recognized the various pejorative connotations associated with the term "subject," including the subordinate role it seems to imply for investigators and the "objects" of their studies. Historically, subjects were subservient to the divine rights of kings; in medical practice, "subject" is sometimes used to refer to a cadaver for dissection and demonstration. One critic of the labeling of research participants as "subjects" argued that the term undermines "the individual and social qualities of the human being" (Riegel, 1979, p. 88). More recently, it has been suggested that a shift in terminology from "experimenter/subject" to "researcher/participant" can enhance the public character of science and thereby facilitate better research (see Box 2.3). The author agrees that the use of "subject" to refer to research participants can have negative associations in research contexts, and that the reader should be mindful of that possibility. What is more important, however, is that we remember to treat the individuals who assist us in the research process with respect and dignity, no matter how we choose to refer to them. Thus, the terms "subject" and "participant" are used interchangeably throughout this book. At other times, especially in references to surveys and research in other disciplines, the more general term "respondent" is used.

A second point regarding the ethical principles is that, like the 1953 standards, responsibility for ethical decision making was placed on the individual investigator. This emphasis is characteristic of the approach taken throughout the 1973 document, and subsequently has been a source of contention for those who feel that researchers are not always in the best position to evaluate objectively the ethicality of their own investigations. In chapter 9 we shall review some evidence that indeed supports this view. Although principle A recommends that the researcher should seek the advice of others when decisions suggest deviations from a principle, some scientists and laypersons believe that independent review committees should have the responsibility to oversee the ethics of human subject research. This, in fact, is an approach to ethical review that has been

Box 2.3

From "subject" to "participant"

Traditionally, the use of the terms "experimenter" and "subject" in behavioral research has reflected a rigid hierarchical distinction between the researcher as "expert" on human behavior and the research subject as "novice" (Young et al., 1995). This distinction is at the heart of the power differential that typically bestows greater authority upon the researcher in his or her interactions with subjects (Kelman, 1972). That is, generally it is the researcher who is fully informed about the nature of the research procedure, its purpose, and its theoretical underpinnings (the exception being when "blind" experimenters are employed to carry out the data collection but are left unaware of certain aspects of the study). Research subjects are presumed to be naive about these things and, as a result, are left relatively powerless in the situation, passively subject to the researcher's procedures.

In recent years, however, as sensitivities regarding the ethical treatment of others have increased, there have been calls within several disciplines for the establishment of a more egalitarian and respectful relationship between researchers and subjects. One indication of movement in this direction within psychology is apparent in the latest revision of the American Psychological Association's (1994) *Publication Manual,* an influential guide intended to aid authors in the preparation of manu-scripts. The current guide advocates a shift in terminology from "subject" to "participant" when referring to human beings. Such a shift towards eliminating the term "subject" in psychology publications has been noted in two key developmental psychology journals over the past two decades (Young et al., 1995).

According to Young et al. (1995), the move to replace the term "subject" with "participant" is a positive development that can serve to facilitate better research. In their view, "participant" more accurately reflects the ideal that persons involved in research are actively engaged in the scientific process, and wider adoption of the term represents a step in the direction of recognizing them as true collaborators. As a result, researchers may be more apt to establish a collaborative relationship with research participants (e.g. by obtaining their input during the formulation of questionnaires and interview protocols, by having them provide feedback as to how they have been represented in a study, and the like). Young et al. believe that such changes can lead to an enrichment of psychological data and ultimate enhancement of the public image of psychological science. Whether these outcomes actually will result from the elimination of the "subject" in psychological research remains to be seen. Or, rather, could it be that the "participant" will suffer the same fate in the future?

initiated at many research institutions, and is a government requirement for federally funded research in the United States.

A third point regarding the ethical principles, and also a source of much controversy both before and after their approval, is that few of the statements contained within them consist of absolute prohibitions. In some obvious cases, the principles are straightforward; for example, principle G states outright that procedures are not to be used if they pose the risk of serious harm to research participants. But for the most part, it is clear when reading over the principles in Box 2.2 that the APA had not developed a set of "Ten Commandments" for research, prefaced by phrases like "thou shalt not," but instead took an approach which emphasized the *weighing* of considerations that maximize positive outcomes ("benefits" or "gains") relative to negative ones ("costs" or "risks"). This so-called "cost–benefit approach," of course, is consistent with placing responsibility for ethical decision making on the investigator, who is expected to evaluate prior to carrying out an investigation whether the anticipated benefits of the study outweigh the potential costs.

A cost-benefit approach to ethical decision making is consistent with a moral theory known as *utilitarianism* which, in short, asserts that an action or rule of conduct is right if it produces the greatest well-being of everyone affected by it (Bentham, 1789). Based in part on the nineteenth-century British philosopher John Stuart Mill's (1957) "principle of utility," this perspective suggests that an individual ought to take the course of action that promotes the greatest good, happiness, or satisfaction for the most people. Utilitarians thus maintain that when a person is considering alternative moral choices, the net consequences of each of the available alternative actions must be considered.

In the context of behavioral research, utilitarians would argue that scientific ends sometimes justify the use of means that would necessarily sacrifice individual subjects' welfare, particularly when the research promises the "greatest good for the greatest number." For example, a research procedure that poses serious risk of harm to a group of human subjects would be justified from a utilitarian perspective if the study shows promise for developing a medical treatment that could save thousands of lives. This philosophical approach, essentially a variation of the moral suggestion that the ends justify the means, can be contrasted with *deontological* ethical theories, which hold that at least some basic actions are good in and of themselves without regard to whether they maximize positive consequences. "Deontologism" is a term that stems from a Greek word for duty or bindingness and thus is used to refer to duty-based moral theories (Atwell, 1981). Such theories suggest that general principles can be formulated that describe right behavior, and that individuals must strive not to violate these general principles or allow for exceptions. Although deontologists often differ in their views about what is good or right, they generally minimize the role of consequences in ethical decision making. Thus, an action might be viewed as morally right or wrong even though it does not maximize the greatest balance of good over evil. A deontological principle such as "do no harm" would suggest that the study described above should not be carried out, even though many individuals might benefit from its results.

Given the wide range of views espoused by psychologists during the two-year process of developing the profession's ethical principles for research, the Cook Committee, in the spirit of compromise, chose not to propose ethical statements in the form of absolute (or deontological) prohibitions. Instead, the research code that emerged was a utilitarian one, consisting of so-called "qualified" principles phrased in such a way as to guide rather than to direct individual decision making. Consider, for example, the concept of voluntary informed consent which appears as a central concept in psychology's research principles. Principle D first states that the investigator must inform participants of all aspects of the research that might influence their decision to participate. However, this principle is then "qualified" by the next sentence which begins with the phrase, "Failure to make full disclosure . . . requires additional safeguards." In short, while the code promoted the importance of informed consent, it allowed for exceptions. A researcher thus might choose *not* to obtain the informed consent of participants if he or she believes that there are good reasons for withholding information or misleading subjects about certain aspects of the study. In so doing, the researcher incurs additional ethical responsibilities to safeguard the well-being of the research participants; nevertheless, the point is that informed consent was not required in all circumstances. Other principles, such as those pertaining to the use of concealment or deception (principle E) and debriefing (principle H) are similarly qualified. Exceptions to these ethical principles often are warranted in psychological research on certain topics, such as conformity and attitude change, which could not meaningfully be studied in the laboratory with subjects' complete awareness. We consider this and other methodological justifications for deception in detail in chapter 4. As is discussed below, the latest revision of the APA code does appear to offer clearer language regarding adherence to the basic principles and for clarifying the research situations in which exceptions might be justified.

The weighing of benefits and costs

In deciding whether or not an exception to an ethical principle is justified, the investigator must consider the benefits and costs of carrying out the study in the intended way. From a utilitarian perspective, the study would be deemed ethical if the benefits outweighed the costs (to participants, science, or society), assuming the potential costs were not too extreme. Among the possible benefits of conducting a study are monetary or other incentives for participants, the discovery of new information, the replication of results obtained from previous studies, etc. The costs of a research project could include physical or psychological discomfort for participants, invasion of privacy, expenditures of money or time, reduction of trust or respect for science, etc.

As an illustration of cost–benefit decision making, we can imagine the sort of decision making Stanley Milgram might have engaged in to evaluate the ethicality of his obedience experiments. (The reader should keep in mind that the 1973 principles were not in place at the time Milgram conducted his obedience research, and that the available ethical guidelines were vague at best. There is good reason to believe, based on some of his subsequent writings, that Milgram did ruminate at length about the benefits and costs of deceiving his subjects prior to actually carrying out the studies.)

Among the possible costs or disadvantages of a deceptive research procedure are that subjects might be offended by the deception, their self-esteem could be lowered, their relationships with others could be impaired, their subsequent actions might be modeled after the experimenter's deceptive behavior, and their perception of behavioral research might be so negatively affected as to give psychologists and other researchers a bad name (APA, 1982). Additionally, in regard to the deception in Milgram's experiments, subjects' trust in authority figures could be diminished and subjects might be more prone to behave disobediently when confronted by legitimate authorities in the future.

Among the possible benefits or advantages of a deceptive procedure are the theoretical or social advances gained from the research findings, the avoidance of misleading findings that might have resulted from the study had subjects not been deceived, and the fact that subjects would have an opportunity to learn something about psychology, to perhaps gain some positive insight about their own behavior, and to feel good about their contribution to the research enterprise (APA, 1982). Milgram's deceptive procedure provided the necessary "technical illusions" for studying the power of a malevolent authority to elicit obedience. He later suggested that his results provided tangible evidence of the psychological potential in human nature to perpetrate horrendous acts such as those committed by the Nazis during World War 2, although it is unclear whether he anticipated such a gain in understanding prior to conducting the research.

Of course, once a listing of costs and benefits is developed, the decision must be made as to whether the particular study should be carried out. Unfortunately, the APA code does not provide any sort of simple equation or formula for arriving at such a decision. The 1982 *Ethical Principles* handbook states that "The investigator must judge the likelihood and seriousness of the costs, the probability and importance of the gains, and the number of people who will be affected" (p. 19), and recommends that the investigator seek the assistance of others in making such judgments in order to avoid personal bias. After weighing the considerations listed above, and after consulting several experts, Milgram might have decided that the potential theoretical and social advances far outweighed the potential risks involved; that without the use of deception, it would not have been possible to obtain new knowledge about the psychological mechanisms underlying obedience to authority; and that the probabilities of the costs were not great enough to justify abandoning the research as designed.

It is not difficult to imagine some of the criticisms that the APA research principles generated in psychology. Some individuals believed that the cost–benefit approach and the qualified nature of the principles would simply make it easier for researchers to justify ethically questionable methods in light of worthwhile research goals. For example, Waterman (1974) and Bok (1978) expressed dissatisfaction with the explicit exceptions contained within the code's basic principles. Bok interpreted the qualification to the informed consent principle as suggesting that when investigators can persuade themselves that a procedure which lacks the fully informed consent of subjects is for a good purpose and presents no harm, they can proceed with secrecy and lies. In this view, the personal rights of subjects are seen as violable at the researcher's discretion (Waterman, 1974).

Baumrind (1975) also objected to cost–benefit calculations for resolving ethical conflicts, asserting that such analyses inevitably lead to moral dilemmas. Since in her view the function of a system of moral philosophy is precisely to avoid such dilemmas, she attacked the code for failing to set forth clearly what is right or wrong in order to provide the researcher with guidance in dealing with specific ethical questions. Moreover, Baumrind argued that a code based on a system of moral evaluation in which principles of informed consent, protection from psychological harm, and deception are qualified represents a move towards legitimizing improper conduct and desensitizes, rather than elevates, ethical sensibilities. Similar complaints were voiced by Aitkenhead and Dordoy (1983) in response to the British Psychological Society's 1978 ethical principles for human subject research (see appendix 2).

Others critics wrote extensively about the various practical problems that a researcher could encounter in using the principles as a guide to ethical decision making. Diener and Crandall (1978) have summarized these practical problems by pointing to five serious drawbacks to cost–benefit analyses (see Box 2.4). First, costs and benefits are impossible to predict accurately prior to conducting a study. It probably is safe to say that a poorly designed or otherwise methodologically flawed investigation likely will yield no benefits, and most probably will have negative consequences for those individuals involved. However, how does one predict the extent to which a study will have positive consequences? After all, as Diener and Crandall imply, if the outcomes of an investigation were completely

---

Box 2.4

Cost–benefit approach: drawbacks

1   Costs and benefits are impossible to predict accurately prior to conducting a study.
2   Potential research outcomes are impossible to measure or quantify with precision.
3   There often are different recipients of the costs and benefits of an investigation.
4   The assessment typically must be completed by the very person who has a vested interest in a favorable decision.
5   Cost–benefit analyses tend to ignore the substantive rights of individuals in favor of practical, utilitarian considerations.

Source: Adapted from Diener and Crandall (1978).

predictable, why conduct the study in the first place? As a further complication, some researchers (e.g. Haywood, 1976; Rosenthal and Rosnow, 1984) have argued that ethical decision making should involve a consideration not only of the costs and benefits of *doing* a study, but also the costs and benefits of *not doing* the study (see Table 2.1). How does one effectively predict the benefits of *not* doing something?

Second, potential research outcomes are impossible to measure or quantify with any degree of precision. Without some sort of scale by which costs and benefits can be numerically measured and weighted, the balancing of the two against each other cannot be managed objectively. For example, a college student who volunteers for an experiment in order to earn some desperately needed cash (e.g. a US$15 incentive for participation) may be willing to risk a loss of self-esteem in the process. On the other hand, a psychologist might view a student's loss of self-esteem as the outcome having the greater consequence. In this case, the two individuals obviously would be using different weights for quantifying the possible consequences of the experiment. Even behavioral scientists tend to disagree among themselves about the value of many studies that already have been completed.

A third problem with cost–benefit decisions is that there often are different recipients of the costs and benefits of a behavioral science investigation; for many studies, individual subjects incur the costs while society in general stands to reap the benefits. This sort of outcome is contrary to the spirit of cost–benefit decision making, which had its roots in medical situations in which the individual patient faced both the risks and potential gains of a treatment. It is more difficult to justify exposing research participants to certain risks in studies that promise positive outcomes for others only, unless the subjects voluntarily agree with full awareness of the potential dangers (Kimble, 1976; Smith, 1976).

A fourth drawback to cost–benefit analysis pertains to the fact that the assessment typically must be completed by the very person who has a vested interest in a favorable decision. Because of the researcher's conflict of interest in the decision-making process, objective cost–benefit analyses may be impossible to achieve. Many scientists are apt to overrate the importance of their work and to underestimate potential harms as they move closer to a decision to proceed with a proposed study (Baumrind, 1975; Kimmel, 1991).

The final problem cited by Diener and Crandall is that cost–benefit analyses tend to ignore the substantive rights of individuals in favor

Table 2.1 Possible costs and benefits of behavioral research decisions

Recipient	Benefits	Costs
A. *Decision "to do" research*		
Subjects	Self-insight Extrinsic rewards (e.g. course credit or monetary payment) Increased understanding of science Feeling of having contributed to science	Mental anguish (e.g. stress or frustration) Physical discomfort Inflicted insight Loss of time Image of science lowered Mistrust of others Embarrassment
Researcher	Satisfaction of desire for knowledge or scientific interest Professional success and career benefits Recognition Altruistic satisfaction	Investment of time and effort Forgoing other activities Legal sanctions (e.g. if confidentiality breached) Undermines integrity and commitment to truth (e.g. if deception used)
Profession	Empirical tests of theories Attracts qualified researchers Provides for replications of earlier research	Exhausts pool of naive subjects Jeopardizes community support for the research enterprise
Society	Scientific advancement and progress Applications toward the betterment of humanity Increased understanding of behavior and interpersonal relations	Undermines trust in expert authorities and science Increased suspiciousness (e.g. self-consciousness in public places) Desensitizes people to needs of others

B. *Decision "not to do" research* (in general, the benefits and costs are the reverse of those for the decision to do research. The following list is thus simplified by providing only a few of the possible benefits and costs that could have been listed)

Subjects	Avoidance of physical or mental harm More time for other activities	Loss of opportunity for self-insight, monetary incentives, understanding of science, etc.
Researcher	Increased opportunities for other activities Avoidance of ethical dilemmas involving the treatment of subjects	Fewer opportunities for professional advancement and status among peers

Table 2.1 *cont.*

Recipient	Benefits	Costs
Profession	Use of available funds for other purposes Less likely to have image tarnished by controversial studies	Greater difficulties in determining the validity of theories, previous research, and assessment instruments
Society	Reduces opportunity for invading the privacy of others Maintains trust in expert authorities	Loss of opportunity to better society (e.g. by studying ways to reduce violence, prejudice, or mental illness)

Source: Baumrind, 1985; Forsyth, 1987; Rosenthal and Rosnow, 1991; Schuler, 1982.

of practical, utilitarian considerations. The notion of substantive rights suggests that all persons have certain inalienable rights, such as the rights to free speech and religious freedom. These substantive rights are considered to be inviolable, even if benefits to society could be maximized if such rights were eliminated. The problem here is that cost–benefit analyses take a more pragmatic orientation and may fail to recognize the inviolable nature of human rights.

When taken collectively, these problems indicate that a cost–benefit approach to ethical decision making alone cannot justify potentially harmful or risky investigations (Diener and Crandall, 1978). In fact, a set of ethical principles represents only one important part of the overall machinery needed to effect self-regulation in the research process. The criticisms described above have not gone without recognition by the APA, which alludes to most of the drawbacks of cost–benefit assessments in its 1982 *Ethical Principles* booklet. An attempt was also made to address several of the criticisms in the latest revision of the principles. However, the original intent in developing the principles was not to mandate a set of constricting, absolute rules that researchers would have to adhere to without regard to the complexities and uniqueness of each research situation. Despite the various problems inherent in the cost–benefit approach, the ethical principles provide at least a workable framework within which psychologists who seek guidance can analyze the issues relevant to the demands of their research. This is a primary function of professional ethical codes.

It also is noteworthy that the preamble to psychology's 1973 research principles included the call for consideration of "alternative directions" when planning a research investigation. This reflects the recognition that most research questions in science can be pursued in more than one manner and that the ethical researcher is one who selects the methodological approach that is most likely to satisfy his or her research goals while minimizing potentially negative consequences. This suggests that the appropriate question for researchers may not be the all-or-none one, between doing and not doing the research, but between doing the research in one way and using the same resources to do a different research project.

Subsequent modifications of the APA principles were made prior to the formidable 1992 revision, including a distinction between "subjects at risk" and "subjects at minimal risk" (which represents one of the ways that professional and governmental guidelines have become more consistent in recent years); a greater emphasis on protecting subjects' right to withdraw from a study, especially when research participants come from "special" populations with limited abilities to understand aspects of the research situation (such as children and the mentally infirm); and a stipulation that investigators should inform their subjects of procedures for contacting them if any stresses or questions emerge after participation (APA, 1981). A discussion of these and other changes in the 1973 document made prior to the 1992 revision can be found in Schuler (1982, pp. 183–5).

A new code of ethics for psychology

The latest revision of the APA ethics code began in 1986, as previously mentioned, leading to limited changes in the form of an amendment of the principles in 1989. Three successive draft versions of a substantially rewritten document were subsequently presented to the APA membership for comment, and a newly structured version of the code was adopted by the APA's Council of Representatives in August 1992. This revised code, known as *Ethical Principles of Psychologists and Code of Conduct* or APA Ethics Code has been in effect since December 1, 1992 (APA, 1992). Included in the code are an introduction, a preamble, six general principles (A–F, pertaining, respectively, to competence, integrity, professional and scientific responsibility, respect for people's rights and dignity, concern for others' welfare, and social responsibility), and eight areas of specific ethical standards. In distinguishing between the principles and standards, the code explains that the former are *aspirational* (and non-enforceable) goals, intended to guide psychologists towards the highest ideals of their profession, whereas the latter are *enforceable* rules for conduct as professional psychologists. One of the problems with earlier versions of the code is that they mixed aspirational and enforceable language (Bales, 1992).

The 14 standards that are most directly relevant to the conduct of research with human participants are found in section 6 of the ethical standards, entitled "Teaching, training supervision, research, and publishing" (see appendix 1). These standards describe appropriate conduct in reference to the following: (1) planning research; (2) responsibility; (3) compliance with law and standards; (4) institutional approval; (5) research responsibilities; (6) informed consent to research; (7) dispensing with informed consent; (8) informed consent in research filming or recording; (9) offering inducements for research participants; (10) deception in research; (11) sharing and utilizing data; (12) minimizing invasiveness; (13) providing participants with information about the study; and (14) honoring commitments. Other research-related standards included in the code focus on the care and use of animals in research, the ethical publication of research results, and privacy and confidentiality.

At first glance, the organization of the research standards is substantially different from that of previous versions of the code; but a closer scrutiny reveals that the basic ethical components have not fundamentally changed. The important ingredients for ethical conduct continue to be informed consent, cost–benefit assessments, protection from harm, and debriefing. However, some of the vague language evident in earlier versions has been replaced by clearer and more precise rules. For example, separate standards have been added which clarify the investigator's responsibilities regarding informed consent (standard 6.12) and the use of deceptive research techniques (standard 6.15).

As an illustration of the greater precision in the current code, one need look no farther than the statements regarding debriefing in studies where deception is an integral feature of the research design. Principle 9.e. of the 1981 standards states that the investigator has a responsibility to provide the participants "with sufficient explanation as soon as possible." Contrast this with the statement in standard 6.15. of the 1992 code, which maintains that the deception "must be explained to participants as early as is feasible, preferably at the conclusion of their participation, but no later than at the conclusion of the research." It is also recognized that informed consent may not be required in research "involving only anonymous questionnaires, naturalistic observations, or certain kinds of archival research" (previously referred to as "minimal risk research"). Greater attention also is paid in the new code to the confidentiality of research data, especially in terms of the protection of research participants from personal identification. Throughout, the importance of consulting with colleagues and review boards and adhering to federal and state law and regulations is emphasized. A practical resource for applying the APA code is presently available (Canter et al., 1994). Further clarification regarding the application of the current research standards will become available to psychologists when a newly formed APA task force completes its rewriting of the 1982 *Ethical Principles* booklet. Unlike the previous publication, which included a format that explicated the ethical principles, the revision will be organized according to specific content areas (e.g., planning research, informed consent, confidentiality), with the revised ethical principles referred to throughout. This shift from principles to content areas is intended to create a "user friendly" handbook for investigators by discussing the issues they would be faced with in conducting research (S. C. Cunningham, personal communication, December 8, 1994).

The 1992 APA Ethics Code is an impressive document, and one is impressed by its breadth and by the apparent care and seriousness that went into its preparation. As described above, the code has largely evolved into its present form as a result of nearly 40 years of diligent work by members of the psychology profession. While recognizing that "most of the Ethical Standards are written broadly," and that their application "may vary depending on the context," the APA took into account earlier criticisms of its research principles and issued a set of statements that are clearer about what psychologists must and must not do in specific circumstances. In so doing, the APA has moved closer in the direction of establishing minimum standards in the form of "thou shalts" and "thou shalt nots." The implications of this shift in terms of the rights of behavioral researchers to pursue scientific knowledge are discussed in subsequent chapters.

Ethics Codes in Other Behavioral Science Disciplines

Now that we have reviewed psychology's ethical principles in detail, we can turn our attention briefly to the ethical activities of professionals in related behavioral science disciplines. As illustrations, we shall focus here on the ethical principles for the conduct of sociological and anthropological research in the United States.

Sociology

The initial attempts to develop codes of research ethics in the fields of sociology and anthropology occurred during the 1960s. The American Sociological Association's (ASA) first ethics code was put into effect in 1971, following several years of study and debate. A standing committee on professional ethics also was appointed at that time. Unlike the initial guidelines in psychology, which focused mainly on professional issues, the early ASA ethical standards almost entirely concerned research, including such issues as objectivity and integrity in research, respect of the research subject's rights to privacy and dignity, protection of subjects from personal harm, confidentiality of research data, honesty in the reporting of research findings, and acknowledgment of research collaboration (ASA, 1971).

Sociology's ethics code was first revised in 1984 following concern that it emphasized research issues to the exclusion of other professional activities performed by sociologists. Thus, beginning with the 1984 code, sections are included that deal with ethical guidelines relating to research, publication and review processes, teaching and the rights of students, and relationships among sociologists. The most recent revision of sociology's ethical principles for research have been in effect since August 1989. The current ethical principles for sociological research are contained within the profession's generic *Code of Ethics* (ASA, 1989). The ten standards that are listed under the heading "Disclosure and respect for the rights of research populations" (see appendix 1) highlight many of the ethical concerns that we already have identified. The concepts of informed consent and confidentiality of subjects' responses are incorporated as critical elements of ethical research practice in sociology. However, because sociologists commonly use non-experimental methodologies as alternatives to controlled laboratory research, the ASA position on informed consent recognizes that the procedure may not be appropriate in many cases. In this view, there are potential problems with the automatic use of informed consent and that, in contrast to accepted practice in medical research, it should not be an absolute requirement for all behavioral science research.

Perhaps as a result of the focus of sociological research on social groups, organizations, communities, and societies, the research principles emphasize the importance of protecting the confidentiality and privacy of the individuals within these contexts who may serve as research participants. An introductory paragraph that precedes the research principles warns that procedures requiring full disclosure "may entail adverse consequences or personal risks for individuals and groups," and principle 6 advises sociologists to anticipate threats to confidentiality and to use appropriate procedures to protect it whenever possible (such as removal of subject identifiers).

Unfortunately, the ASA ethics code does not contain illustrative examples or further details regarding how the principles are to be interpreted and applied. Thus, researchers must use their own criteria for judging whether the potential risks imposed by a study exceed the risks of everyday situations; they must determine what constitutes "culturally appropriate steps" for securing informed consent, and so on. The guidelines do specify, however, that research design and data collection procedures should conform to the regulations outlined by the American Association of University Professors (AAUP) (1981), and should be consistent with appropriate federal and institutional research regulations. In addition to concerns about the lack of specifics regarding interpretation and application of the principles to research situations, the ASA revised code has been criticized on other grounds, including its apparent focus on requirements that are not firmly grounded in moral principle (Mitchell, 1990). At the time of this writing, an ASA committee had begun meeting to propose changes in the current code (B. D. Melber, personal communication, October 24, 1994).

Anthropology

In the discipline of anthropology, position statements on ethical problems in research were first made available by the American Anthropological Association (AAA) in 1948, and were more fully articulated in 1967. Based on these early statements, a set of ethical standards detailing the responsibilities of professional anthropologists was adopted by the AAA in May 1971 in a document entitled *Professional Ethics: Statements and Procedures of the American Anthropological Association*. Reissued in June 1983, these standards presently serve as the primary ethics statement of the AAA, as well as the American Association of Physical Anthropologists and the American Ethnological Association. The current standards specify that the responsibilities of anthropologists extend to those individuals studied, the public, the discipline, students, research sponsors, and their own government and host governments. The standards also acknowledge that conflicts between the various responsibilities of anthropologists are inevitable and that choices between conflicting values will have to made. Toward assisting anthropologists in successfully resolving these conflicts, the standards emphasize that priority should be placed on the well-being of research subjects and the integrity of the profession, and that when these conditions cannot be followed, it is best that the research not be carried out.

Because anthropological investigations in foreign areas are more likely to give rise to ethical dilemmas than other forms of research in the discipline, the AAA research-related principles focus on such concerns as freedom of research, support and sponsorship of research, and the involvement of anthropologists in government service. Emphasis is placed on potential threats to governmental and international relations as a function of questionable research practices involving "constraint, deception, and secrecy," which are viewed as having "no place in science." The AAA ethics statement also views as unacceptable the involvement of anthropologists in clandestine activities, and the restrictive imposition of external security regulations which are contrary to the best interests of scientific research. While the AAA statement may raise sensitivities to the important ethical issues in the discipline, like sociology's ethics code, it lacks specific recommendations that can be readily applied by a researcher who needs to resolve a research dilemma. This is a complaint that is often voiced in reaction to professional codes of ethics.

US Governmental Regulations for Behavioral Research

Much of the professional activity and concern about ethical issues in research was stimulated as a consequence of the increasing scope of governmental regulation, policies of the larger scientific community, and new research support requirements. As was also true within the scientific professions, some of the political impetus for regulation of human research came about as a result of widely publicized accounts of ethical misuse and abuse in biomedical research (e.g. Beecher, 1959).

Details of studies like those described in chapter 1 led to the development of governmental investigatory committees in the United States, and this activity played a role in the eventual formulation of American federal regulations for human research. Investigations into controversial studies like the New York cancer cell study typically resulted in the same conclusions – that established policies for reviewing experimental procedures were virtually non-existent in governmental agencies, and that more effective controls were needed to protect the rights of research participants. Federal policy also was stimulated by the following (Frankel, 1975; McCarthy, 1981):

1 widespread acceptance of the Nuremberg Code;
2 publication of the World Medical Association's Declaration of Helsinki in 1964, permitting therapeutic experiments with patients' consent;
3 enactment of the Kefauver–Harris Amendments of the Food, Drug and Cosmetic Act in 1962, which followed hearings into human subject abuses in drug research and resulted in regulations governing the testing of new drugs;
4 recommendations of the National Advisory Health Council (NACH), urging the establishment of governmental standards on human subject research.

Federal safeguards concerning the rights and welfare of human participants in research have been in place as part of US Public Health Service (PHS) policy since 1966, although the initial focus was limited to clinical research in the medical fields (McCarthy, 1981; Seiler and Murtha, 1980). In part, this policy required that a clinical investigation could be funded by the Public Health Service only after it was first subjected to an institutional ethics review. This review, conducted by a committee consisting of members of the principal investigator's institution, was to include a determination of the rights and welfare of subjects, the adequacy of the procedure for obtaining informed consent, and the risks and potential medical benefits of the proposed study (Frankel, 1975).

Following some confusion as to whether the initial federal policy applied to research in the behavioral and social sciences, Surgeon General Philip Lee issued a statement in May 1969 which extended Public Health Service policy to all research involving human participants, including biomedical and behavioral investigations. In addition to assuring greater consistency in interpretation and implementation of the provisions, the revised policy also provided greater detail as to what constituted "consent" and acknowledged that certain non-medical benefits could justify a review committee in permitting an informed subject to accept possible risks (Frankel, 1975). The revised policy was amended in 1971 to cover all research activities supported by grant or contract within the Department of Health, Education and Welfare (DHEW, now the Department of Health and Human Services, or DHHS), if the research placed human subjects at risk.

At that time, however, there were several calls for a reevaluation of

DHHS policy, especially following revelations of research abuses which came to light in hearings before the Senate Health Subcommittee chaired by Senator Edward Kennedy. One of the studies that received particular attention was the Tuskegee syphilis experiment (see chapter 1). Another matter to receive scrutiny was the alleged use of psychosurgery and behavioral modification techniques to control violent prisoners. The upshot of these discussions was for Congress to sign into law in July 1974 the National Research Act, which created the National Commission for the Protection of Human Subjects of Biomedical and Behavioral Research.

The primary purpose of the commission was to propose governmental regulations for ensuring the protection of human participants in virtually all behavioral, social, and biomedical research. The commission was composed of eleven members from several fields, including law, medicine, philosophy, the humanities, and theology, who were nominated by their professional organizations and appointed by the Secretary of the DHHS, to whom they were to directly submit their recommendations. Among the several specific issues considered by the commission were those involving the selection of subjects; conditions under which research participants were capable of giving fully informed consent; appropriate research procedures for investigations involving prisoners or children; and the proper functioning of ethical review boards (Schuler, 1982; see Seiler and Murtha, 1980, for a critical assessment of the early work of the National Commission).

In addition to creating the National Commission, another major outcome of the National Research Act was to issue regulations requiring the ethical review of human subject research by institutional review boards (IRBs). At first, this requirement for IRB review was applied in an uneven manner. Only DHHS-funded research was to be covered by the regulation, representing a small percentage of the behavioral research carried out in the country. However, if an institution received considerable support from the DHHS for human subject research, even non-funded research carried out at the institution was subject to IRB review (McCarthy, 1981). Non-funded research at institutions receiving only a small amount of DHHS funding were not required to undergo IRB review. This approach was further muddied in 1978 when the National Commission finalized its report on IRBs, which ultimately led to the extension of the requirement for IRB review to all research involving human subjects carried out at an institution receiving DHHS funds. This requirement provoked substantial criticism from those who argued that it was inappropriate for the government to extend its authority to research not directly funded by federal agencies.

These federal developments for regulating research actually were attacked by behavioral and social scientists on a number of grounds, only one of which involved the confusion over IRB coverage. Each argument tended to reflect fears that the government was exceeding its bounds by implementing increasingly stringent policies to control the country's human subject research. For example, many researchers were wary that investigations would be delayed as project proposals were reviewed by individuals who lacked an awareness of research problems outside their own particular disciplines. Some researchers contended that separate forms of regulation should have been developed for biomedical and behavioral research, and that the application of federal policies for certain categories of research, such as studies involving questionnaires and surveys, threatened the academic freedom of university researchers or otherwise violated the First Amendment of the US constitution. Others argued that federal coverage should extend only to human research for which ethical abuses actually could be documented. The National Commission instead had based its recommendations on a consideration of whether a given research strategy could *possibly* cause harm to human subjects (McCarthy, 1981; Seiler and Murtha, 1980). The proposed regulations for research involving prisoners, which essentially limited such research to studies offering direct benefits to the participants, also met with disfavor as being overly restrictive. Researchers have viewed the prisoner population as a rich source of willing participants who happen to live under controlled conditions essential for conducting a wide variety of investigations (Schuler, 1982).

Some of these fears that the research process was becoming increasingly enmeshed in governmental bureaucracy may have been unjustified. McCarthy (1981) has maintained that the National Commission's proposed rules actually would have resulted in substantial deregulation from the 1974 regulations, which had been broadly implemented. The National Commission had recommended that regulations be applied in a flexible way, taking into account the specifics of each case. In fact, in its report on IRBs, the commission called for the determination of categories of low-risk research that would be exempt from coverage. On the other hand, there were those who felt that all human participants were entitled to similar protection and that all human subject research should be subject to ethical review, regardless of the source of funding.

The National Commission presented its final recommendations in 1978 along with the two-volume Belmont Report, which included testimony from various experts (National Commission, 1979, appendixes 1 and 2). Hearings then were held on the proposed regulations by the President's Commission for the Study of Ethical Problems in Medicine and Biomedical and Behavioral Research. In drafting its new regulations, the DHHS responded to recommendations of both Commissions, public comment, and advice obtained from IRB members and research investigators, breaking sharply with the earlier trend toward greater mandatory IRB coverage of human subject research. The regulations, which first appeared in the January 26, 1981 issue of the *Federal Register*, apply directly only to research projects receiving DHHS funds and no longer apply to unfunded projects. (The Department stressed, however, that institutions seeking DHHS funds should demonstrate their commitment to the protection of human research participants in all non-funded research as well.) Moreover, the regulations substantially reduced the scope of DHHS coverage by exempting broad categories of research that present little or no risk of harm to subjects (so-called minimal risk research). Among the exempted research categories are certain forms of educational research; survey and interview research involving non-sensitive issues; observational studies of public behavior; and research involving the collection or study of existing databases, documents, and records. Additionally, the regulations specify that expedited review (an abbreviated procedure conducted by one IRB member) is appropriate for certain categories of minimal risk research, such as research requiring moderate exercise by healthy volunteers and non-manipulative, stress-free studies of individual or group behavior or characteristics.

For the approval of research covered by the federal regulations, an IRB must be established at the institution where the research has been proposed, which is to determine whether the following conditions are met: (1) risks to subjects are minimized by sound research procedures that do not unnecessarily expose subjects to risks; (2) risks to subjects are outweighed sufficiently by anticipated benefits to the subjects and the importance of the knowledge to be gained; (3) the rights and welfare of subjects are adequately protected; (4) the research will be periodically reviewed; and (5) informed consent has been obtained and appropriately documented. (The basic elements of informed consent, as defined in the regulations, are described in chapter 3.) These federal regulations have been in effect since July 27, 1981, and were revised as of March 8, 1983.

The impact of governmental regulation on human subject research has been substantial. Prior to the establishment of federal regulations, there was very little evidence of any type of committee review of research projects in university departments of medicine (Curran, 1969; Welt, 1961), and probably none at all in social and behavioral science areas (Reynolds, 1979). Today, ethical review boards are commonplace in most research-oriented institutions. While unfunded studies no longer require institutional review according to current federal regulations, it is likely that most universities, hospitals, and other research settings are requiring some form of review for the approval and monitoring of all human research conducted at those institutions (e.g. Institute for Social Research, 1976).

The increasing prevalence of ethical review boards, which often are composed of individuals lacking expertise in the areas proposed for investigation, has given rise to serious concerns among investigators. For example, one issue is the apparent lack of consistency of review board decisions, such that a study might or might not receive approval depending upon the particular composition of the committee (Kimmel, 1991; Rosnow et al., 1993). This issue and others related to the ethical review process represent a major focus of chapter 9.

Distinguishing Between Government Regulation and Professional Ethics

As we have seen, present federal policy effectively removes much behavioral and social research from IRB scrutiny and shifts the burden of ethical responsibility onto individual investigators and institution officials. Of course, it is generally assumed that investigators will apply their profession's ethical standards in conducting exempted or expedited research or studies not falling under federal coverage. Before concluding this chapter, it might be helpful to distinguish more clearly between professional ethics codes and governmental ethics regulations, clarifying the functions and goals of each.

The development of professional ethical standards and other ethical activities reflect the willingness of a profession to self-regulate the behavior of its members on behalf of the public and the profession. In the context of behavioral research, professional ethics refer to the rules that define the rights and responsibilities of behavioral scientists in their relationships with research subjects, clients, employers, and each other (Chalk et al., 1980). Because researchers may work in a variety of scientific disciplines, each having developed its own separate set of standards for guiding and controlling human subject research, no single set of professional rules of conduct presently is available.

By contrast, federal and state laws have emerged as requirements for scientists to take certain precautions or to refrain from engaging in certain research activities likely to increase the risk of harm to an individual or to society. Government regulations like those imposed by the DHHS are designed to protect or advance the interests of society and the individuals who make up that society. The mechanism for the enforcement of such regulations is the established system of law in the society.

While sharing the goal of laws and regulations to protect society and its individual members from harm, professional ethics exist primarily to instruct members of the profession in behavior considered appropriate by their peers. In fact, ethical standards can be viewed as a basic means of socializing those entering the profession to normative rules defining acceptable and unacceptable conduct.

Summary

In every decision-making situation, a researcher is armed with the ethical maxims of his or her scientific community, which are intended to provide guidance for dealing with ethical problems. However, unlike governmental regulations, which must be adhered to as directives, an ethical decision may or may not be consistent with the standards of one's profession. Rather, an investigator is free to exercise judgment regarding the appropriate procedures, allowing for exceptions to the ethical standards of the profession if they seem justified by the uniqueness of the situation (Lowman and Soule, 1981).

Professional codes of ethics thus are not guarantees for the full protection of research participants, society, and scientific integrity, but might instead be regarded as workable frameworks which provide researchers with guidance in order to analyze the issues relevant to their investigations. In chapter 3 we begin to consider some of these issues in depth by focusing on the ethical dilemmas that arise in the conduct of laboratory research.

3

Ethical Issues in the Conduct of Human Subject Research I: Laboratory Research

I observed a mature and initially poised businessman enter the laboratory smiling and confident. Within 20 minutes he was reduced to a twitching, stuttering wreck, who was rapidly approaching a point of nervous collapse. He constantly pulled on his earlobe, and twisted his hands. At one point he pushed his fist into his forehead and muttered: "Oh God, let's stop it." And yet he continued to respond to every word of the experimenter, and obeyed to the end.
Stanley Milgram, Behavioral study of obedience

It was not long after publication of Milgram's (1963) "Behavioral study of obedience" in the *Journal of Abnormal and Social Psychology*, in which the above passage appears, that psychologists began to give serious attention to the appropriateness of experimental procedures routinely used in the investigation of human behavior. Psychology experiments in past decades rarely elicited the sorts of intense reactions in subjects as those reported by Milgram. Nevertheless, by the mid-1960s, a growing number of critics had begun to question the proliferation of deceptive and potentially harmful manipulations which more and more investigators seemed to be using as a matter of course.

As early as 1954, social psychologist W. Edgar Vinacke took issue with psychology experiments in which research participants were deceived and sometimes exposed to "painful, embarrassing, or worse, experiences." Few, if any, psychologists were ready to deal with Vinacke's concerns at the time, probably because the use of deceptive procedures by psychologists was not particularly wide-spread (Christensen, 1988). Further, this was the dawn of an increasingly fruitful period for scientific psychology. An experimental research tradition had emerged that many psychologists hoped would rival progress in the more established physical sciences (Rosnow, 1981). A decade later, however, Vinacke's questions about the "proper balance between the interests of science and the thoughtful treatment of the persons who, innocently, supply the data" (p. 155) were raised anew by Diana Baumrind (1964) in response to Milgram's obedience research (see chapter 1) and a rapidly growing number of other studies involving deception. In her view, Milgram's deceptive procedure, which involved the bogus delivery of electric shocks to a hapless victim, subjected research participants to certain psychological risks. These risks included loss of self-esteem and dignity, and a possible loss of trust in rational authority. Baumrind additionally argued that because of methodological drawbacks, including the questionable representativeness of results obtained from voluntary participants, the research held little potential for serious benefit. She concluded that Milgram should have terminated the experiment as soon as he realized how stressful it was for his subjects.

Milgram (1964) responded point by point to Baumrind's criticisms, and we will return to some of their arguments in chapter 4. However, the exchange between the two psychologists over the ethical implications of the obedience research provided a focus for those who had begun to scrutinize the ethical bounds of an experimental approach that had become so prevalent at the time (Rosnow, 1981).

The experimental approach in psychology often involves the observation of human participants in the controlled environment of a psychology laboratory, an artificial setting that has been devised for the sole purpose of studying behavior. In the psychology laboratory, variables of interest to researchers typically can be more carefully manipulated, and behavioral responses more precisely measured, than in non-laboratory (or "natural") settings. However, unlike experimental research in the physical sciences, the investigation of human behavior by psychologists involves complexities inherent in the study of conscious human beings in a social setting defined as a "psychology experiment." A common assumption, described in greater detail below, is that human research subjects often would act differently if they were told the full truth about the experiment (e.g. Silverman, 1977). Since it is often desirable to keep these participants naive, psychologists frequently rely on the elaborate use of deceit in order to conceal the true purpose and conditions of their experiments.

This chapter and the next focus on the ethical issues stemming from laboratory experimentation in psychology, giving special emphasis to problems arising from the use of deception. As we have seen in chapter 2, present codes of ethics in psychology do not rule out the use of deception in certain cases, assuming that steps are taken to further protect research participants from harm and that there are no alternative procedures for investigating the research problem under consideration. Below, the various types of deception used in behavioral research are distinguished, along with the justifications for their use. Additionally, the nature of informed consent and debriefing are considered. In chapter 4 we turn our attention to the potential threats to research validity imposed by these important ethical procedures and evaluate some suggested alternatives to the use of deception.

The Problem of Deception in Psychology Experiments

In a paper presented at the 1965 American Psychological Association meetings in Chicago, social psychologist Herbert Kelman (1967) helped bring to light just how widespread the use of deception had become in psychology and related fields. Kelman lamented the growing use of deception in psychological research and, in addition to discussing Milgram's study, he mentioned several cases of investigations that he felt warranted ethical concern.

Among the studies described by Kelman were experiments by Bramel (1962, 1963) and Bergin (1962), involving undergraduates who were given discrepant information about their sexuality. In the studies by Bramel, heterosexual males were led to believe that they had become sexually aroused by a series of photographs depicting other men. In order to accomplish the manipulation in his studies, Bramel ingeniously used a bogus "psychogalvanic skin response apparatus," which subjects were led to believe was an extremely effective and accurate measure of sexual arousal. Subjects were instructed to record their arousal response level to each photograph projected onto a screen by noting the needle indication on the apparatus, which they were informed would indicate their homosexual arousal to the photograph. Unknown to them, however, was the fact that the experimenter had complete control of the galvanometer readings. Each photograph had been assigned a needle reading in advance, so that photographs portraying "handsome men in states of undress" received higher readings than did those of unattractive and fully clothed men.

It should be noted that Bramel took certain precautions to protect subjects from undue harm in his studies. He informed the participants that their privacy and anonymity would be protected in the experimental situation and that very strong homosexual tendencies would be indicated by consistently extreme readings on the galvanometer. Nonetheless, Bramel (1962, p. 123) claimed that the procedure was impressive enough for subjects "that denial of the fact that homosexual arousal was being indicated would be very difficult."

In Bergin's (1962) study, both male and female students were given feedback about their sexuality after completing an extensive series of psychological tests. Perceptions of sexuality were manipulated by falsely informing subjects that their scores on the masculinity factor of a personality test were indicative of the opposite sex. As Bramel had done in his studies, Bergin informed the student participants about the true nature of the research prior to their leaving the laboratory. However, the very fact of having received the false feedback during these studies may have caused lasting emotional turmoil in subjects who were particularly sensitive about their sexual identity (Kelman, 1967).

Other, more dramatic illustrations of the potentially harmful effects of deception and the lack of fully informed consent were identified by Kelman. Some of these cases occurred outside the laboratory, such as a series of investigations involving naive army recruits (see Box 5.1). (The implications of using deception in natural settings, where subjects typically are unaware that they are being observed, are described in chapter 5.) In one laboratory study that is often mentioned in discussions of research ethics in psychology, Campbell et al. (1964) tested the effects of the classical conditioning of a traumatic fear response in male alcoholic patients who had volunteered to participate in an experiment "connected with a possible therapy for alcoholism." The procedure used by the researchers involved the pairing of a neutral tone with an intense fear response brought on by an injection of the drug Scoline, which produces temporary motor paralysis, causing an inability to move or breathe. Although the effects of the drug were known to have no permanent physical consequences, it was assumed that the temporary interruption of respiration induced by the drug would cause severe stress in the unsuspecting patients. Not surprisingly, the inability to breathe (for an average of about two minutes) was reported by the subjects as a terrifying experience, to the extent that most believed that they were dying. None of the patients had been warned in advance about the effect of the drug, and there is no reason to believe that the researchers expected to learn anything about its treatment potential for alcoholism (Diener and Crandall, 1978).

As predicted, by injecting the drug immediately after subjects had heard the neutral tone, Campbell and his colleagues found that the tone by itself had become capable of inducing an extreme stress reaction in the subjects. In fact, the conditioning procedure had been so effective that the researchers later had difficulty eliminating the traumatic reaction to the tone in some patients. Thus, in this study, the researchers recruited volunteers and obtained their consent under false pretenses, withheld important information from them which surely might have influenced their decision to participate, and then subjected them to various risks, including extreme discomfort and possible long-term harm. In designing their study, the researchers must have decided that deception regarding the effects of the drug was essential in order to test the hypothesis; however, some physicians who were familiar with the drug actually had participated in the study and they similarly described the experience as an "extremely harrowing" one (p. 632). This suggests that the researchers could have successfully performed the investigation even if the patients had been informed about the effects of the drug, a procedure that would have reduced some of the study's ethical problems. This fact reflects one of the main points in Kelman's early discussion about the use of deception in research. In his view, deception had become a commonplace research procedure that increasingly was being employed even in situations where it was unnecessary.

In psychology, Kelman (1967, p. 2) and others believed, the use of deceptive manipulations had become like a game for many researchers, "often played with great skill and virtuosity." Consistent with this view, the design of some studies sometimes involves a series of deceptions in which one falsehood is built upon another. "Second-order" deceptions, for example, are procedures in which subjects are observed for the purposes of a study only after they are falsely led to believe that the experiment is over, or that they have been enlisted as the experimenter's accomplice. Both of these sorts of deceptions were utilized in a classic early study on cognitive dissonance theory conducted by Festinger and Carlsmith (1959). The researchers attempted to determine if attitudes towards a boring research task (turning wooden pegs on a board for one hour) would become more favorable if subjects were induced to convey to future subjects that the experiment was interesting and enjoyable. To test this hypothesis, the procedure was designed in such a way that participants were led to believe the study was over after they had completed the boring task. Depending on the experimental condition, subjects then were offered either US$1 or US$20 to misrepresent the task to the next "subject," who really was the experimenter's assistant. The researchers found that subjects who received only US$1 (and thus lacked justification for lying about the experiment) later rated it more positively than subjects who received US$20 (a sum assumed to have provided sufficient justification for the lie).

Festinger and Carlsmith believed that the results of their study were consistent with the tenets of cognitive dissonance theory. Although this conclusion was contested by other psychologists (e.g. Bem, 1965; Linder et al., 1967), one can readily understand why the study also raised ethical concerns about the various deceptions used. The participants were deceived about the purpose of the study, were falsely informed that the study was over when in fact it was not, and were encouraged to lie to someone they were unaware was in on the experiment in order to receive money. The final insult was that subjects were asked to return the money they had earned for "helping out" the researcher! In Kelman's (1967) view, such multiple-level deceptions have the potential to undermine the relationship between experimenter and subject, eliminating the possibility of mutual trust and weakening researchers' respect for the very individuals whom they depend on in order to carry out their studies.

Kelman was not alone in criticizing the excessive use of deception in psychological experimentation (e.g. Bok, 1978; Seeman, 1969; Warwick, 1975). Ring (1967, p. 117), for example, similarly attacked the "fun-and-games approach" taken by many psychologists who, in their attempts to create increasingly elaborate deceptions, appeared to be engaged in "a game of 'can you top this?'" Indicative of this tendency is an extreme case cited by Carlson (1971) in which researchers employed 18 deceptions and three additional manipulations in a single experiment (Kiesler et al., 1968). Baumrind (1964, 1975, 1985), another outspoken critic of deception, focused her arguments on the issue of trust. In her view, all cases of deception in human relationships represent violations of human rights which inevitably weaken people's trust in future interactions. When researchers deceive their subjects, according to Baumrind, they not only weaken their subjects' confidence in the integrity of others, but they also may serve as models for encouraging dishonest behavior.

Whatever the specific complaint, the various criticisms leveled against deception often suggest that, because it involves lying and deceit, its use in psychological research is morally reprehensible and may have potentially negative effects on subjects, the profession, and on society (Christensen, 1988). Of course, how one feels about research deception ultimately reflects the ethical theory one espouses. Thus, there are those who believe that deception can be ethically justified from a cost–benefit (or "act-utilitarian") perspective, given the methodological requirements of a particular investigation, and provided that certain safeguards are followed, such as debriefing and protection of research participants from harm. This is the position that has been adopted by several professional societies, such as the American Psychological Association (1992). Others, arguing from a deontological (or "rule-utilitarian") ethical position, believe that deception should be eliminated from research settings because its use represents a clear violation of the subject's basic right to informed consent (Baumrind, 1985; Goldstein, 1981). Some general examples of ethical and methodological arguments for and against the use of research deception appear in Box 3.1.

Why deceive?

As we have seen, strict adherence to the letter of the law regarding ethical standards in the behavioral sciences requires that researchers first obtain voluntary informed consent from their subjects. In the current American Psychological Association (1992) regulations (see appendix 1), for example, the basic elements of informed consent maintain that prospective research participants be informed of (1) the nature of the research; (2) their freedom to decline participation or withdraw at any time during the investigation; (3) the possible consequences of non-participation; (4) the factors that may be expected to influence their decision to participate, such as risks, discomforts, and the like; and (5) other aspects of the study about which they inquire. (For a more complete listing of the "ingredients" that make up fully informed consent, see Box 3.2.).

The carefully worded criteria that define informed consent in the APA guidelines do not stipulate that researchers must provide subjects with full information about their experimental plans, procedures, and hypotheses. In fact, it is certain of these aspects of their investigations that researchers often choose to withhold from their

Box 3.1

Sample arguments for and against deception

In a study of the effects of deception on subjects' trust of psychologists, Sharpe et al. (1992) found that participation in research did not appear to result in the development of negative attitudes in college students, regardless of whether they had been exposed to deception. The students also were generally accepting of arguments justifying the use of deception as a research tool. In their investigation, Sharpe et al. presented their subjects with brief methodological and ethical arguments for and against the use of deception (p. 587). These arguments, presented below, are typical of those that appear in the literature on the ethics of deception in research.

An affirmative methodological argument for deception

Deception is a good idea, because the only way to study "natural" behavior is to catch people off guard. If they know that some specific aspect of their behavior is being watched, then they'll change it so that they look as good as possible.

A methodological argument against deception

Deception is a bad idea, because it only allows us to study spontaneous and "spur of the moment" reactions. Much behavior is deliberate and well thought out. To study such intentional behavior, psychologists must provide participants with all relevant information about a situation before observing behavior.

An ethical rationale for conducting deception research

Deception is ethical and should be permitted if the information being sought outweighs the costs to subjects of being deceived, if some other viable procedure cannot be found, and if subjects leave the experiment in a positive psychological state.

An ethical argument for not conducting research involving deception

Deception is unethical. Experimenters should not be allowed to mislead subjects about the study's purpose, deny subjects the opportunity to give their informed consent, or invade their privacy without their knowledge.

Source: Sharpe, D., Adair, J. G. and Roese, N. J. (1992). Twenty years of deception research: a decline in subjects' trust? *Personality and Social Psychology Bulletin, 18,* 585–90. © 1992 by Sage Publications, Inc. Reprinted by permission.

Box 3.2

Informed consent: what is sufficient information?

In a paper prepared for the National Commission for the Protection of Human Subjects of Biomedical and Behavioral Research, Robert Levine (1975) identified 11 types of information that should be communicated to the prospective research participant during the informed consent procedure.

1 Describing the overall purpose of the research.
2 Telling the participant his or her role in the study.
3 Stating why the subject has been chosen (and how).
4 Explaining the procedures, including the time required, the setting, and those with whom the participant will interact.
5 Clearly stating the risks and discomforts.
6 Describing the benefits of the research to the participant, including the inducements.
7 Disclosing alternative procedures, where applicable. (This is a requirement more clearly applicable to medical research.)
8 Offering to answer any questions.
9 Suggesting that the subject may want to discuss participation with others, especially where substantial research dangers are involved.
10 Stating that the participant may withdraw at any time and that exercising this right will have no negative consequences.
11 Stating that the initial information is incomplete and that further information will be given after the experiment, where applicable.

Levine recognized that in most cases some of these elements will be inappropriate or unnecessary to obtain (pp. 3–72):

Each negotiation for informed consent must be adjusted to meet the requirements of the specific proposed activity and, more particularly, be sufficiently flexible to meet the needs of the individual prospective subject.

Even Diana Baumrind (1971), one of the more outspoken defenders of subjects' rights, similarly posited that certain information need not be divulged in order to obtain sufficiently informed consent to participate in an investigation (p. 891):

I do not agree that the researcher has a moral obligation, in order for consent to be informed, to tell a subject why he has been selected; the investigator need only not misinform the subject as to why he has been selected and to let him know what will be expected of him and what may happen to him as a consequence of the experiment.

For some illustrative samples of informed consent forms that have been used in academic settings, see appendix 3.

subjects or else misinform them about. If they did not, the resulting psychology of human behavior might very well be based on biased research findings. Thus, despite the ethically questionable nature of deception as a research technique, there is widespread agreement that its use in psychological experimentation is often a methodological necessity for some investigations (e.g. Adair et al., 1985; Schuler, 1982; Sharpe et al., 1992; Weber and Cook, 1972).

In order to clarify what is meant by "methodological necessity" and how research results can be biased by fully informed consent, consider the following hypothetical research examples suggested by Baron and Byrne (1987). Imagine a study in which subjects are correctly informed at the outset that they are participants in an investigation of the effects of flattery on their liking of strangers. Aware of the purpose of the investigation, it would be difficult to imagine that subjects' feelings towards the strangers from whom they received compliments would be the same if they were unaware of the study's purpose. Similarly, we might expect many subjects to "bend over backwards" to show how accepting they are of members of other races if the subjects are aware that they are participants in a study of racial prejudice. The message in both of these cases should be clear: in revealing the exact substance of the research, the psychologist runs the risk of distorting the reactions of his or her subjects and ultimately limiting the applicability of the research findings.

Let us next consider an example that pertains to a series of laboratory investigations that actually were carried out. In order to test certain hypotheses about the conditions affecting whether people will intervene during emergencies, Bibb Latané and John Darley (1970) created research situations in which naive subjects were led to believe that a room next to the laboratory was on fire, that another subject was suffering an epileptic seizure, or that a secretary had had a serious fall from a stepladder in an adjoining room (see Box 3.3). Rather than waiting for such unpredictable events to occur and then to observe bystanders' reactions, it was more practical for Latané and Darley to create the events themselves under the more controlled conditions of a psychology laboratory. This, of course, required the researchers to deceive their participants about the real nature of the situation and the purpose of the investigation. If such facts had been made known to the subjects – that they were participants in a helping experiment or that the situations were not actual emergencies – the researchers clearly would not have been able to learn much about bystander intervention during actual emergencies.

Box 3.3

Creating emergencies in the laboratory

During the mid-1960s, social psychologists Bibb Latané and John Darley began a series of investigations in order to determine the conditions leading to prosocial behavior during emergencies. Like other behavioral scientists at that time, their interest in this subject matter was sparked by actual incidents involving groups of people who idly stood by, failing to come to the aid of victims in distress. Perhaps the best known example of the perplexing phenomenon of "bystander apathy" was the case involving Catherine "Kitty" Genovese, a 28-year-old bar manager who was brutally murdered near her apartment in Queens, New York, after returning home from work late during the early morning hours of March 13, 1964. Thirty-eight bystanders in the neighborhood witnessed the crime but nevertheless failed to take effective action to help her.

Given the number of factors that might underlie such complex behavior, Latané and Darley conducted their initial experiments in the laboratory, where they were able to isolate certain variables in order to study their effects on the behavior of interest. Because of the large number of witnesses present during several of the reported cases of bystander apathy, they reasoned that people may be less apt to intervene when others are present in the situation. But how might an emergency be arranged in a laboratory setting in order to test this hypothesis? One strategy employed by Darley and Latané (1968) was to have college students show up for a study allegedly pertaining to personal problems associated with campus life. Upon arrival in the laboratory, each participant was taken to a small room equipped with an intercom system which would enable communication with other participants (who were actually "figments of the tape recorder"). The subject then was informed that each "participant" in turn would be given two minutes to discuss their school-related adjustment problems, followed by an opportunity to comment on what others had said. The individual comments then would be followed by a free discussion. The experimenter also explained that in order to ensure confidentiality, he would not listen to the discussion and that participants would remain anonymous to each other. During the initial round of comments, each prerecorded participant in the discussion, as well as the one naive subject, took their turns and described common adjustment problems. However, the future victim added to his remarks that he was prone to seizures, especially when under stress. When it came time for that individual to speak again, his comments became progressively disoriented and he seemed to be experiencing great distress. Although somewhat incoherent, he then complained that he was having a seizure and needed immediate help. Finally, he choked loudly and then became silent.

In order to manipulate the number of bystanders thought to be present during the emergency, the actual subject was led to believe that the discussion group involved either one other person (the victim), two other persons (the victim and another bystander), or five other persons (the victim and four other bystanders). The results of the study were consistent with expectations that the more people present during an emergency, the less likely they would come to the aid of the victim. Whereas 85 percent of the subjects who believed they alone knew of the "victim's" plight reported the seizure before the victim became silent, only 31 percent of subjects who thought that four other bystanders were present did the same.

Was the use of deception justified in Darley and Latané's study? We might expect critics of deception in research to argue that the research paradigm exposed subjects to psychological risks, including guilt and a threat to their self-esteem for not helping, stress during the emergency itself, and embarrassment at being duped by the researchers. On the other hand, others might defend the procedure by contending that such risks were far outweighed by the importance of the subject matter and potential gain in knowledge about helping behavior. What do you think?

In the examples just described, we can identify some of the more common methodological justifications for the use of deception in behavioral research. Foremost among these is the recognition that if research participants were aware of the true purpose of an investigation, their resulting behavior might be influenced by such knowledge (see Rosenthal and Rosnow, 1969). Research findings may be systematically biased by altruistic subjects who attempt to confirm the research hypotheses in order to help in the cause of science (the so-called "good subject effect") or by subjects who attempt to put their best foot forward in order to make a good impression on the researcher (a bias known as "evaluation apprehension"). An experimenter's openness may also produce boomerang effects from more suspicious subjects who seek ulterior motives and alternate explanations. These potential threats to validity that stem from the social nature of the experimenter–subject interaction collectively are referred to as research artifacts.

Confronted by such problems, it is easy to understand why an investigator may feel compelled to mislead his or her subjects about the true nature of the experiment. By doing so, the researcher gains confidence that his or her subjects have responded to the experimental variables under investigation and that the observed behavior has not been influenced by certain motivations associated with participation in the study, such as the good subject effect. In this sense, we would say that deception has the capacity to improve the internal validity of an investigation.

A related methodological advantage of deception is that it can elicit more spontaneous behavior from subjects than otherwise might be the case in a laboratory setting. This possibility can increase the generalizability of the research findings to more natural, everyday situations, thereby improving the external validity of the investigation. Finally, as we have seen in the example of the bystander intervention studies, deception can increase the researcher's degree of methodological control over the experimental situation. In their initial investigations, Latané and Darley (1970) were particularly interested in determining whether the number of bystanders present during an emergency would influence the likelihood of intervention by any one bystander. Clearly, they could not have expected an emergency to occur repeatedly in a natural setting under precisely the same circumstances with a different number of bystanders present during each occurrence. It was more feasible for them to conduct their studies in the controlled setting of the laboratory, where the number of bystanders present could be systematically manipulated during a series of carefully contrived "emergencies." By creating such "fictional environments" in the laboratory, investigators can manipulate and control the variables of interest with much greater facility than if deception is not used.

In sum, with regard to the use of deception in human subject research, a balance often must be struck between methodological and ethical considerations. Realistically, ethical research procedures such as informed consent are not always the most methodologically sound procedures. In other words, what is the most ethical is not necessarily the most effective. Thus, under certain circumstances, deceiving research participants about the true nature of an investigation may be the only feasible way to collect the necessary data. An absolute rule prohibiting the use of deception in all research would make it impossible for researchers to carry out a wide range of important studies.

Types of research deception

As is apparent in the research examples previously described, various kinds of research deceptions have been utilized by behavioral scientists. Although the focus of this chapter is on laboratory research, one must bear in mind that deception may be practiced by researchers within both laboratory and field settings. In laboratory studies, subjects are at least aware that they are participating in a study, whereas participants in field studies may or may not be aware of that fact. In either setting, deception may be of the active or passive sort (Arellano-Galdames, 1972). "Active deception" is deception by commission, as when the researcher blatantly misleads the subject about some aspect of the investigation (e.g. the purpose of a study or the procedures to be followed). "Passive deception" is deception by omission, as when the researcher purposely withholds relevant information from the subject. With passive deception, a lie is not told; rather, a truth is left unspoken (Rosenthal and Rosnow, 1991).

Given this basic distinction between deception by commission and deception by omission, it is possible to classify a variety of deceptive research practices (Arellano-Galdames, 1972). Among the research procedures that represent examples of active deceptions are the following:

1 misrepresentation of the research purpose;
2 untrue statements about the researcher's identity;
3 use of research assistants (called "confederates," "stooges," or "pseudosubjects") who act out predetermined roles;
4 false promises;
5 violation of the promise of anonymity;
6 incorrect information about research procedures and instructions;
7 false explanations of scientific equipment and other measurement instruments (including "paper-and-pencil tests");
8 false diagnoses and other reports;
9 use of placebos (i.e. inactive medication) and secret application of medications and drugs;
10 misleading settings or timing for the investigation (when the study actually begins and ends, its duration, etc.) and corresponding behavior by the experimenter.

Deceptions by omission include the following research practices:

1 concealed observation;
2 provocation and secret recording of negatively evaluated behavior;
3 unrecognized participant observation;

4   use of projective techniques and other personality tests;
5   unrecognized conditioning of subjects' behavior.

Most of the examples listed in this taxonomy of deceptive research practices are more likely to be practiced in the laboratory context than in field settings. The list of deceptive procedures is not exhaustive and some deceptive methods might be classified in more than one of the 15 categories (Schuler, 1982). However, the taxonomy is helpful in pointing out that research deception varies along several dimensions. A survey of research reports conducted by Gross and Fleming (1982) provides an indication of the frequency with which these types of deceptions have been employed by researchers (see Box 3.4).

Additional typologies of research deception have been suggested

---

Box 3.4

Types and frequency of research deceptions

In an extensive survey of 1,188 research studies appearing in leading social psychology journals over a twenty-year period, Gross and Fleming (1982) found that more than 60 percent involved at least one of the kinds of deception techniques listed below. Milgram's obedience experiments, for example, utilized four of the deceptions included in the table: a false cover story, incorrect information concerning materials (i.e. the shock generator), use of a confederate, and false feedback about the "victim's" pain (Forsyth, 1987).

Type of deception	Percentage of studies
False purpose or cover story	81.5
Incorrect information concerning materials	42.0
Use of confederate or actor	29.2
False information (feedback) given to participant	17.5
Participant unaware of being in a study	13.7
Two related studies presented as unrelated	9.1
False information about a confederate or other person given	5.3
Participant unaware study is in progress at time of manipulation or measurement	4.1

Source: Gross and Fleming (1982).

by other researchers. Sieber (1982a) has identified several varieties of deception, including self-deception, third-person deception, offering false information, and offering no information. In addition, she distinguishes between three types of "forewarning" (i.e. limited informed consent) by the researcher: (1) consent to the fact that a deception will be employed; (2) consent to be deceived; and (3) obtaining waivers to inform. According to Sieber, whether these forms of deception will result in harm or pose serious risks to research participants depends in large part on the nature of the actual experiment. For example, the offering of no information implies that participants are not aware that they are being observed for research purposes. While the appropriateness of this form of deception depends on *what* behavior is being observed, the setting and the purpose of the observation also are critical factors. A research technique in which "lost letters" are dropped in public settings (such as shopping center parking areas) to see how many will be returned poses no risk to participants and the magnitude of the deception is relatively small. In the laboratory, the omission of certain information irrelevant to the subject in a verbal learning experiment likewise would hardly represent a serious breach of ethical standards.

Another important distinction pertaining to kinds of deception has to do with the recognition that deception in the research context often may be unintentional (Keith-Spiegel and Koocher, 1985). Thus, Baumrind (1981, 1985) has chosen to classify deception as non-intentional or intentional. Non-intentional deception, which cannot be entirely avoided, includes absence of full disclosure, failure to inform, and misunderstanding. Intentional deception, on the other hand, includes the withholding of information in order to obtain participation, concealment and staged manipulations in field settings, and deceptive instructions and confederate manipulations in laboratory research. According to Baumrind, absence of full disclosure does not constitute intentional deception, so long as participants agree to the postponement of full disclosure of the research purpose. But when the investigator's purpose is "to take the person unaware by trickery" or to "cause the person to believe the false" for whatever reason (e.g. to induce subjects into agreeing to participate or to reduce threats to causal inference), the study invariably involves the intentional deception of research participants (Baumrind, 1985, p. 165).

Baumrind's notion of non-intentional deception suggests that while full disclosure of all information that may affect an individual's willingness to participate in a study is a worthy ideal, it is not a realistic possibility. Even the investigator with a sincere desire to disclose all relevant aspects of a study's purpose and procedure to subjects nonetheless may fail to reveal certain information. In part, this may be a function of the researcher taking for granted that subjects have a basic level of knowledge about scientific procedures, testing materials, and the research equipment or apparatus used. In other cases, information provided to subjects may not be fully understood. Experimental research procedures can be quite complex and difficult to explain in a language that is understandable to the typical research participant. Further, young children and the mentally impaired have certain cognitive limitations that seriously limit the extent to which fully informed consent can be obtained from these groups (see chapter 7). Finally, due to the imperfect nature of human communication, there is likely to be some degree of misunderstanding between the subject and researcher during the informed consent procedure (Baumrind, 1985).

In sum, deception experiments differ so much in the nature and degree of deception used that even the harshest critic would be hard pressed to state unequivocally that all deception has potentially harmful effects. Indeed, simple generalization about the benefits or harms of deception research is not possible (Sieber, 1982a). Since experimenters can never convey everything about a study to their subjects, it may be that all behavioral research is disguised in one respect or another and that the range of ethical questions related to the use of deception must be thought of as falling on a continuum. The style of the deception is not so much the issue as is its probable effect on the subject.

Frequency and magnitude of deception

Several attempts have been made to gauge the prevalence of intentional deception in human subject research (e.g. Adair et al., 1985; Carlson, 1971; McNamara and Woods, 1977; Menges, 1973; Seeman, 1969; Smith and Richardson, 1983; Stricker, 1967). Consistent with more casual observation, these reviews of the published literature have revealed that deceptive procedures have been regularly used in psychological research and that the frequency of deception has not diminished from its peak in the 1970s (Sharpe et al., 1992).

It is not very surprising to find that a relatively high frequency of deception was evident in psychological research during the 1960s and early 1970s (see Table 3.1). During this period, as previously discussed, psychologists used deceptive procedures as a primary means of coping with research artifacts. One must also bear in mind that separate sets of ethical guidelines for human subject research did not exist until the early 1970s in the behavioral sciences, although there is evidence that ethical regulations have not led to a reduction in the use of deception (Adair et al., 1985).

Taken together, reviews by Stricker (1967), Seeman (1969), Carlson (1971), and Menges (1973) revealed that nearly one-fifth of the studies appearing in major psychological journals involved incidents of overt deception by experimenters who deliberately told untruths to their subjects. In fact, deceptions involving direct lying to subjects

Table 3.1 The extent of psychological studies using deception

Source	Year(s) surveyed	Journals	Studies using deception (%)
Stricker (1967)	1964	*Journal of Abnormal and Social Psychology, Journal of Psychology, Journal of Social Psychology, Sociometry*	19.3
Seeman (1969)	1948, 1963	*Journal of Personality, Journal of Abnormal and Social Psychology, Journal of Consulting Psychology, Journal of Experimental Psychology*	18.5 (1948) 38.2 (1963)
Carlson (1971)	1968	*Journal of Personality, Journal of Personality and Social Psychology*	57
Menges (1973)	1971	*Journal of Personality and Social Psychology, Journal of Educational Psychology, Journal of Consulting Psychology, Journal of Experimental Psychology, Journal of Abnormal and Social Psychology*	17 (false information) 80 (incomplete information)
Gross and Fleming (1982)	1978–9	*Journal of Personality and Social Psychology, Journal of Experimental Social Psychology, Social Psychology, Journal of Social Psychology*	59

Source: From A. E. Gross and I. Fleming, "Twenty years of deception in social psychology," *Personality and Social Psychology Bulletin*, 8, 402–8. Copyright © 1982 by Sage Publications Inc. Reprinted by permission.

characterized between 19 and 44 percent of the studies published in the areas of social psychology and personality during the period surveyed (1961–71). For example, in an early examination of 457 studies published in the *Journal of Abnormal and Social Psychology*, the *Journal of Personality*, the *Journal of Social Psychology*, and *Sociometry* during 1964, Stricker (1967) reported that subjects were intentionally misled about some aspect of the investigation in 19.3 percent (88) of the investigations. Specifically, subjects were given false information about instruments, stimuli, tasks, and other conditions in 76.1 percent of the deception studies, and about other participants' performance and behavior in 60.2 percent of the studies. The research purpose was misstated in slightly more than half of the investigations.

Stricker also found that the frequency of deception varied greatly according to the specific topical area studied. The use of deception was most pervasive in studies of conformity (81 percent) and cognitive dissonance and balance theory (72 percent). Similarly, Seeman (1969) found that journals emphasizing personality and social psychology had a relatively high incidence of deception in comparison with experimental and clinical psychology publications. Further, Seeman noted a significant rise in the use of deception in the personality and social areas, from 18.5 percent in 1948 to 38.2 percent in 1963. In another survey, Carlson (1971) focused specifically on the frequency of deception in personality and social psychology journals. She found that subjects were given complete and accurate information in less than 5 percent of 226 studies published in the 1968 issues of the *Journal of Personality* and the *Journal of Personality and Social Psychology*. The deceptions took a variety of forms, involving cover stories, miscommunication of purpose, the use of confederates, and false interpretations of test performances. Similar findings were obtained in a more recent survey conducted by Gross and Fleming (1982). It is interesting to note that when psychological journal editors were asked to nominate studies they considered to be empirical landmarks in social psychology, the five most frequently nominated (Asch, 1955; Festinger and Carlsmith, 1959; Milgram, 1963; Schachter and Singer, 1962; Darley and Latané, 1968) involved elaborate laboratory deceptions of the studies' purpose and procedures, and the sixth (Sherif et al., 1961) involved the observation and manipulation in the field of unsuspecting youngsters (Diamond and Morton, 1978).

The high frequency of deception in personality and social psychology research is understandable, given the nature of the subject matter and the greater potential for threats to validity posed by research artifacts. Many studies involving social and personality topics such as conformity, impression formation, self-esteem, attitude change, helping behavior, and aggression simply could not be carried out without the use of some form of deception. However, deception has not been limited to these areas of psychology. Menges (1973) examined nearly one thousand studies published in 1971 in a variety of psychology journals, including those in abnormal, counseling, and educational psychology, and found that the use of deception was widespread. Subjects received incomplete information in 80 percent of the studies, inaccurate information about the independent variable in 17 percent, and complete information in only 3 percent of the studies.

Given the growing attention that has been paid to ethical issues in research since the early 1970s, one might imagine that the use of deceptive procedures has declined in recent years. Yet there is evidence that this is far from the case. McNamara and Woods (1977), for example, reported a rise in the use of deception to 57 percent in studies published during the years 1971 to 1974. In a more recent survey, Smith and Richardson (1983) asked 464 psychology undergraduates to describe the experiments in which they had participated and found that approximately half reported that the study had used deception.

In order to assess the extent to which psychologists' research practices have been altered by ethical regulations, Adair et al. (1985) conducted a survey of methodological and ethical practices reported in all empirical studies appearing in the *Journal of Personality and Social Psychology* for 1979. Among the research practices considered was the frequency of the use of deception, defined as providing information that actively misled subjects about some aspect of the study. It was found that deception was utilized in 58.5 percent of the examined studies, a percentage which, according to Adair et al. p. 63), represents a "monotonically increasing function" over previous decades and suggests that deception has become a normative practice, especially for research in social psychology (see Table 3.2).

In addition to the fact that the frequency of deception has not declined in recent years despite the implementation of more stringent ethical regulations, there is also some evidence that the nature and the intensity of the deceptions have also remained relatively unchanged. As one indication of the consistency over time in the magnitude of deceptions used, Adair et al. (1985) reported that more

Table 3.2 Percentage of empirical studies in social psychology employing deception 1948–79

Journal of Abnormal and Social Psychology		
1948	(Seeman, 1969)	14.3
1961	(Menges, 1973)	16.3
1963	(Seeman, 1969)	36.8
Journal of Personality and Social Psychology		
1971	(Menges, 1973)	47.2
1971–74	(McNamara and Woods, 1977)	56.8
1979	(Adair et al., 1985)	58.5

Source: Adapted from Adair et al. (1985, p. 63).

than two-thirds of the 1979 studies they assessed employed multiple types of deception. Further, the same authors found that it was relatively easy to identify comparable examples of deception studies in their 1979 sample that matched some of the extreme laboratory experiments cited as problematic by Kelman some fifteen years earlier.

In one study, White (1979) used a typical paradigm for studying the link between aggression and sexual stimuli. Subjects were angered by a confederate who evaluated them in a highly negative and personally insulting way (in order to manipulate subjects' frustration levels); they next were assigned to the role of having to give electric shocks to the confederate, following a rigged draw, after having been given a false cover story as to the purpose of the experiment. Prior to administering the shocks, the angered subjects were asked to rate a series of slides depicting various sexual acts, including some which were intended to create unpleasant emotional reactions. At this point, the reader should readily recognize some similarities between this approach and Milgram's obedience procedure. However, had Adair et al. selected from studies published in 1977 they would have found an investigation that even more closely matched Milgram's research – a duplication of the obedience procedure carried out by Shanab and Yahya (1977), whose subjects were children as young as six years old.

Recalling certain aspects of Bergin's (1962) sexual identity investigation is a study by Baumeister et al. (1979) in which female subjects were given false feedback informing them that they had scored high on a bogus trait called "surgency." Half of the subjects were led to believe that their score indicated that they were immature, ill adjusted, and possessed a variety of additional negative characteristics. As part of the manipulation, the researchers went on to explain that these traits had resulted from an unusual developmental background which had caused the subjects to be insecure about their femininity. So informed, half of these women then had their surgery score and its presumed meaning publicly disclosed in the presence of a male confederate.

A third study cited by Adair et al. parallels (and may even exceed) the interruption-of-respiration manipulation employed in the 1964 experiment by Campbell and colleagues, at least in terms of the multiple, high-magnitude deceptions employed. The study, conducted by Marshall and Zimbardo (1979), was intended to determine the emotional consequences of inadequately explained physiological arousal. As in Campbell's investigation, subjects were misinformed about the purpose of the experiment and the cause of their experimentally manipulated physiological state. Marshall and Zimbardo informed their participants that they would receive a vitamin injection when instead they were injected with epinephrine or an inactive placebo; the subjects also were misinformed about the somatic effects of the drug. Fake equipment was used to mislead subjects into thinking that their physiological responses were no longer being monitored when in fact they were. A doctor then pretended to administer the drug to a confederate who proceeded to behave in a bizarrely "euphoric" manner (in order to provide a context for the real subject's interpretation of the effects of the epinephrine). As if this scenario was not enough, complete debriefing of subjects was postponed for up to six weeks until all had been tested.

The close match between these experiments and the extreme deception studies described by Kelman and others more than a decade earlier suggests that the intensity of some deceptive procedures, at least as practiced by social psychologists, has not changed and in fact may have increased (Adair et al., 1985). Of course, the studies considered by Adair and colleagues were conducted more than fifteen years ago, and are themselves now somewhat dated. However, extreme cases of deceptive research are still to be found in the psychological literature, as illustrated by the following two examples. In one study, Zimbardo et al. (1981) induced partial deafness in a sample of university students through post-hypnotic suggestion. This procedure was based on the assumption that many elderly individuals become paranoid and fearful because they have hearing problems, not because they are actually suffering from psychological deficits. In the investigation by Zimbardo and colleagues, subjects were misinformed about the purpose of the experiment and the experiences they would undergo. As predicted, it was found that the experience of partial deafness, without awareness of its source, induced changes in the cognitive, emotional, and behavioral functioning of subjects. Specifically, subjects in the deafness-without-awareness treatment condition became more paranoid and were rated by themselves and by others as more irritated, agitated, hostile, and unfriendly than control subjects. Although the experimental sessions were followed by extensive debriefing and attempts to remove any tension or confusion in subjects, lingering effects of the deceptive manipulation cannot be completely ruled out.

A more recent example of extreme deception involved a variation of Milgram's obedience procedure. Dutch social psychologists Meeus and Raaijmakers (1986) carried out an experiment on "administrative obedience" which, rather than commanding subjects to inflict physical harm on another person, required them to engage in behavior intended to cause apparent psychological harm. Subjects were led to believe that they were participating in a study of people's ability to work under stress and that their task was to administer a series of test questions to a job applicant. Supposedly, the applicant – actually a confederate – would get the job only if he passed the test. However, during the test, subjects were ordered to make an escalating series of fifteen insulting "stress remarks" to the applicant over a microphone from an adjoining room (such as "This job is much too difficult for you. You are more suited for lower functions").

As in Milgram's procedure, the confederate pleaded with the subject to stop, and then angrily refused to tolerate the treatment. The confederate finally pretended to fall into a state of despair, thus apparently losing out on the job for which he was applying. As in Milgram's research, complete obedience was obtained from a majority of subjects (92 percent) in the condition that included an experimenter who ordered them to continue with the procedure. No subject persisted in a control condition which lacked a prodding experimenter. In a follow up to this study, Meeus and Raaijmakers (1987) sent another group of subjects a letter prior to their participation forewarning them about the research procedure. Despite the fact that these subjects had ample opportunity to consider the conflict posed by their eventual experimental role, all were completely obedient in the study.

Although Meeus and Raaijmakers eliminated from their obedience studies the use of ominous shock generators and any suggestions that intense physical pain was being experienced by the unsuspecting victim, the risks to participants probably were no less than those posed by the original Milgram procedure. Yet despite an increase in sensitivities to ethical issues since Milgram's studies, little attention seems to have been paid to the ethicality of the use of deception in the Dutch replications. However, as Adair et al. (1985, p. 65) correctly concluded following their discussion of contemporary deception experiments, even the most extreme deceptions "are controversial and subject to differing interpretations in a discipline that remains divided on the issue of deception."

Safeguards Against Deception

In view of the various ethical and methodological problems posed by deception, it is essential for the researcher who uses this procedure to take special precautions to minimize potentially negative effects. In addition, before deciding to employ deception in a study, investigators should give serious consideration to the full range of alternatives that might be employed instead. In the remainder of this chapter we turn our attention to the nature and implications of the important procedure of debriefing as a means of protecting subjects from negative research consequences. Alternatives to the use of deception are considered in chapter 4.

Debriefing

Debriefing is perhaps the most commonly used safeguard when deception studies are carried out in laboratory settings. In its basic form, debriefing (also sometimes referred to as "dehoaxing," "desensitization," or "disabusing") describes the procedure that occurs once data have been collected from the research participant. The investigator engages the participant in a discussion to clarify various aspects of the study, including its true nature and purpose. It is during the debriefing session that the researcher attempts to correct any of the subject's misconceptions about the study, answer the subject's questions honestly, and supply additional information that may have been purposely withheld. It also is a good idea to inform subjects how they can contact the researcher if later they have additional concerns or experience negative effects from having participated in the study. All of this information is to be provided sensitively and with the education of the subject in mind.

Debriefing is considered an important safeguard against the effects of deception because it is during this formal post-experimental interaction that the adverse effects of deception may be detected, including negative reactions (such as embarrassment, upset, and stress), resentment at having been duped, suspiciousness, and the like. From an ethical point of view, the most important function of debriefing is to ensure that no permanent harm to subjects will result from their involvement in the study (Schuler, 1982). As Kelman (1968, p. 222) has aptly stated, "A good rule is that the subject ought not leave the laboratory with greater anxiety or lower self-esteem than he came in with." Others have reiterated this point and added that participants should feel even better about themselves or derive some educational benefit from their laboratory experience. It can be difficult to detect whether these goals have been met and, as described below, in some situations debriefing might do more harm than good. As Tesch (1977, p. 218) has suggested, ethical guidelines that stress the researcher's obligation to detect and remove (or correct) undesirable research consequences are based on the hidden assumption "of magical undoing, of using debriefing as an eraser for emotional and behavioral residues." However, there appear to be at least three necessary steps for maximizing the effectiveness of debriefing in deception studies: (1) a careful debriefing immediately following the subject's participation in the study; (2) a clear explanation of why the deception was necessary; and (3) an expression by the experimenter of regret for having deceived the subject (Rosenthal and Rosnow, 1991; Tesch, 1977).

To be sure, debriefing may serve other functions for the researcher in addition to the ethical one of identifying and eliminating the harmful effects of deception. Thus, debriefing is a procedure that is used in non-deceptive studies as well as deceptive ones. For example, a subject may have experienced stress or tension as a result of the experimental manipulations, even if he or she had been informed about them and agreed to undergo the risk prior to participation. Debriefing in such cases should be oriented toward assessing any lingering effects from these negative reactions, helping the subject to better understand and cope with them.

In addition to the ethical functions of debriefing, there are two methodological functions that can be identified (Schuler, 1982; Tesch, 1977). The first has to do with ascertaining whether the experimental manipulations were successful. As suggested below, one way for the researcher to assess whether an investigation has high experimental realism is to have subjects reflect on the study and describe their thoughts about it. In addition to serving as manipulation checks, such subject reports can determine the extent and accuracy of their suspicions, and can verify whether they construed the research situation as intended and were involved with it (Aronson and Carlsmith, 1968). Unfortunately, the extent to which debriefing is capable of detecting procedural problems and subject suspiciousness is questionable in light of evidence suggesting that some subjects are unwilling to confess their prior awareness or to let on that they did not take the procedure seriously (Golding and Lichtenstein, 1970; Newberry, 1973). What may underlie the detection inadequacy of many debriefings is the difficulty for participants to clearly separate the debriefing from the experiment *per se* (Orne, 1962; Tesch, 1977). That is, subjects may construe the debriefing procedure as merely a continuation of the experiment and continue to play whatever subject role they adopted during the experiment.

A second methodological function of debriefing is that it can operate as a safeguard against the participants' sharing of information about the study with potential subjects. As described in our discussion of subject suspiciousness in chapter 4, some researchers have found that pledging subjects to secrecy during the debriefing procedure can limit the extent to which relevant information is passed on to others. This outcome is especially likely when subjects are given insight as to the problems involved with the loss of naivete (Aronson and Carlsmith, 1968). Other researchers are less optimistic about the likelihood that participants will keep their knowledge about an experiment confidential, even when they have entered an agreement with the investigator (e.g. Altemeyer, 1972).

There is widespread consensus that debriefing should fulfill an educational function in addition to those functions already described. That is, research subjects should receive some sort of an educational benefit in return for their participation. This might take the form of acquired insight and knowledge of a psychological nature (such as better understanding of the conditions that influence bystander intervention during emergencies), a better appreciation for the scientific process, and self-knowledge. Consider, for example, a study on shyness in which some subjects receive feedback that merely confirms their fears (that they are shy). This sort of self-knowledge could conceivably serve to exacerbate their social anxiety. A more complete and potentially beneficial debriefing would be one in which the researcher provides the subject with some insight into the causes of shyness, helps the subject recognize that it is a common but treatable problem, and suggests some steps that the subject might take to overcome it. This is not to suggest that the researcher should act as a psychotherapist during the debriefing session; such an approach would represent a conflict of interest and would likely be viewed as unethical conduct.

There are potential pitfalls in using debriefing as an educational tool. For example, Tesch (1977) has described the possibility that experimenters might use debriefings as a means to manage their subjects' impressions of the experiment and its value, redefining the situation in such as way as to promote their own explanations. Consistent with this argument is the fear that researchers often provide subjects with only a cursory explanation about the study and instead devote most of the debriefing session to convincing them that they were not unconsciously manipulated or misinformed. Because explicit knowledge about concealed aspects of the study can be more detrimental to the research results than vague suspicions, some researchers choose to limit the information they provide to subjects or else postpone the debriefing until all subjects have been tested and the study has been completed.

The extent to which subjects attend to the information provided at the end of a study represents another concern that may limit the educational utility of debriefing. Previous surveys have revealed that subjects did not highly value the debriefings they received from researchers (Gerdes, 1979; Sharpe et al., 1992; Smith and Richardson, 1983). In addition to the possibility that some debriefings are poorly conducted, it may also be the case that participants give minimal attention to the debriefing message, assuming that their obligation to the researcher is over. In other words, once subjects have gone through the experimental procedure they are likely to be more interested in obtaining the promised incentive (such as course credit or payment) than in learning about behavioral research (Sharpe et al., 1992). One interesting approach for overcoming this potential problem has been recommended by Davis and Fernald (1975). They suggest that successful learning and assimilation of experiments can be maximized by requiring student subjects to compose a laboratory report based on the experiments in which they participate, with the report containing all of the critical information that would appear in a scientific publication. They could also be asked to evaluate the experiments and provide their personal experiences. One limitation to this approach, in addition to the fact that it only applies to participants who are students, is that a common alternative to the research participation requirement in many college courses is for students to write a term paper (an option that is rarely selected). Because Davis and Fernald's recommendation would result in a more time-consuming activity for students than participation (or perhaps the term paper) alone, there is the possibility that the university subject pool could be depleted by students who opt for seemingly less demanding alternative options for satisfying course requirements. (Ethical issues pertaining to the research participation requirement are considered in chapter 7.)

There is some concern that the educational functions of debriefing are ignored by researchers, such as when they do not follow through with their promises to subjects that debriefing will occur once all of the data have been collected. Especially in academic settings, there is much gossiping about ongoing research investigations among student members of the university subject pool; as a result, researchers simply choose to withhold the debriefing of subjects until the study is completed. However, as Carroll et al. (1985) have observed, there are certain factors that work against this actually happening. Intact classes of students disperse at the end of the semester and the make up of the subject pool changes, or the study reveals nothing of significance and as a result the subjects do not receive feedback. It is Carroll and colleagues' belief that this is one area in which self-regulation on the part of researchers clearly has failed, and that some form of institutional regulation may be necessary.

Despite the important functions of debriefing and the emphasis placed on its use in current codes of ethics, there is not strong evidence that it is practiced by most laboratory researchers. For instance, Carlson (1971) noted that approximately only one-third of the investigations in her review of deception experiments indicated that debriefing procedures were carried out. Similarly, Menges (1973) observed that whereas 18.6 percent of the experiments he reviewed utilized deception procedures, only 10 percent of all of the studies mentioned the inclusion of debriefing. However, in their survey of studies published in the *Journal of Personality and Social Psychology* in 1979, Adair et al. (1985) found that 66 percent of all of the deception studies and 45 percent of the studies overall (including non-deceptive investigations) reported the use of debriefing, suggesting that the reporting of debriefing perhaps has increased over the years.

Because most professional ethics codes now require debriefing as a necessary procedure in studies that utilize deceptive manipulations (e.g. APA, 1992), it may be that these figures represent an under-estimation of the actual amount of debriefing that actually does occur (Adair et al., 1985). Adair and Lindsay (1983) found some support for this possibility in a mail survey of the authors of empirical articles in the 1979 volume of the *Journal of Personality and Social Psychology.* Whereas 81 percent of the authors claimed to have included some form of debriefing in their studies, the practice was reported in less than half of their published studies. However, as Adair et al. (1985) found, descriptions of the nature of the debriefing procedure in published research reports typically are uninformative and often are limited to vague phrases such as "all subjects were fully debriefed." Further, students of research methods are unlikely to find discussions about debriefing in their experimental psychology textbooks, despite the fact that the coverage of research ethics in general has increased over the years (Adair et al., 1983).

Evaluating debriefing: drawbacks and effectiveness

In chapter 4 we discuss the possibility that debriefing can create certain problems for researchers by influencing the naivete of subjects and thus posing threats to research validity. Other possible drawbacks to the use of debriefing have been identified, which may account for the reluctance of some researchers to include the procedure in their studies. Some concerns have been raised that debriefing may itself cause harm in situations where participants were selected because of some deficit, such as low self-esteem, or if their performance in a deception study revealed embarrassing behavior (Keith-Spiegel and Koocher, 1985). Such "inflicted insight," of course, might not be welcomed by research subjects (Baumrind, 1976). In such cases, it might be a wiser and more ethical course of action for researchers to provide general information about the study instead of a thorough debriefing (Sieber, 1982b). Experimenters might also convey to participants who respond in ways that are socially perceived as negative that their behavior was not atypical and that it is not uncommon for other subjects to respond in kind. It is important that researchers take care not to lie to participants when using this approach, as this would undermine the very purpose of debriefing and exacerbate the ethical problems.

There is always the possibility that subjects may not believe that researchers are telling them the truth during the debriefing process. Subjects may simply conjecture that the debriefing is another part of the experiment that involves new deceptions or rather is a therapeutic attempt to relieve any stress, embarrassment, or loss of self-esteem that they have experienced (Carroll et al., 1985; Schuler, 1982). Imagine what might be going on in the mind of a participant who has just left an experiment in which he or she received false feedback about performance on an intelligence test. This participant might think, "I know I was lied to during the experiment, but I'm not sure when: was the poor score I received on the test really a fake, as the researcher confessed at the end of the study, or did the researcher simply make up a story about the test before I left to make me feel better about having received such a horrible score?" The very fact that the subject has these concerns after participation would suggest that the researcher's debriefing was not completely effective.

As to the effectiveness of debriefing in eliminating negative consequences, the research is mixed, with some studies revealing that it often does eliminate experimental effects and others suggesting that sometimes it does not (e.g. Holmes, 1976; Ross et al., 1975; Walster et al., 1967). In one study, Walster et al. (1967) found that a thorough debriefing procedure was not enough to reverse the experimental impairment of self-esteem in subjects prone to low self-evaluations. Ring et al.'s (1970) modified replication of Milgram's obedience experiment demonstrated that certain modes of debriefing apparently are more effective than others. Three post-experimental procedures were compared, using only obedient participants: an accurate discussion of the procedure and a rationale for obedience, an accurate discussion of the procedure and a justification for defiance, and a polite "thank you" with no discussion of the true nature of the procedure. The results revealed that the debriefing which provided subjects with justification for their (obedient) behavior lowered tension more than the debriefing that did not. Overall, careful debriefing of the participants substantially reduced their negative feelings about the use of deception in the study.

Some researchers have maintained that because question marks remain regarding the effectiveness of debriefing in completely removing the negative consequences of experiments, it cannot be used to justify every study. Debriefing should not be overrelied on to eliminate negative effects and it is inappropriate to condone ethically problematic manipulations simply by reference to debriefing (Diener and Crandall, 1978; Schuler, 1982).

Whether or not the goals of debriefing are achieved no doubt depends on the seriousness and care taken by the researcher in conducting the procedure. Investigators would be wise to devote as much time to planning their debriefing procedure as they do in designing other aspects of their research, such as elaborate deceptive manipulations. Although the specific procedure used to debrief participants will depend on the nature of the study itself, some of the elements that can provide a starting point for a debriefing protocol are presented in Box 3.5 (see also Mills, 1976).

Removing negative consequences

Researchers are ethically obligated to remove or ameliorate whatever negative consequences arise as a result of research-related participation, including those that are unforeseen and unintended. At the very least, some attempt should be made to assess whether subjects are experiencing negative effects from having participated in a study and, in cases where potential harm is a major concern, subjects should be

---

Box 3.5

Sample debriefing procedure

Although the debriefing procedure typically involves an oral discussion with one's research participants, it is a good idea for the researcher to plan what will be said to them and what areas will be covered. At the least, a thorough debriefing should present, in non-technical language, the purpose of the experiment, various aspects of the procedure that would be helpful to the participants in understanding their role in the experiment (such the nature of the treatment conditions and the reasons for any deceptions that were included), and the expected results of the investigation. Subjects also should be informed about how they can contact the investigator if they have any questions or concerns about their participation after leaving the laboratory.

A useful exercise for providing a starting point in developing a more complete debriefing protocol is to finish the following paragraph lead ins.

"The purpose of my experiment is to. ... "

"The reasons why I misled you or withheld certain information about my study prior to your participation are. ... "

"The treatment conditions are. ... "

"I expect to find. ... "

"Do you have any questions about the experiment?

"For future contacts, I can be reached by. ... "

"Thank you for participating."

Source: Adapted from Kiess, H.O. (undated) *Experimental psychology learning guide*, unpublished manuscript, Framingham State College, Framingham, MA. Reprinted by permission.

followed up at some later point in time. Post-experimental checks and long-term follow ups are essential for ensuring that negative consequences truly have been removed. Researchers must be particularly sensitive to the possibility that participants from vulnerable populations might experience long-term effects.

Another action that can be taken by researchers to assist participants in overcoming the negative effects of deception or other research procedures is to make arrangements for referrals to individuals who are qualified to assist disturbed participants. Monetary compensation mechanisms also have been proposed (Silverstein, 1974; President's Commission for the Study of Ethical Problems in Medicine and Biomedical and Behavioral Research, 1982) and Keith-Spiegel and Koocher (1985) have suggested that "malresearch insurance" may eventually become a reality. Researchers in applied areas also bear an ethical responsibility to untreated control group subjects who have been deprived of a potentially beneficial treatment and who experience this deprivation as a loss. In studies of known positive treatments, these subjects could be offered the treatment once the research is completed, as long as they do not suffer substantially in the meantime as a result of withholding a required treatment. Many researchers no doubt will feel an obligation to aid untreated controls who have assisted them in their research upon completion of a study. (For a more complete discussion of untreated control groups, see Kimmel, 1988a.)

A related consideration regarding the responsibility of researchers to take steps to assist participants has to do with the possibility that some subjects may be found to have been in a state of preexisting distress prior to their involvement in the research. That is, during the course of an investigation, a researcher might identify potentially depressed or suicidal subjects whose distress had nothing to do with methods used during the study. It is widely acknowledged that researchers have a responsibility to intervene when serious distress is discovered in such cases (e.g. APA, 1992; Burbach et al., 1986; Stanton and New, 1988), by initiating contact with the subject and a significant other or through some other form of intervention. Despite the researcher's duty to promote the welfare of distressed subjects, post-experimental intervention might be viewed by subjects as a violation of their privacy and autonomy rights. One way of allaying this reaction, however, would be to inform subjects before they actually participate in a study of the possibility of follow-up contact by the experimenter. This suggestion, however, may have certain methodological consequences that also must be weighed by the researcher. For example, Stanton et al. (1991) found that when subjects were presented with the possibility of intrusive follow up (i.e. experimenter contact with the participant and a significant other) on an informed consent form, they were less likely to report depressive symptoms than subjects expecting a less intrusive follow up. In light of this finding, Stanton et al. recommend that one way to satisfy both ethical and methodological concerns would be merely to provide all research participants with a list of treatment sources once they have completed their participation in a study.

Summary

When the methodological requirements of a laboratory investigation lead the researcher to conclude that the only way a study can be carried out is by employing deceptive research tactics, the decision to deceive necessarily results in additional ethical responsibilities for the researcher. Deception was once accepted without comment, but it is now accompanied by elaborate justifications and extensive debriefing procedures are required by current ethical codes. Most behavioral scientists, when caught up in situations involving conflicting values, such as whether or not to use deception, are willing to weigh and measure their sins, judging some to be larger than others. Complicating the decision-making process in the context of laboratory research are the various methodological issues related to the use of deception, which we consider next in chapter 4.

4

Methodological Issues in the Use of Deception

Psychologists always lie!
Anonymous student cited by M. T. Orne, On the social
psychology of the psychological experiment

This chapter represents a continuation of our discussion of the issues that emerge in the conduct of laboratory research, with a closer look at the methodological implications of deception. Several methodological issues relating to the use of deception have emerged along with the gradual increase of studies that have employed deception over the years. As a result, the issues themselves have become the subject of countless research studies. In short, an analysis of the methodological issues requires some insight into the subject's point of view and an understanding of how research participants are affected by deceptive practices. Does deception make them more cynical and suspicious of researchers, as the above quotation implies? Do they develop negative feelings about behavioral science research in general? What are the effects of suspiciousness on research results? Are there some alternatives to the use of deception that are both methodologically and ethically sound? These are the questions that are considered in detail below.

Evaluating Deception From a Methodological Perspective

The use of deception procedures in behavioral research depends upon certain important methodological assumptions, foremost of which are that: (1) the level of naivete among research subjects is high; (2) the experimental procedure does not produce cues that are interpretable by participants as indicating that deception is taking place; and (3) subjects' suspiciousness of deception does not alter the experimental effect (Golding and Lichtenstein, 1970). In addition to the ethical implications of using deception for research purposes, concern has focused on these methodological assumptions and the potential shortcomings and limitations of deception as a laboratory research tool. Deception has thus been evaluated in terms of the degree of subject naivete concerning its use, the consequences of suspicion on experimental validity, and the effects of deception on subsequent experimental performance. Some critics of deception in fact have argued that the scientific costs of deception are considerable. For example, Baumrind (1985) suggested that deception can lead to the depletion of the pool of naive subjects for future research, a reduction in community support for the research enterprise, and an undermining of researchers' commitment to truth. We begin our consideration of the potential effects of deception from a methodological point of view in the following section.

Assessing the effectiveness of deception

The use of deception procedures in laboratory research is largely predicated on the principle that an experimental effect is interpretable only with a naive subject population. Accordingly, most researchers believe that the detection of any effects solely attributable to subject awareness about the study's true nature and purpose should be treated as artifacts of the research setting rather than as sources of meaningful data (Golding and Lichtenstein, 1970; Forward et al., 1976).

An obvious criterion for the effectiveness of deception is the extent to which the deception duplicates the essential conditions it is intended to portray (Orne and Holland, 1968). Two important dimensions for evaluating research methods are relevant in this regard: experimental realism and mundane realism. "Experimental realism" refers to the extent to which the experimental situation is realistic to subjects and taken seriously by them. "Mundane realism" refers to the degree of similarity between the events surrounding the research participant in the experimental setting and those that occur in the natural environment. The extent of mundane realism typically varies according to the research setting. An investigation conducted in a field setting in which participants are not aware that their behavior is being investigated is said to have a very high degree of mundane realism (Geller, 1982). At the other extreme are studies conducted in the artifical setting of the laboratory in which subjects are placed in situations that bear little resemblance to their lives outside the research context.

A deception procedure typically will have high experimental realism if it is successfully implemented and the psychological impact of the experimental manipulation on the research subjects is high (Geller, 1982; Rosenthal and Rosnow, 1991). Some critics of Milgram's studies on obedience have argued that the experiments lacked experimental realism because several subjects may have been aware of the deception (e.g. Patten, 1977). Evidence supporting this claim was provided by Orne and Holland (1968) in their description of an unpublished replication of Milgram's research. In post-experimental interviews, participants in the replication claimed that they did not really believe the deception. Further, prior to their participating in the study, two additional groups of subjects were led to believe that things were not what they seemed in the research. In order to raise suspicions, one group was informed that there would be something "fishy" about the experiment but not to "let on" and the other group was prebriefed that the administered level of shock was really only one-tenth the intensity indicated on the shock machine.

Despite these manipulations, the level of obedience for the two groups was not significantly different from Milgram's original study and a "blind" experimenter was unable to ascertain the experimental condition to which subjects belonged. Thus, although Milgram's subjects reported in post-experimental interviews and questionnaires that they believed that victims were receiving painful shocks, they may have been aware of the deception but still behaved as if they were naive. This sort of behavior is consistent with the good subject role described earlier. This role requires that participants pretend to be naive even when they are aware that the researcher is playing tricks on them (Orne and Holland, 1968).

One difficulty in assessing the degree to which an investigation possesses experimental realism has to do with a "pact of ignorance" that sometimes develops between the experimenter and the subject (Orne, 1959). That is, subjects may prefer to conceal the fact that they saw through a deception because it might compromise the value of their participation; in turn, the experimenter may not forcefully attempt to discover compromising information that would invalidate subjects' data and delay completion of the investigation. According to Patten (1977), a philosopher with serious reservations about the effectiveness of Milgram's deception, Milgram's disobedient subjects may have been reluctant to say they saw through the hoax because it would have meant giving up the hero status gained by standing up to the malevolent authority and refusing to continue with the experiment.

It is unclear to what extent these methodological points accurately apply to the behavior exhibited by Milgram's subjects. In responding to his critics, Milgram claimed that the obvious signs of tension and stress among a majority of his subjects during the procedure (as exemplified in the opening passage of chapter 3) clearly suggest that his method was quite real for the participants. But can researchers better detect whether an experimental manipulation has actually deceived the participants in a study? One proposal is to use quasi-controls, subjects who are asked to step out of their traditional roles and to serve as "coinvestigators" in the research process (Orne, 1969).

In essence, quasi-control subjects are asked to reflect "clinically" on what is happening during a study and to describe how they think they might be affected by the research context and the experimental events. The intent is to carefully elicit from these subjects their perceptions and beliefs about the research situation, without raising suspicions or prompting their responses. This technique can involve having experimental subjects function as their own quasi-controls. In this case the experimenter would engage the subjects in an exhaustive inquiry following their participation in the experiment or a pilot study. Another approach, which is referred to as preinquiry, employs quasi-control subjects who are asked to imagine that they are real subjects in the experiment. During the preinquiry, subjects are provided with all of the significant details of the experiment and are asked to predict how they might behave in such a study. A high degree of similarity between their responses and those obtained from real subjects implies that the experimental results could have been caused by subjects' guesses about how they should have responded (Strohmetz and Rosnow, 1995).

There is not much evidence that researchers have made use of quasi-control techniques to assess the adequacy of their deceptive procedures; however, much useful feedback can be obtained from subjects during a judiciously carried out debriefing session. Some of the characteristics of an effective debriefing were described in chapter 3. Without making some attempt to ascertain subjects' reactions to their participation in a study, it is appropriate to ask the question that has been posed by some critics of deception: who is actually being deceived – the subject or the researcher?

Subject suspiciousness

"Psychologists always lie!" This remark, purportedly made to Martin Orne by one of his student subjects, may have been repeated over the years by countless other individuals who have experienced research deceptions firsthand or else have learned about such techniques in university courses or from other sources, including friends and the mass media. Orne (1962) aptly observed that such a comment might have some support in reality and reflects the concern that subjects' growing sophistication about deceptive practices may cause them to behave unnaturally in behavioral science investigations. As early as the 1960s researchers (e.g. Jourard, 1968; McGuire, 1969; Orne and Holland, 1968) suggested that subjects participating in psychology experiments have considerable awareness of the implicit rules that govern the situation, and have learned to distrust the experimenter because they know that the true purpose of the experiment may be disguised. Of course, if the prevalence of deception over the years has decreased naivete among prospective subjects, this would diminish the methodological value of using deception in the first place (Kelman, 1967). It also implies that even honest investigators might not be trusted by subjects who view apparent "openness" with suspicion (Orne, 1962). An anecdote related by MacCoun and Kerr (1987) is enlightening in this regard. During a study in which subjects were asked to serve as members of a mock jury, one of the participants suffered a seizure. Fortunately for this individual, another subject who was trained in emergency procedures offered quick assistance to the victim. What was unsettling, however, was the number of subjects who reported that they initially doubted that the emergency was real. Reflecting a growing suspiciousness among research participants, these subjects misinterpreted the seizure as a staged incident contrived by the experimenter and victim.

There has not been a great deal of research on the extent of subjects' suspiciousness in the research setting and the existing studies present something of a mixed bag. In an early review of deception experiments, Stricker (1967) found that appraisals of the effectiveness of deception were uncommon, and that those that were made tended to be superficial. Only 23.9 percent of the 88 studies utilizing deceptive procedures in 1964 included any attempt to gain some information about suspicion and awareness of deception in the subject samples. The amount of suspicion among subjects in the 21 studies that assessed it was small, ranging from zero to 23 percent, with an average percentage of only 3.7. However, in contrast to Stricker's reassuring report, estimates from experiments investigating conformity to group pressure revealed higher percentages of distrustful subjects. In one study, Stricker et al. (1967) assessed the extent of subjects' suspicions about deceptions in two simulated-group conformity situations and on a questionnaire bearing fictitious norms. As a measure of suspicion of deception, subjects completed an open-ended questionnaire following the experimental procedure in order to ascertain their perceptions of the study's purpose and the behavior of other subjects. The proportion of subjects classified as suspicious about the purpose of each of the two conformity procedures and the methods employed ranged from 9 percent for subjects who were suspicious about the simulated group to 61 percent for subjects who were suspicious about the questionnaire. These findings were surprisingly high given the fact that Stricker et al.'s participants were young (high school) students who were expected to be naive about psychological research procedures.

Other investigators have reported high rates of suspiciousness among subjects in conformity studies (as indicated by subjects' comments during post-experimental inquiries), ranging from 50 percent to nearly 90 percent (Gallo et al., 1973; Glinski et al., 1970; Willis and Willis, 1970). Additionally, in an unpublished study, Stang found a positive correlation of 0.76 between the amount of reported suspicion in published research and the year, indicating a steady rise in suspicion over a two-decade period (cf. Stang, 1976). These findings suggest that as the common deception paradigms used to investigate certain topical areas become progressively known among subject populations, research may have to be varied or reinterpreted as dealing with something other than naive behavior. Kenneth Gergen (1973b) has theorized that the dissemination of psychological knowledge can alter previous patterns of behavior which then become difficult to test in an uncontaminated way. Just as powerful drugs to combat virulent diseases can lead to hardier strains of the diseases which then become immune to the drugs, people may become "immune" to deception as higher rates of suspicion are found among new subject populations (Rosnow, 1978). Thus, researchers might be inclined to increase the number and degree of deceptions in their studies as subject sophistication increases. On the other hand, subject suspiciousness may motivate investigators to orient their attention to new subject populations and to research settings outside the laboratory. Eventually, however, suspicions can be expected to spread throughout the new subject populations as well (Diener and Crandall, 1978).

Sources of suspiciousness

Several potential sources of subject suspiciousness have been identified. Some subjects come to the laboratory setting unaware that deception may take place or else have only a vague knowledge that it is a possibility. In such cases, so-called demand characteristics of the experimental situation can cause them to become suspicious. Demand characteristics comprise the mixture of various hints and cues that govern a subject's perceptions of his or her role, the experimenter's hypothesis, and other aspects of the study (Orne, 1962). For example, certain comments in the instructions to subjects or cues in the laboratory setting itself might suggest that something is going on that is inconsistent with the version given by the experimenter. A subject who is informed that he or she is participating in an investigation of verbal learning naturally may become suspicious after happening to notice on the top of a paper held by the researcher a heading which reads "Conformity Experiment."

Subjects may come to the research situation already harboring certain suspicions as a result of demand characteristics encountered in other settings. Cues in the original appeals used to recruit participants, information about behavioral science studies conveyed in college courses, and campus scuttlebutt (i.e. rumors) about ongoing studies represent common sources of demand characteristics outside the laboratory context (see Box 4.1). The author can recall how, when he was a student in an introductory psychology course at the University of Maryland, he had to select from among a large number of experiments being carried out on campus in order to fulfill a course requirement for research participation. There was much gossiping about the various studies among his fellow classmates, as they attempted to find out which would be the most interesting (and least time consuming) research experiences. This sort of information transmission within student subject pools apparently is commonplace; in fact, one researcher who has greatly advanced our understanding of the psychology of rumor recently admitted that his interest in the topic originally was piqued by such campus scuttlebutt (Rosnow, 1991).

One established principle that has emerged from the research on rumors clearly is applicable here – that uncertainty and anxiety provide optimal conditions for the transmission and reception of rumors (Kimmel and Keefer, 1991; Rosnow et al., 1988). These two psychological factors are likely to be high among potential research participants, thus motivating them to gather as much information as possible prior to their participation in a study. Such information often might be in the form of suspicion-arousing rumors.

There is evidence that research participants who have been debriefed sometimes communicate the true purpose and other details of the studies in which they have participated to future subjects, a tendency referred to as "leakage" (Diener et al., 1972; Farrow et al.,

---

Box 4.1

Demand characteristics and subject performance

An early study by Orne (1959) effectively demonstrated how subjects' behavior in an experiment can be affected by earlier knowledge obtained outside the laboratory. Orne's study involved the presentation of false information to college students about a presumably novel characteristic of hypnosis. He first informed a large college class during a lecture on hypnosis that upon entering a "trance," a hypnotized person will manifest "catalepsy of the dominant hand" – that is, the right hand of a right-handed individual will take on a waxen flexibility or rigidity; for left-handed persons, this effect will be experienced in the left hand. This characteristic, which was fully concocted by Orne, was demonstrated to the class using volunteer students. As a control condition, Orne gave the same lecture and demonstration to another class section, but this time did not include any mention or display of catalepsy of the dominant hand.

Some weeks later, students from both lecture sections were invited to participate as subjects in a study of hypnosis. When they were hypnotized during the study, Orne found that catalepsy was present for nearly all of the students who had learned about it in the previous lecture. By contrast, none of the control subjects exhibited the fictitious characteristic. Apparently, subjects acted on the demand characteristics received prior to their participation in such a way as to confirm what they believed to be the experimenter's hypothesis or to provide the "correct" response to hypnosis.

1975). Wuebben (1967) provided some indication of the extent of leakage by reporting that 64 percent of 113 subjects who had agreed to secrecy following a detailed debriefing session revealed the nature of the research deception to other potential subjects within one week.

Other investigators have found that large proportions of subjects leak crucial information into the subject pool, to an extent that could substantially affect the experimental outcome (Glinski et al., 1970; Lichtenstein, 1970). Apparently, however, information is much less likely to be divulged when research participants are cautioned not to talk about an experiment to others or are asked to make an oral or written agreement with the researcher (e.g. Aronson, 1966; Walsh and Stillman, 1974). Aronson (1966) provided nine subjects with an unusually thorough debriefing following their participation in a deception study, including a vivid description of the consequences of testing sophisticated subjects. None of the subjects revealed the true nature of the experiment when approached by a confederate who attempted to obtain information about the study, suggesting that the nature of the debriefing can offset the possibility of leakage. In their surveys of the *Journal of Personality and Social Psychology*, however, Adair et al. (1985) found that the practice of establishing secrecy agreements with debriefed subjects was infrequent, having been reported in only 13 percent of the deception studies.

Another interesting finding that emerged from Adair et al.'s review of deception research that bears mentioning has to do with the number of types of deception included within a study. They observed that the percentage of studies reporting suspicious subjects and the number of suspicious subjects identified tended to increase systematically along with the number of deceptions employed. This seems to suggest that the more complex the deceptive procedures utilized in an investigation, the less successful they will be in achieving their purpose. However, Adair et al. added that even for studies involving highly complex deceptions, the average percentage of subjects identified as suspicious rarely rose much above 6 percent. The method used to assess subject suspiciousness may have something to do with this low percentage: researchers have found that when they assessed subject suspiciousness by asking their subjects up to three post-experimental questions, only about 5 percent were classified as suspicious (Page, 1973; Spielberger, 1962). This percentage increased to about 40 percent when more extended questioning was utilized.

One additional potential source of subject suspiciousness that bears mentioning is prior experience in deceptive research. That is, when subjects who participate in multiple investigations are deceived in an early study, they may become less trusting in subsequent studies. Prior experience may operate in such a way as to influence the subject's anticipation of what is going to happen in another study and perhaps influence his or her performance as well (see below). Research on the effects of prior research experience has been somewhat contradictory, although there is some evidence that suspicions aroused in one study may not generalize to another unless the two have something in common (Brock and Becker, 1966; Cook et al., 1970; Fillenbaum, 1966; Stricker et al., 1967).

Effects of subjects' suspicions

Perhaps the issue that has raised the greatest concern among researchers regarding the problem of subject suspiciousness is its potential for influencing research performance. Subjects' mistrust of the experimental procedure may alter their reactions enough to cause misleading results or invalidation of the experiment altogether.

The effects of suspicion on behavior in an experiment are apt to be varied, depending on the subject's perceptions of the situation and motivations for acting on the suspicions. For example, subjects who are aware that they are being deceived by a researcher may resent that fact and thus try to respond in a manner that is counter-compliant with what they believe are the experimenter's expectations. Research participants who approach an investigation with an uncooperative attitude are sometimes referred to as "negativistic subjects" (Masling, 1966). By contrast, subjects who are motivated to cooperate with the researcher might actively seek out additional demand characteristics in order to more successfully play the good subject role (cf. Strohmetz and Rosnow, 1995). Other motivations may be operative in suspicious subjects' minds as they approach the research situation. Awareness that the research procedure is not all that it appears to be may increase subjects' fears about being poorly evaluated; in turn, this can lead to a desire to be more alert and observant in the research situation than under conditions of greater trust.

Research results on the effects of subject distrust (e.g. Cook et al., 1970; Epstein et al., 1973; Stricker et al., 1967) are somewhat inconsistent and have led some behavioral scientists to conclude that in general there are not major differences between the data of suspicious and reportedly naive subjects (Schuler, 1982). Some studies have shown that the results produced by suspicious or pre-informed subjects at times differ markedly from those of naive subjects (Allen, 1966; Golding and Lichtenstein, 1970; Levy, 1967; Newberry, 1973). In one conformity study, subjects who were aware of the deception conformed less than those who were deceived, but more than participants in a no deception control group (Allen, 1966). Newberry (1973) reported the results of two experiments in which subjects who had received prior information from a confederate consistently used it, whether intentionally or not, to improve their performance on various problems-solving tasks. Golding and Lichtenstein (1970) used a bogus feedback procedure and found that subjects who admitted awareness of the deception tended not to show the hypothesized effect, whereas subjects who were either not aware or who did not admit awareness showed a substantial effect.

In addition to the possibility that prior experience in deception studies can serve to increase subject suspicions, there also is some evidence that it can affect performance in subsequent experiments. Silverman et al. (1970) reported that subjects who had been previously deceived in a preliminary experiment differed significantly from naive controls on all measures in a later experiment. The direction of the obtained differences seemed to indicate that prior deception increases the tendency for favorable self-presentation and decreases compliance with demand characteristics. In other words, evaluation apprehension (i.e. "looking good") is more likely to emerge as the predominant motivation of previously deceived subjects, as opposed to the desire to cooperate with the researcher in the cause of science (i.e. "doing good"). Silverman et al.'s results are consistent with other studies that have investigated these two subject artifacts (e.g. Rosnow et al., 1973: Sigall et al., 1970).

Although it appears that previous research experience can play a role in later studies, some researchers have been unable to obtain performance effects from subjects with foreknowledge about an investigation (e.g. Brock and Becker, 1966; Fillenbaum, 1966). Several methodological explanations have been posited for this lack of predicted effects. There is the possibility that if a prior deception is seen as mild and legitimate, subjects will accept it and proceed faithfully, adhering to the instructions in later studies despite their suspicions. Some studies on the effects of suspiciousness have also relied on two immediately consecutive experiments to manipulate degree of experimental naivete. That is, subjects are either debriefed or not debriefed in a first experiment and their performance is then measured in a second experiment. The problem with this sort of procedure is that it does not ensure a great difference in naivete between the comparison groups (Cook et al., 1970).

Diener and Crandall (1978) have argued that there is a strong probability that many suspicious subjects in studies which have failed to find performance effects simply have not been identified. They also suggest that when subjects who have not been taken in by the deception act the same as subjects who have, problems arise when one attempts to interpret the findings – the results possibly may have been caused by an extraneous factor that was not under study.

Another serious problem with suspicion pertains to the possibility that it can operate differentially across different conditions of a study (Diener and Crandall, 1978). Because different treatments are likely to cause different levels of suspicion, apparent treatment effects in some studies may have been caused by the differential suspicion rates. This point raises certain problems for the researcher in terms of the most appropriate means for dealing with data obtained from suspicious subjects (assuming such subjects are identified during a post-experimental interview). The researcher may choose to discard subjects who did not see the experiment as it was intended or simply decide to analyze the data obtained from suspicious and unaware subjects separately (Adair et al., 1985). If the rate of suspicion is not distributed randomly across treatment conditions, omitting the data obtained from suspicious subjects can serve to increase the apparent treatment effects. On the other hand, the decision to eliminate suspicious subjects from the study could lead to the loss of truly random assignment of subjects to conditions. This would make it difficult for the researcher to rule out the possibility that treatment effects were caused by selection biases or, for that matter, undetected suspicion that exists differentially across conditions (Diener and Crandall, 1978). Although there are no easy solutions to these problems, at the very least researchers should attempt to assess the effectiveness of each deception used in an investigation and the extent to which subjects idiosyncratically perceived the experiment and its rationale (Adair et al., 1985).

Before leaving the topic of subject suspicions, an intriguing theoretical issue should be mentioned. This has to do with the possibility that disclosure of information to subjects may turn out to be more stressful than non-disclosure in certain instances (Rosenthal and Rosnow, 1991). Resnick and Schwartz (1973) similarly alluded to a possible drawback of informed consent when they posed the question: "Does being ethical trigger paranoid ideation in otherwise nonsuspicious subjects?" It is possible in some cases that informing subjects will cause more stress and raise more suspicions than leaving subjects in the dark about the nature of the study and its purpose.

Effects of deception

In the preceding section we briefly considered the effects of prior deception on research participants' suspicions and performance in further studies. But what about more general effects? Are subjects harmed by research deceptions? Do deceived subjects become resentful about having been fooled by a researcher? Does deception have an impact on subjects' perceptions about psychology or attitudes about science in general?

Much of the anecdotal and empirical evidence on these and related questions has been encouraging – the negative effects of deception appear to be minimal. For example, Milgram (1964) defended his obedience experiments in part by citing thank you letters he received from some of his subjects and data from a follow-up questionnaire revealing that 84 percent of the subjects were glad they had participated; fewer than 1 percent indicated that they regretted having participated; and 15 percent were neutral or ambivalent. Moreover, thorough interviews were conducted one year after the study and there was no evidence of apparent permanent damage. Short-term effects experienced by some of the subjects were readily dispelled and any remaining discomfort was eliminated through the mailing of an explanatory letter and some preliminary findings to subjects (Errera, 1972).

The results of a replication of Milgram's experiment by Ring et al. (1970) produced a similar picture regarding the lack of negative after-effects resulting from deception. In a follow-up interview, most subjects described the experiment as having been a positive experience for them and claimed that they did not resent the deception. Only 4 percent of the subjects mentioned that they had regretted participating in the experiment. However, some did state that they were experiencing difficulty in trusting adult authorities, and others reported persistent disappointment with themselves or expressed self-doubts.

Subjects' reactions following participation in a bystander intervention experiment also have been assessed (Clark and Word, 1974). A majority of the participants in the experiment viewed the research as valuable (95 percent) and considered the deception unavoidable (94 percent). When surveyed several months later, 92 percent of the subjects claimed that they did not believe that their rights had been violated or that they would rather not have participated in the study.

Based on a review of studies that have assessed subjects' reactions to deception experiments (e.g. Pihl et al. 1981; Smith, 1981; Smith and Richardson, 1983), Christensen (1988) concluded that subjects who have participated in deception experiments versus non-deception experiments in psychology tend to report that they did not mind being deceived, enjoyed the experience more, received more educational benefit from it, and did not perceive that their privacy had been invaded. These results may argue for a "situation ethics" approach to solving ethical dilemmas, whereby participants are asked whether they are concerned that certain ethically questionable practices exist after being told the reasons for utilizing them.

The data obtained from the deception after-effects studies have not been readily accepted by some critics of deception. For example, Baumrind (1964) has argued that the positive reactions from those who have participated in deception studies may reflect attempts to resolve cognitive dissonance resulting from the deception experience. That is, participation in a deceptive experiment like Milgram's, for example, could have generated negative self-perceptions in subjects as well as unpleasant feelings associated with having been used and tricked by the researcher. In order to cope with these negative outcomes, subjects may have reinterpreted the experiment as being valuable, an interesting learning experience, and so on. Baumrind (1985) also has discounted the after-effects evidence as being a product of experimental demands and inadequately conducted post-experimental inquiries. She maintains that well-trained clinical interviewers are required to uncover true after-effects among research participants. In her view, self-report questionnaires typically are added to a study as an afterthought and the instruments themselves tend to be poorly constructed.

Despite these criticisms, the results of surveys intended to gauge reactions to deception consistently have shown that most individuals in the general population apparently do not have serious objections to its use for research purposes (Collins et al., 1979; Epstein, et al., 1973; Rugg, 1975; Sullivan and Deiker, 1973). Rugg (1975) found this to be the case for several groups of individuals, including college professors and lawyers. Epstein et al., (1973) reported that college students – the very persons who comprise the typical research population and who are likely to experience harm from its use – were generally accepting of deception. A majority of those questioned expected to be deceived as participants in psychological studies.

Although they viewed this as personally undesirable, they also believed that deception in research is permissible and appropriate. Similarly, Sullivan and Deiker (1973) found that psychologists tend to have more serious reservations about the use of deception than do college students. When presented with hypothetical experiments that differed in the amount of stress, physical pain, or threat to self-esteem inflicted on subjects, it was the psychologists who turned out to be far more negative toward the practices and the propriety of using deception in each instance than were the students.

Another question regarding the use of deception has to do with the belief that such practices reflect poorly on the discipline and cast suspicion on the motives of behavioral scientists. For some time, critics of deception have warned that its frequent and continued use would have an adverse effect on subjects' trust of psychologists and would lead them to adopt a negative attitude toward behavioral research in general (Kelman, 1967; MacCoun and Kerr, 1987). The results of a recent study by Sharpe et al. (1992) provide little support for these fears. Questionnaire responses obtained from students at the beginning of the 1970 academic year were compared with those of students surveyed at a comparable time in 1989. Data were obtained from both subject samples through the use of the Psychological Research Survey (PRS), a scale that measures attitudes toward psychological research. Scores on the PRS for the two subject samples were found to be similar, suggesting that there has not been a predicted increase in negative attitudes toward psychological research in the subject population as a result of the continued use of deception during the past 20 years.

Sharpe et al. also administered the PRS to a third sample of students in 1990, shortly after the students had participated in a number of experiments during the academic year. Although these subjects were accepting of arguments justifying the use of deception, their attitude toward psychological research was somewhat more negative than subjects in the other samples. According to the researchers, the finding that participants in experiments did not possess a more positive attitude toward research suggests that participation does not facilitate appreciation for the discipline, and thus contradicts a common justification for requiring student research participation (Lindsay and Holden, 1987) (see chapter 7). Sharpe et al. reasoned that previous reports of subjects' positive evaluations of deception experiments may be due to a contrast effect – the unfolding deception scenario may represent a more engaging experience for subjects, a welcome relief from the more typically mundane and less interesting studies conducted in many university settings.

Alternatives to Deception

Various alternatives to the use of deception techniques have been proposed, foremost of which are simulations and role playing. Other recommended techniques, such as quasi-control groups and fore-warning, are not "alternatives" in the strict sense of the word. Rather, they are used in conjunction with deception in order to minimize its potential for harmful effects. Each of these approaches are described below.

Quasi-controls

Previously, we considered the usefulness of quasi-controls as a technique for detecting demand characteristics, thereby providing the researcher with some indication of the effectiveness of a deception manipulation. Quasi-control subjects can also be used in some studies to minimize the level of deception. According to Rosenthal and Rosnow (1991), this can be accomplished by using the following procedure. The researcher would first employ a quasi-control group to check for demand characteristics. If none were uncovered, rather than immediately proceeding with the actual experiment, the researcher would develop a less deceptive manipulation and have the quasi-control subjects once again reflect on the study. If they remained unaware of the demands of the study the experimenter could then use this lower level of deception to carry out the intended investigation.

Although quasi-control techniques are time consuming and often more costly to carry out, they can be useful to the researcher who is concerned about balancing ethics and artifacts; that is, deception can be minimized without risking a corresponding increase in demand cues (Suls and Rosnow, 1981).

Forewarning

In laboratory studies where deception is deemed necessary, the informed consent process can include a forewarning to prospective participants that some information will be withheld or that certain manipulations will be carried out that cannot be divulged until later. In essence, forewarning is a procedure whereby subjects are informed before they participate that deception may be involved in the investigation. So informed, the researcher then can ask the subjects if they are willing to participate in research that involves deception and include in the sample only those who answer affirmatively (Geller, 1982).

As a form of limited informed consent, forewarning is viewed as ethically preferable to not obtaining consent at all because subjects essentially agree to be deceived and the researcher will not have directly misled them (Diener and Crandall, 1978). There is some concern, however, that forewarning will sensitize participants to demand characteristics (Geller, 1982). Subjects may be more inclined to engage in problem-solving behavior in order to determine the nature of the deception than if they had not been forewarned. Despite this possibility, evidence seems to suggest that forewarning subjects does not appear to influence the accuracy of data (Holmes and Bennett, 1974; Horowitz and Rothschild, 1970).

Simulations

A simulation is an experimental strategy that is sometimes used as an alternative to deception in laboratory settings. This technique involves the creation of conditions that mimic the natural environment in some definite way, and participants are asked to pretend or act as if the mock situation were real. Simulations require the scaling down of the natural environment to a size that is conducive to analysis in the laboratory but which preserves the key elements thought to underlie the dynamics of the the real-world phenomenon under study (Rosenthal and Rosnow, 1991).

Geller (1982) has distinguished between three basic types of simulations: game simulations, field simulations, and role playing. *Game simulations* consist of participants who take on roles in staged situations lasting until a desired outcome has been attained or a specified length of time has passed. Such simulations, which often involve participants working in teams, are popular in research in the areas of business, international relations (including conflict resolution and negotiation), community planning, and education. International simulations or war games, for example, can be used to predict future world events from present national characteristics or conditions. Another popular version of a game simulation is the mock jury study. This approach is intended to closely simulate actual jury procedures and sometimes involves actual cases and real jury candidates and judges. The minimal requirements of such studies are that subjects be brought together in groups which read, hear, or see the proceedings of a court case and deliberate as a jury (Bermant et al., 1974; Lamberth and Kimmel, 1981). Each subject indicates his or her opinion concerning the guilt or culpability of the defendant and often is asked to assess punishment or award damages.

Field simulations are distinguished by their setting. Such simulations are characterized by highly realistic staged settings that encourage participants to believe they are participating in natural situations. A well-known example of a field simulation in psychology is the Stanford prison study (Haney et al., 1973), a mock prison simulation which itself was the subject of much ethical criticism (see Box 4.2). That field simulations can be highly realistic and capable of eliciting spontaneous behavior from participants was effectively demonstrated in an investigation known as the Grindstone experiment (Olson and Christiansen, 1966). Sponsored by the Canadian Friends Service Committee, the study involved the defense of an island under simulated attack. The participants who were to defend the island consisted of Quakers who were committed to pacifist ethics. Although the objective was for the participants to repel the simulated attack through non-violent methods, the study had to be terminated after 31 hours due to the breakdown of behavioral norms of non-violence, which resulted in 13 mock "deaths" (Mixon, 1971).

The *role-playing simulation* has become increasingly used in recent years. According to Geller (1982), role playing has the most developed potential as an alternative to deception and it can be adapted to a wide variety of laboratory settings. Because there has been much scientific debate about the merits of role playing, details pertaining to this approach are presented separately in the following section.

As the Stanford prison study illustrates, the simulation alternative to deception is not free of ethical problems. When effectively carried out, such studies can cause intense emotional involvement and distress in participants, risks that some say should not be condoned for research purposes (Baumrind, 1964; Savin, 1973). Additionally, simulations are not always free of intrinsic deceptions (Geller, 1982). This was true of the Stanford prison study in the sense that although the subjects had volunteered to role play, they were not told which role they would play (prisoner or guard) until the onset of the study. The participants also were unaware that they would be dramatically "arrested" by local police as part of the study they had volunteered for several weeks earlier.

Another concern about simulations is that their mundane realism is sometimes suspect (Rosenthal and Rosnow, 1991). This was

Box 4.2

The Stanford prison study

The Stanford prison study represents one of the most effective and controversial field simulations in the behavioral sciences. Conducted by social psychologist Philip Zimbardo and two of his graduate students (Craig Haney and W. Curtis Banks) at Stanford University, the simulation was intended as an intensive investigation of the behavioral and psychological effects of imprisonment on the guards who maintain and administer the prison as well as those who experience it as inmates (Haney et al., 1973; Zimbardo et al., 1973). The researchers had hoped to obtain a better understanding of the processes by which prisoners adapt to this novel and alienating environment and guards derive social power from their status in controlling and managing the lives of others.

The researchers decided to conduct their investigation in the context of a mock prison rather than an actual one for several reasons, including the lack of outsiders' access to the prison system, the impossibility of separating out to what extent an individual's behavior is a function of pre-existing personality characteristics or attributed to some aspect of the prison situation, and the obvious ethical implications of placing innocent people inside a real prison for the purpose of studying them. As a result, a mock prison (the Stanford County Prison) was constructed in the basement of the psychology department building on Stanford's campus. The mock prison was equipped with iron-barred cells, a solitary-confinement closet, and a recreation area.

The participants for the prison study consisted of 21 college-aged males who were selected from over 75 volunteers who had answered area newspaper ads promising US$15 a day for a two-week study of prison life (see Box 7.1). After being tested for psychological well-being – all applicants underwent extensive psychological testing and psychiatric interviews – 11 subjects were randomly designated to role play being guards and 10 to be prisoners. At the outset of the study, the prisoners were unexpectedly "arrested" at their homes and were booked, fingerprinted, and taken to the simulated prison by local police officers. Once inside the prison, they were stripped, searched, and given a loose-fitting smock to wear; they also were required to wear a stocking over their hair, were told to refer to themselves by identification number only, and wore a light chain and padlock around one ankle. These steps were taken in order to minimize each prisoner's uniqueness and prior identity and to heighten the prisoners' sense of powerlessness and emasculation. Guards were outfitted with khaki uniforms, including mirror sunglasses and nightsticks, and were told to do anything necessary to protect the security of the prison. All of these details were intended to give as much mundane realism to the situation as possible.

Although scheduled to run for a two-week period, the study had to be terminated after only six days because of the extreme psychological effects experienced by the participants. The guards progressively began to exploit their power by behaving abusively toward the prisoners. They woke prisoners in the middle of the night and arbitrarily forced them to do pushups or stand at attention for hours, required them to clean toilets with their bare hands, locked them in solitary confinement, refused to grant them permission to use the bathroom, and cruelly imposed point-less rules. After an initial rebellion, the prisoners became progressively passive, submissive, and depressed. Several prisoners had to be pre-maturely released – the first after only 36 hours – as a result of suffering severe hysterical reactions, such as uncontrollable crying and disordered thought. Zimbardo himself confessed that he had become seriously affected by his role as warden, autocratically controlling visiting proce-dures and worrying about rumored "prison breaks." By the sixth day, the remaining prisoners were so obviously shaken by the experience that the researchers concluded that the study had to be prematurely terminated. A series of individual and group debriefing sessions and discussions were conducted and apparently none of the participants showed signs of lasting distress.

Not surprisingly, the prison simulation was subject to much ethical criticism, similar in intensity and kind to those that were directed at Milgram's obedience experiments (see Savin, 1973). Zimbardo (1973a) countered his critics by pointing out some of the steps that were taken to protect the well-being of his participants, including pre-screening sub-jects, early termination of the simulation, and extensive debriefing ses-sions. Further, he seemed to suggest that the study was justified from a cost–benefit perspective (p. 243):

> While acknowledging that the subjects in the prison experiment did suffer pain and humiliation, data are presented indicating that the subjects learned a number of things about themselves and that there were no persisting negative reactions.

Whether or not an increase in self-knowledge can be said to justify "pain and humiliation" suffered in the research context is, of course, open to ethical debate.

According to some, the prison study does not tell us much about prisoners and prison life because of the discrepancies between the simulated prison and real prisons (Forsyth, 1987). However, the results of the study provide an impressive illustration of the power of the norms of a situation and of pressures to conform in social life. In a span of only a few days, a group of emotionally stable, law-abiding students were dramatically transformed by their simulated institutional roles. From the Stanford prison study one point is particularly clear: under certain circumstances, a role-playing simulation can become a totally involving life experience for its participants.

evident in Allport and Postman's (1947) early laboratory simulations of the rumor transmission process. Allport and Postman were interested in the ways rumors change as they are transmitted. Their rumor simulation studies involved a serial transmission chain (similar to the popular game of Telephone) in which a description of an innocuous stimulus (typically a story or picture) is passed along from one person to another and distortions and eliminations in the communication are observed. For example, the experimenter or another participant would describe a picture projected on a screen to a subject who could not see the picture. That subject would then relate everything he or she could recall about the picture to a second subject; the second subject would then convey the description to a third, and so on until the description was transmitted six or seven times. When the accounts were analyzed it was found, consistent with processes of memory, that subjects tended to omit certain details of the communication to simplify the structure (so-called "leveling") and to accentuate or point up other details (so-called "sharpening").

Although Allport and Postman's findings were generally accepted over the years, it is now clear that many rumors do not act that way at all in natural settings; in fact, rumors tend to expand with a growing number of invented facts and details as they circulate (Rosnow, 1991). Communications may indeed shrink and become distorted as they are serially reproduced in laboratory simulations; however, such demonstrations may be of limited value for studying rumor transmission (Koenig, 1985). When rumors spread outside the controlled setting of a laboratory many more people are involved who are likely to be emotionally involved in the message content and intrinsically motivated to pass it accurately to others. In everyday interactions there are also opportunities for discussion and clarification of the rumor. Allport and Postman (1947) did, in fact, recognize the limitations inherent in their simulation studies, as indicated by the following quote (p. 64):

> Laboratory control can be achieved, we admit, only at the expense of oversimplification. By forcing serial reproduction into an artificial setting we sacrifice the spontaneity and naturalness of the rumor situation.

Thus, Allport and Postman (1947) implied that the mundane realism of their simulations was limited (and in turn their external validity), but apparently decided that it was the best available procedure for their early investigations. Contemporary rumor researchers are more likely to utilize other research strategies, typically in natural settings, in their attempts to uncover what happens to rumors as they are transmitted through interpersonal communication networks (e.g. DiFonzo et al., 1994).

Role playing

Some researchers who object to the use of deception in any form have recommended the use of an experimental alternative known as role play. This alternative to deception in the laboratory involves the diametrically opposed method of enlisting the subject as an active collaborator in the investigation. The subject is told what the experiment is about and is then asked to play a role as if he or she were actually participating in the experiment.

As a method, role playing is quite flexible in terms of setting, participants, and procedure. The essence of a role-playing study is for the participant to pretend that a situation is real when in fact it is not (Geller, 1982). For example, subjects may be asked to passively read a description of an experiment provided by the researcher and to predict how they would respond (similar to the preinquiry method described earlier). In other cases, subjects are asked to play a more active role, from the totally improvised to one in which the experimental scenario is acted out with the aid of a script provided by the researcher. In contrast to deception studies, which require that subjects must be kept unaware, participants in role-playing studies must be fully informed. According to proponents of role playing, the strategy allows for a wide latitude of response, is capable of exploring complex behavior, and is a more humanistic alternative to many experiments that typically rely on deception (Forward et al., 1976; Geller, 1982; Mixon, 1972, 1974). One early advocate of the method argued that the behavior of role-playing subjects would more closely approximate life situations than that of deceived subjects (Brown, 1962).

In one of the first attempts to design a role-playing study as an alternative to deception, Greenberg (1967) utilized this technique in a replication of Stanley Schachter's (1959) early affiliation studies. In order to test the hypothesis that fear leads to affiliation, Schachter's initial experiments were designed such that some subjects (the "high-anxiety" group) were led to believe that the experimental procedure would involve painful electric shocks. The deception was accomplished by having a confederate describe the supposed procedure in a frightening way and by prominently displaying ominous-looking equipment in the laboratory. For other subjects (the "low-anxiety" group) the experiment was presented in a non-threatening way as involving no physical discomfort. When the subjects were given the opportunity to wait with others or alone until the experiment actually would get underway, a significantly greater number of high-anxiety subjects chose to wait with others in a similar plight, in support of Schachter's expectation that "misery loves (miserable) company." In a later experiment, Schachter obtained evidence that this effect was more likely to hold for anxious firstborns and only children than for later-borns.

In Greenberg's role-playing experiment, subjects were told to imagine that they were performing in an important experiment and were instructed to "act as if the situation were real." They then were exposed to a scenario that was very closely modeled after Schachter's original design; the threatening apparatus was present for some of the role players and subjects similarly were told that they would soon undergo a painful (or non-painful) procedure. The major results of Greenberg's role-play experiment were consistent with Schachter's findings but not statistically significant. For example, the induction of fear had a lesser effect on affiliation than was expected, but the extent to which high-anxiety subjects chose to affiliate with others was in the predicted direction. Similar results were obtained for the birth-order effect. Supporters of the role-playing approach have interpreted the qualified success of this experiment as evidence that role play can replicate more traditional deception experiments. However, because Greenberg had regrouped his data according to perceived fear rather than by experimental manipulation, some critics maintain that his results are suspect and open to alternative interpretations.

Several methodological studies on role play have been carried out since Greenberg's early experiment and they have met with some success, particularly in research on conformity (Horowitz and Rothschild, 1970; Wahl, 1972; Willis and Willis, 1970) and obedience (Geller, 1978; Mixon, 1971; O'Leary et al., 1970). However, despite these limited successes and the ethical advantages of role playing, there has been much criticism of it as a research tool (e.g. Freedman, 1969; Miller, 1972), which may account for the paucity of studies that have utilized this approach in recent years. Critics have claimed that the expectation that role playing would replicate general relationships has not been adequately demonstrated. In the same vein, it is argued that role players often cannot duplicate the most interesting findings that come from true experiments, such as counter-intuitive ones. It also has been suggested that this technique lacks the capacity to induce involvement in role players when motivational or emotional variables are studied (West and Gunn, 1978). Some critics have gone so far as to assert that role playing is unscientific because it lacks adequate methodological controls. For example, Freedman (1969) alleged that role playing represented a return to the pre-scientific days of intuition and "psychology-by-consensus."

Opinions continue to be sharply divided over the adequacy of role playing as an alternative to more traditional experimental procedures. Those who have defended its use emphasize the apparent successes of role-play studies and have argued that there is some confusion among critics of the approach as to what actually constitutes role playing. For example, Coutu (1951) distinguished between role playing, which is actual role enactment on the part of the subject in a given situation, from role taking, which is imagining what someone else would do in a particular situation. According to Geller (1982), Freedman's view of role playing as psychology-by-consensus really only applies to role taking, as exemplified by Milgram's (1963) finding that a group of psychiatrists could not accurately predict the outcome of his obedience experiments.

One troubling aspect of role-play studies for its critics is that it is not possible to predict in advance whether the results would have corresponded with the results of non-role-play studies; thus, the best one can conclude after conducting a role-play investigation is that the results might have occurred had truly experimental studies been used (cf. Rosenthal and Rosnow, 1991). According to others, however, discrepancies between role-playing and traditional experimental results are problematic only if one assumes that the latter produce the truth. Indeed, there are enough methodological problems with deception experiments – in addition to concerns that laboratory data often cannot be generalized to everyday behavior – that perhaps deception results should not be used as the criterion against which the value of role playing is considered (Diener and Crandall, 1978). As we have learned from the research on subject artifacts, much role playing already seems to occur unintentionally in laboratory experiments.

Given the pros and cons of the role-playing alternative, perhaps the best advice that can be offered to investigators who are considering its use is the suggestion offered by Rosenthal and Rosnow (1991, p. 152) that "researchers must go into this methodology with their eyes open." Role playing can be a useful research technique in certain situations and appears to be an efficient aid to theory development and hypothesis generation (Cooper, 1976). Creative combinations of role playing with the classical experimental approach might be worth exploring, as has been suggested elsewhere (Schuler, 1982). For example, subjects can be required to play various roles while other aspects of the study correspond to usual experimental procedures, such as the withholding of information about certain manipulations. In the long run, few would argue that the role-play approach will completely replace the use of deception in experiments. Nearly twenty years ago, Diener and Crandall (1978) concluded that more information was needed to assess the value of experimentation by means of deception and role playing, in addition to obtaining greater understanding about the roles subjects play in all sorts of investigations. Unfortunately, it appears that Diener and Crandall's conclusion still holds true today.

Summary

This concludes our two-chapter coverage of the ethical issues involved in the conduct of laboratory research and the arguments for and against the use of deception in the laboratory setting. In addition to the numerous ethical implications inherent in the practice of deceiving research subjects, we have seen that there also are a number of difficult methodological issues that need to be weighed by the researcher. Although viewed as morally reprehensible by its critics, there is mounting evidence that deception is not considered to be an unacceptable or aversive methodology from the research participant's point of view and does not appear to have the negative effects that its critics have claimed. However, it is also clear that deception has been (and no doubt continues to be) used more out of convention than necessity by experimental behavioral scientists and that the days of experimentally naive subjects are long past. Considering the variety and breadth of existing research alternatives, one only hopes that investigators will progressively explore these potentially rich sources of data.

One alternative to laboratory experimentation that we have not yet considered in depth is for researchers simply to leave the laboratory setting and conduct their investigations "in the streets" – that is, in so-called "real-life" or natural settings. While some of the ethical issues pertinent to research in the laboratory setting also apply to research in the field, there are other ethical dilemmas that present themselves in field settings which are not relevant to traditional research settings. In chapter 5, our focus turns to the ethical issues involved in the conduct of field research.

5

Ethical Issues in the Conduct of Human Subject Research II: Field Research

In one sense, field experiments are more problematic, more troubling, than lab experiments for the reason that in a lab experiment at least the person knows that he or she is coming for an experiment. It's like going to the theater ... you're not going to know exactly what's happening and, by having decided to participate, you have in a sense consented in a very general way to being subjected to certain strange things that you won't know about. But if you're just walking on the street and, like in some of the helping behavior experiments, some things happen in front of you ... you have not been warned and you have not agreed to participate in an experiment. I think that is a more severe violation of the rights of the subject than the laboratory situation.

Herbert Kelman, cited in Allan J. Kimmel,
Herbert Kelman and the ethics of social-psychological research

Among the helping behavior experiments that Kelman may have had in mind when he referred to the ethical problems of field research during an informal talk at the 1987 Eastern Psychological Association meeting (cf. Kimmel, 1988b) was a series of studies conducted by Jane and Irving Piliavin during the early 1970s. The studies were designed to test a cost–reward model of bystander helping behavior based on the assumption that the decision to come to the assistance of a victim is the result of a consideration of the anticipated costs and rewards of helping as well as the costs and rewards of choosing not to help. In one of the Piliavins' experiments a confederate pretended to collapse in a moving subway train (Piliavin and Piliavin, 1972). As the confederate fell, he released a trickle of "blood" from an eyedropper in his mouth; in a control condition, no blood was released. The results of the study supported the expectation that fewer bystanders would come to the aid of the bloody victim because of the greater potential costs involved in assisting an apparently severely injured person. In a later variation of this research, the experimenters had another confederate who pretended to be just another passenger stand near the "victim." However, this second confederate wore either an intern's jacket, priest's attire, or ordinary street clothes in order to determine how bystanders' reactions would be affected by the presence of a doctor or clergyman.

While such "subway victim" studies represent ingenious alternatives to the staged laboratory emergencies that were described in chapter 3, they indeed may be more problematic from an ethical perspective, as Kelman has speculated. For example, because such unexpected events are not common in most persons' lives, they pose certain risks to unsuspecting bystanders. The sight of a victim who appears to be suffering from internal bleeding could cause substantial distress in some onlookers; although unlikely, it is entirely possible that a bystander could faint or even have a heart attack. Those persons who rush to the assistance of the "victim" could be injured in the attempt, and there is the additional possibility that they would experience some embarrassment upon learning that they were fooled for the purpose of a research investigation. Those who choose not to help might experience certain negative effects later, such as guilt or a reduction of self-esteem, when they reflect on their apathetic response.

There are societal implications as well from this type of research. When news spreads that bloody victims are merely research stooges, the public is provided with simply one additional reason for not helping, thus adding to the other forces in society that cause people to be cynical and able to rationalize their failure to assist others (Baumrind, 1977; Jung, 1975; Warwick, 1975). Of course, in defense of the research, there are some who would argue that the studies were specifically intended with the opposite outcome in mind – to provide a better understanding of the forces that work against helping during emergencies – and that awareness of this type of research sensitizes people to the apathetic bystander phenomenon. Even in light of these potential benefits it is clear that this research also creates certain problems.

Twenty years prior to his remarks at the Eastern Psychological Association meeting, from which this chapter's opening quotation was taken, Kelman brought to public attention some examples of even more dramatic and ethically problematic field investigations in his seminal 1967 paper on deception (see chapter 3). One of these examples involved a series of studies conducted under the auspices of the United States army to investigate the psychological effects of .stress on performance (Berkun et al., 1962). When brought to public attention, the studies aroused considerable furor because of the powerful stresses that were created in young soldiers (see Box 5.1)

It is important to note, as will become evident throughout this chapter, that most behavioral research in natural settings does not involve the staging of emergencies and other horrific events to see how unsuspecting subjects will react. Other field research procedures are much more benign, such as those that involve the observation and description of naturally occurring public behavior or analyses of the by-products of human behavior (e.g. public records and mass media content). However, the bystander intervention experiments and the army stress studies serve to illustrate some of the pervasive ethical issues that emerge in research conducted outside of the laboratory. These issues serve as the focus of the current chapter.

Field Research in Perspective

A substantial amount of behavioral science research is conducted "in the field": that is, in naturalistic settings outside of the artificial setting of the laboratory. This is especially true in disciplines such as sociology and anthropology. While laboratory research continues to be the preferred choice of a majority of psychologists, greater attention has been focused on the potential value of field studies for testing psychological hypotheses (e.g. Webb et al., 1981; Yin, 1989) and there is some indication that field research has gradually increased in psychology during recent decades (Adair et al., 1985).

A growing interest in field research as an alternative to laboratory experimentation can be attributed to several factors, largely stemming from some of the methodological problems that were identified in chapters 3 and 4. The most serious problem that confronted laboratory researchers was the recognition that their "test tubes" were dirty, in the sense that laboratory settings are subject to numerous artifacts that can serve to reduce the validity of the research findings. In short, all of the needs, fears, and intentions that result from subjects' awareness that their behavior is being scrutinized in a scientific investigation represent potential biases in the

---

Box 5.1

The army stress experiments

During the early 1960s, a series of experiments was carried out by a group of United States army scientists in order to assess the effects of psychological stress in military settings. Specifically, the researchers were interested in determining whether there is a deterioration of performance during combat as a result of battle stress (Berkun et al., 1962). The subjects in these studies were young army recruits who had not given their prior consent to undergo stress and who were completely unaware that they were being deceived for research purposes. In each of the studies, the subjects were exposed to elaborate and convincing experimental hoaxes which made it appear to them that their lives were in danger or that they were responsible for injuring a fellow soldier.

In one harrowing study, 66 recruits in their first eight weeks of basic training were aboard a military aircraft which they were led to believe was disabled and about to crash-land. This deception was accomplished when, towards the end of the flight, the subjects saw that one of the plane's propellers had stopped turning and they heard about other malfunctions over the intercom. The subjects then were informed directly that there was an emergency, an announcement reinforced by a simulated pilot-to-tower conversation (which could be heard over their earphones) and by the sight of ambulances and fire trucks on the airstrip below in apparent preparation for the crash-landing.

In order to ascertain whether the recruits' performance would be impaired by these terrifying manipulations, the plane steward distributed two questionnaires which were to be quickly completed. One of the questionnaires was an "emergency form" consisting of deliberately complicated instructions asking about disposition of the soldiers' personal possessions in case of death. The second form included a series of multiple-choice questions intended to test the soldiers' retention of airborne emergency procedures, the results of which supposedly would be furnished to the army's insurance company. Subjects were informed that all of the completed forms would be put in a waterproof container and jettisoned before the crash.

Of course, the plane did not crash after all, but instead landed safely. The subjects then were informed that their intense ordeal was simply part of a research study, and blood and urine samples were collected from all of the participants. Not surprisingly, post-experimental inquiries revealed that the subjects had suffered various degrees of anxiety about the possibility of injury or death.

Other studies designed for the army stress project involved new recruits not yet in basic training who were transported to an isolated mountain area where they were to participate in military maneuvers. There they were informed that new concepts of atomic-age warfare were to be tested which required them to perform individually rather than as a group. Each recruit was taken separately to a remote outpost where, after a short period of time, they were exposed to one of four experimentally implemented emergencies. In three of the emergency situations, the recruit was led to believe that he was in immediate danger of dying or of becoming severely injured. In one situation, the radiation needle on the recruit's equipment began to fluctuate and he received bogus radio messages detailing the accidental release of dangerous nuclear radiation in the area. In another condition, subjects received messages that a forest fire had suddenly broken out in the area and was closing in on them from all sides. To make the emergency appear more realistic, a smoke generator was used to simulate the fire. The third emergency condition involved supposedly misdirected artillery shells. This deception was accomplished via radio transmissions contrived to confirm the hazard, and the subjects heard a series of progressively closer explosions intended to simulate the incoming shells. In this condition, a radio transmission suggested that the shells were being fired by gunners who were unaware that anyone else was in the area. In a final condition, there was no threat of injury to the subject himself, but rather he was deceived into believing that he had caused serious injury to another recruit after setting off an explosion. This condition proved to be the most stressful, perhaps in part because the subject was continually reprimanded by a senior officer for having caused the injury.

In each of these emergency conditions, subjects were led to believe that the only chance of an immediate rescue was to use an elaborate two-way radio to call for help and to report their location. However, in each case, as was pre-arranged, the radio failed to work. The measure of impaired performance in these studies was how long it took the subject to repair the broken radio. After 75 minutes had elapsed, the recruit was picked up at the outpost and blood and urine samples were collected.

While these dramatic studies no doubt contributed to scientific knowledge by shedding some light on the effects of stress in realistic, non-laboratory settings, they raise a number of ethical problems as well. The uninformed participants were exposed to elaborate deceptions that posed undue or unnecessary risks, including the experience of severe stress and other psychological and physiological harms, invasion of privacy, and disrespect for human dignity. Perhaps the unsuspecting recruits also became less trusting of their army officers as a result of being victimized by the elaborate hoaxes.

Because the army experiments were carried out in an environment in which participants naturally expect a loss of individual freedom, there are some who might argue that the studies were less ethically problematic than had they been conducted in civilian settings. One also could contend that the experience of stress during army basic training is a "given" that recruits naturally come to expect (though perhaps not at the same level as the life-threatening stresses described here!). When researchers conduct their investigations in field settings, the extent to which the risks posed are similar to those that individuals experience in their everyday lives is a particularly relevant consideration in evaluating the ethical acceptability of the research.

laboratory. Faced with these problems of "observational reactivity" – a term used to suggest that the act of observing affects that which is being observed or measured – it is easy to understand why many behavioral scientists may simply choose to leave the laboratory setting in order to study naturally occurring behavior in field settings. In the field, individuals can be studied without their awareness that they are research participants.

A second problem with laboratory research that led to an increased interest in field studies is linked to the first. As the methodological problems of experimentation were exposed, elaborate deception manipulations became commonplace as a strategy for coping with potential subject biases. This, in turn, led to a serious consideration of the specific effects of deception (e.g. whether subjects accept the researcher's cover story or instead create their own interpretation of the situation that deviates from the intended manipulation) and the ethical implications of using deception (e.g. whether the research technique represents an abuse of subjects' rights).

At a more general level, ethical issues were linked to the realization that the interaction between experimenter and subject in the laboratory situation is a human relationship that is governed by some of the same characteristics that define other human interactions. As soon as the experimental situation began to be viewed as a human interaction situation and researchers began to address the methodological issues created thereby, it became relatively easy to suggest that norms that apply in other human relationships ought also to apply to the research relationship.

Field research offered researchers a ready alternative to some of the ethical concerns that arose in the laboratory context. Many behavioral research hypotheses can be studied in everyday reality without necessitating the deception or manipulation of subjects. Schuler (1982) provides the example of a study intended to investigate the effects on behavior of frustrating experiences. In the laboratory, frustration no doubt would have to be induced by the researcher through the use of some complex and deceptive manipulation, thereby exposing subjects to risk of harm. (In fact, there is a tradition of such research in the empirical literature on the frustration–aggression hypothesis; see, for example, Berkowitz, 1969.) Schuler suggests that there is no need to induce frustration when one can study many individuals who have been frustrated by events in their natural environments, such as university students who were unable to obtain preferred courses during registration, high-school students who scored poorly on college entrance exams, or shoppers who missed out on a desired sale item.

Another reason that some researchers have become increasingly disenchanted with laboratory research has to do with questions about the "ecological validity" of laboratory experiments, a term that is generally used to refer to whether an experimental situation reflects the outside world (Rosnow and Rosenthal, 1993). Experiments that are set up in psychology laboratories often bear little resemblance to the real-world settings they are intended to reflect. As a result, the obtained findings only may pertain to subjects' responses in relatively artificial environments.

Some researchers maintain that field research represents one possible solution to the limited ecological validity of laboratory research (see Box 5.2). Along these lines, Ellsworth (1977) has argued that experimentation in the field actually offers better prospects for revealing complex causal relationships than laboratory experimentation, where so many variables are kept constant that the complex network of interactions operating in everyday life cannot possibly be revealed. The problem with this argument is that it presumes that the possibilities of experimental control and precision of measurement are not seriously diminished in field settings. Much of the early enthusiasm for field research in fact has decreased, largely because of its lack of manipulatory control and constancy of conditions (Adair et al., 1985; cf. Cook and Campbell, 1979).

The recognition that various behaviors are not amenable to investigation in laboratory settings also may have led some researchers to turn their attention to field research (Keith-Spiegel and Koocher, 1985). Behavior as varied as responses to litter, reactions in doctors' waiting rooms, race relations in a community following a highly publicized police beating of a minority group member, and coping behaviors in the aftermath of natural disasters are more appropriately studied in the field than in the laboratory (Sommer and Sommer, 1991). Similarly, the rise of interest in socially relevant research since the early 1970s seems to have encouraged many researchers to leave the laboratory and to study social problems

---

| Box 5.2 |

Studying obedience in the field

In addition to the ethical concerns raised over Milgram's obedience research, there also was a difference in interpretation of the significance of the research, especially in terms of its ecological validity (cf. Rosnow, 1981). Milgram, of course, chose to create a miniature social world within the confines of his laboratory to study obedience to authority. But how might one study such behavior in a more natural setting? An experiment by Leonard Bickman (1974) provides a novel illustration.

In his field study of obedience Bickman manipulated the authority of three male experimenters who gave orders to 153 randomly chosen pedestrians on a street in Brooklyn, New York. The authority level of the experimenter was varied by style of dress. The experimenters wore either a sports coat and tie, a milk carrier's uniform, or a guard's uniform closely resembling that of a police officer. The procedure involved having an experimenter give one of the following three arbitrary orders to each chosen subject:

1 *Picking up litter*: the experimenter pointed to a bag lying on the ground and commanded, "Pick up this bag for me!"
2 *Dime and meter*: the experimenter nodded in the direction of a confederate standing nearby at a parking meter and stated, "This fellow is overparked at the meter but doesn't have any change. Give him a dime."
3 *Bus stop*: The experimenter approached a subject waiting for a bus and exclaimed. "Don't you know you have to stand on the other side of the pole? This sign says 'No standing.'"

The results of Bickman's experiment revealed that participants were more likely to be obedient when the experimenter was dressed as a guard rather than as a milk carrier or civilian. Whereas Milgram's laboratory experiments sacrificed realism for experimental control, Bickman's field study sacrificed rigorous control for realism. Because field research is embedded in an existing social situation, it is difficult for the researcher to control other variables that might be capable of influencing the results (Forsyth, 1987).

The fact that Bickman's study was conducted in the field does not mean that the results can be generalized to other natural settings or that it is without ethical problems. Different conclusions might have been obtained had he carried out the study in another area, manipulated authority differently, or used female experimenters (Mook, 1983). The use of deception with participants who were unaware that a study was in progress raises certain ethical issues as well, although perhaps in this case it was the experimenters who were subjected to the greatest risk of harm at the hands of disobedient subjects!

where they occur (see chapter 6). Some researchers believe that the investigation of such complex social problems as violence and crime, drug abuse, and race relations in the laboratory would result in a reduction of their complexity, a sort of reduced reality from which generalizations would be seriously restricted (Schuler, 1982). Finally, citing the fact that approximately three-fourths of psychology graduates in the United Kingdom go on to jobs not directly related to psychology, Robson (1993) has implied that field research is both more relevant and more applicable to the "real-world" problems encountered in professional settings such as industry, commerce, health, and education.

It is somewhat ironic that while many early proponents of field research believed it to be an ethical alternative to laboratory deceptions, it now is similarly criticized for the frequent use of that very research technique. Adair et al.'s (1985) survey revealed that deception was introduced as an experimental manipulation in order to provide greater control of variables in over half of the field studies examined in 1979 social psychology journals. On the one hand, the fact that subjects usually are unaware that they are being observed in the field (as opposed to the laboratory setting) can be seen as a methodological strength. On the other hand, the manipulation and observation of unsuspecting persons can be viewed as ethically problematic in many situations. Thus, like their counterparts in the laboratory, field researchers must cope with the dilemma of scientific validity versus the ethics of the research methodology used to advance knowledge (Keith-Spiegel and Koocher, 1985).

Field Research and Ethical Issues

Whether or not one agrees with Kelman's claim (in this chapter's introductory quotation) that field studies may be more ethically troublesome than laboratory investigations is open to debate and depends very much on the nature of the investigation. In some field investigations participants are as aware of being observed as those who take part in laboratory investigations. While there are many ethical issues that emerge in non-laboratory settings, two major areas of concern are especially likely to be encountered and can serve as focal points for a discussion of the various field research methodologies (Webb et al., 1981). The first concern has to do with the privacy rights of subjects. Because the inherent nature of much field research is to elude the awareness of those who are observed, the possibility that participants' privacy will be invaded must be fully considered by field researchers.

The second area of concern involves the informed consent of those who participate in the research. We have seen that this issue also is at the center of ethical decision making in the laboratory setting. As in many laboratory studies, informed consent can be problematic in the field because much research simply could not be carried out with the awareness of the research subjects. Imagine how ludicrous it would have been to carry out either the bystander intervention experiments or the army stress studies described at the beginning of this chapter had the researchers decided first to obtain the participants' informed consent. Exacerbating the problem of informed consent in the field are the greater practical problems in carrying out effective subject debriefings.

Privacy

The central ethical issues in field research are likely to revolve around potential invasions of privacy. Privacy has been defined by Alan Westin (1968, p. 7) as "the claim of individuals, groups, or institutions to determine for themselves when, how, and to what extent information about them is communicated to others." Similarly, Ruebhausen and Brim (1966, p. 426) have defined privacy as "the freedom of the individual to pick and choose for himself the time and circumstances under which, and most importantly, the extent to which, his attitudes, beliefs, behavior and opinions are to be shared with or withheld from others." These definitions regard privacy, when viewed in terms of the individual's relation to social participation, as voluntary and temporary withdrawal from others through physical or psychological means.

There are major international codes of human rights (such as the United Nations' Universal Declaration of Human Rights and the European Convention on Human Rights) that specify the right to privacy, which also is implicit in various European and American ethical codes for research and professional standards. Some common threads running throughout these codes are a fundamental respect for the person, remedies for violations of privacy, the need for protecting anonymity, and the prohibition of breaches of confidentiality. Attempts to apply these formal standards in certain contexts often give rise to difficult choices. Problems of application are especially evident in research and clinical practice where the need for information conflicts with the reluctance to disclose that information.

The circumstances that characterize participation in field research are such that subjects' freedom to determine whether or not to participate frequently is violated. In contrast to the laboratory setting, participants in the field typically are not informed of their role as subjects and are unaware that a study is in progress. No social contract has been established with the researcher; as a result, informed consent cannot be given, subjects cannot choose to refuse to participate or leave once the study is in progress, and in many cases they cannot be debriefed or informed of the results once the study has been completed. Under such circumstances, if researchers obtain or reveal (wittingly or unwittingly) information about attitudes, motivations, or behavior that a subject would prefer not to have revealed, the latter's basic right to privacy will have been compromised and an ethical problem will have been encountered.

An individual's desire for privacy is never absolute, and he or she is continually engaged in a personal adjustment process in which the desire for privacy must be balanced against the desire for disclosure and personal communication with others. This adjustment occurs in the context of various situational forces, including pressures from curious others (such as behavioral scientists), as well as the society's norms and processes of surveillance used to enforce them. In short, under some circumstances or conditions, violations of the right to privacy, when weighed against the right to know, may be viewed as falling within the bounds of ethical propriety.

Webb et al. (1981) have suggested that there is a continuum of situations in which behavior might possibly be observed, anchored at one end by situations in which privacy could not be said to be violated and at the other end by situations in which privacy could be said to be violated. Between these extremes is an area in which it is difficult to draw distinct boundaries and in which the acceptability of violations of privacy will depend upon characteristics of the situation and on individual judgment.

One end of the invasion of privacy continuum can best be characterized by the public behavior of public figures. Politicians and celebrities are well aware that their every move may be observed and reported for public consumption. The nature of what is observed may range from linguistic analyses of political speeches to off-the-cuff remarks made at a dinner party. By nature of their profession, public figures typically recognize that anything they do in public is fair game and reluctantly choose to forgo their right to privacy.

When Warren Beatty, the film actor and director, addressed the American Society of Newspaper Editors in 1983, he stated: "Privacy is a simple matter in my case. I don't have any and I don't really expect any" (Friendly, 1983).

By contrast, privacy problems are more likely to emerge for non-public figures who presumably are unaware of the possibility that they are being observed, even in public situations (Webb et al., 1981). For example, a couple necking on a park bench one sunny day in May, in an open and busy area of the park, surely understand that their behavior can be seen by others. However, they might be offended to learn that strangers were carefully observing them and taking notes of their behavior. This same couple might not mind very much having observers note the amount of time they spend in various aisles of the supermarket while shopping for groceries. Thus, as Webb et al. (1981) warn, while many observations of public behavior are harmless enough and do not raise ethical problems, researchers need to consider specific behaviors and situations carefully.

The other extreme of the continuum is characterized by spying on private behavior in situations where people clearly expect that their privacy will not be violated, as when they are in their own homes, personal offices, physicians' examining rooms, and the like. The privacy within these settings may fluctuate from time to time (e.g. one's office is not private when workers are present to make repairs), but there is general agreement that these are settings in which invasions of privacy would be deemed inappropriate. While surreptitious observations in private settings might be required in certain criminal situations, they would not be considered legitimate for scientific purposes (Webb et al., 1981). Such observations by behavioral scientists are rare, but have been made. For example, the Wichita Jury Study (see chapter 1) created a national uproar in large part because the researchers had "bugged" the jury room, a private setting that is protected from outside intrusion by laws and by tradition.

Falling in the middle of Webb et al.'s privacy continuum are public places in which individuals expect some degree of privacy. That is, in some circumstances, public settings are regarded as socially or psychologically private by the persons who temporarily occupy them. For example, returning to our lovers in the park, if they moved from their highly visible bench to a secluded and unoccupied area of the park they no doubt would expect a far greater degree of privacy than they had earlier. If a researcher were to observe the couple through binoculars from a hidden vantage point for a study on interpersonal attraction, the observation is likely to be viewed as a violation of the couple's privacy and not to be condoned.

In other situations in which private behaviors occur in public settings, violations of privacy would be considered even more problematic. For example, despite the "public" nature of public restrooms, people enter such settings with the understanding that their behavior will be respectfully ignored by others. Such expectations in part led to the strong criticisms of Humphreys's "tearoom trade" observations (see chapter 1). Although his subjects were well aware that their homosexual activity could be noticed by others present, they did not expect to be observed for research purposes or to have their identities eventually revealed to a researcher. The fact that Humphreys obtained the names and addresses of his subjects through devious means and later interviewed them under false pretenses exacerbated the ethical problems created by his private observations in a public place (Webb et al., 1981).

Another highly controversial study involving observations in public restrooms was conducted by Middlemist et al. (1976). In order to study the hypothesis that crowding leads to stress and discomfort, unsuspecting males were observed while urinating in a campus restroom. The researchers arranged to have a confederate stand either near them (at the next urinal) or farther away (at another urinal); other males were observed urinating alone (the control group). The time of delay before urination and the duration of urination were recorded by an observer hidden in a nearby stall through the use of a periscope device. As predicted, subjects took longer to begin to urinate and urinated for a shorter period of time as the interpersonal space between themselves and the confederate decreased. This study was criticized as trivial research that crossed the bounds of propriety, invaded subjects' privacy, and used up valuable space in an important research journal (Koocher, 1977). In defense of their research, Middlemist et al. claimed that the invasion of privacy was of the sort that the subjects experience frequently in their everyday lives and that pilot subjects did not object to the observation when they were informed about it afterwards. Further, the researchers stressed that all data obtained in the study were kept anonymous; that is, at no time in the study were individual subjects' identities or behavior made known to others.

While it might also be argued that Middlemist et al.'s stress hypothesis could have been studied under less invasive conditions, it is important to recognize that their so-called "micturition study" was carried out to test an important theory, there was no intent to demean subjects, safeguards were taken to protect subjects from harm, and no apparent evidence has emerged suggesting that any harm came to the subjects. To a great extent, these points also are relevant in evaluating the ethicality of Humphreys's investigation. As Webb et al. (1981) appropriately observed, the ethical controversies surrounding these studies have involved principles rather than consequences. That is, regardless of the fact that there were no identifiable effects on participants, it has been argued that the studies violated basic ethical principles pertaining to invasion of privacy and should not have been conducted.

In their analysis of privacy issues, Webb et al. (1981) further suggest that four dimensions underlie the placement of different research situations on the invasion of privacy continuum and regularly enter into determinations of the privacy interest (see Table 5.1). First, there is the element of publicness in the location of the behavior under study. People can lay less claim to privacy protections for behavior that takes place in public settings (shopping malls, sports arenas, airports, parks, etc.) as opposed to behavior in private settings such as their own homes. In fact, the value of privacy in the home is recognized by statute and common law. In other settings, as our previous examples illustrate, the distinction between private and public is less clear.

A second dimension underlying privacy considerations is the publicness of the person. As was discussed above, public personalities regularly are subject to observations and reporting that would be considered invasions of privacy by less public individuals.

A third important dimension to be considered in judgments concerning privacy is the degree of anonymity provided, in terms of the explicit identification of the person with particular information. Anonymity pertains to a person's desire not to reveal who he or she is, and the expectation that he or she will not be personally identified or held to the rules of role or behavior that would operate under

Table 5.1 Dimensions underlying privacy judgments

1	Publicness in the location of the behavior under study
2	Publicness of the person studied
3	Degree of explicit identification of the person with particular information (i.e. anonymity)
4	Nature of the information disclosed during the study

Source: Webb et al., 1981.

normal conditions (Westin, 1968). Privacy is clearly maintained when the linkage between the individual and the information obtained for research purposes has been completely severed. This, as mentioned, was something Middlemist et al. (1977) claimed in defense of their micturation study. To the contrary, when information can be linked with a single identifiable person, anonymity will not have been preserved and the risk of privacy invasion is high. Degrees of subject identification falling between these extremes, such as the case in which identification is not immediately available but could be obtained (as in Humphreys's tearoom trade study) also can incur ethical problems.

A final factor in considerations of privacy is the nature of the information disclosed during a study. Certain information is likely to be more sensitive and pose greater risks to subjects than other information. For example, in American society, sensitive topics such as those involving birth control practices, income level, drinking and driving, cheating on income tax, and child and spousal abuse can be expected to raise privacy concerns. Ethical judgments should take into account the possibility that disclosed information, particularly when it can be associated with individual subjects, may be perceived as an invasion of privacy.

These four considerations can be expected to interact in complex ways. While not solely applicable to research conducted in the field, they are more likely to influence ethical judgments relevant to the appropriateness of field observations involving participants who have not willingly entered into an agreement with the investigator.

Informed consent

It is difficult to consider the ethical issues pertaining to privacy fully without taking into account the principle of informed consent. Even the most private settings can be studied without raising ethical concerns if the people within those settings freely consent to observation once informed of the nature of the investigation. For example, college students might willingly open their dormitory rooms to researchers interested in observing how the rooms are decorated for a study on student interests and lifestyles. Those same students doubtless would view their privacy as having been violated if they were to learn after the fact that researchers had gained access to the rooms for whatever research purpose while the students were out attending classes. Some of the problems associated with the recording of individually identifiable data similarly can be overcome by obtaining subjects' prior voluntary agreement.

The problem with informed consent in field settings, of course, is that much field research involves situations in which informed consent is either not feasible or is detrimental to the interests of the research. These are research activities in which subject awareness might defeat the purposes of the research, as in the cases described at the beginning of this chapter. On the other hand, informed consent will be irrelevant for many research activities that involve observations of ongoing public behavior or public records. According to Diener and Crandall (1978), when research procedures used in field settings do not significantly affect subjects' lives and do not pose any risks for them, informed consent procedures often become bothersome and time consuming for all parties involved. As a result, obtaining informed consent in such situations may be both ethically and methodologically undesirable. In short, application of the principle of informed consent in field research depends greatly on the nature of the specific research methodologies and the circumstances under which they are employed.

Research Methods Outside the Laboratory

Behavioral scientists have a wide variety of research tools and techniques to choose from in order to carry out their investigations in field settings. As a viable solution to the awareness problem encountered in laboratory experiments, field researchers are perhaps most apt to select a research approach that utilizes some form of nonreactive observation or measurement. Nonreactive research consists of methods that do not affect what is being studied. In nonreactive (or "unobtrusive") research participants are unaware that they are being observed or measured for research purposes and thus do not "react" to the research; that is, they remain unaffected by it. Nonreactive methods are particularly useful in situations where the researcher is interested in obtaining an accurate measure of behavior as it naturally occurs. Examples of nonreactive field procedures include descriptive studies involving the simple observation of naturally occurring behaviors, surreptitious field experiments involving unknowing participants, and participant-observation studies in which subjects are unaware of the presence of a researcher. Field researchers also utilize nonreactive measures that do not involve the direct observation of subjects at all, but rather physical evidence or by-products (such as archival records) of behavior.

By contrast, reactive research involves obtrusive research procedures that affect what is being observed or measured. Among the
most common reactive procedures utilized in field settings are surveys, structured observations, and participant observations in which
subjects are aware of the presence of the researcher. These techniques
avoid some of the ethical issues raised by studying individuals
without their knowledge, but are likely to encounter other ethical
problems as well as methodological problems associated with observational reactivity.

In the remainder of this chapter we will consider some of these
commonly used field research methods, giving particular emphasis to
ethical issues involved in the study of individuals without their
awareness.

Simple observation

If there is one activity that most people have in common with
behavioral scientists it is "people watching." In scientific jargon, this
activity is more apt to be referred to as simple observation, a research
approach which generally has as its goal the careful mapping out (i.e.
the description) of what happens behaviorally. Like people watching,
simple observation can be fun; it also can be a provocative source of
research ideas. As a means of describing behavioral occurrences,
simple observation may be casual or systematic in nature. In casual
observation, the researcher conducts the observations without pre-
arranged categories or a formal coding or scoring system. For
example, the researcher may simply choose to observe people in a
particular setting and to note anything that occurs of interest. This is
the research approach that most closely approximates everyday,
informal people watching.

Systematic observations consist of a more formalized approach to
observation and measurement, necessitating that the researcher
develop and employ a scoring system with a set of pre-arranged
categories. By "systematic" it is meant that the conditions under
which the observations are made are specified in advance and that
the observations are carried out in a consistent and orderly way.
Typically, this approach requires some sort of a checklist in which
information is recorded or tabulated under appropriate headings.
Checklist categories, often developed from casual observation,
should include examples of observable behavior that one might
expect to occur naturally in the situation.

Simple observations of public behavior typically are innocuous and do not raise ethical problems. For example, Sommer and Sommer (1989) carried out systematic observation research in coffee-houses in order to study a basic principle of behavior known as the "social facilitation effect." Social facilitation involves the enhancement of an individual's performance of simple tasks when that person works in the presence of others. This effect was first noted by Norman Triplett in one of the earliest recorded social psychology experiments. Triplett (1897) found that young children wound fishing reels faster in the laboratory when they were joined by other children performing the same task as opposed to when they performed the task alone. This effect has since been replicated in numerous laboratory studies using both humans and non-human species.

In order to test whether social facilitation effects occur among people in their everyday activities, Sommer and Sommer recruited and trained student observers who were assigned to three different coffee-houses – one that was frequented by office workers and business persons, a second that was in a shopping center, and a third located in the college student union. The observers sat in their respective coffee-houses unobtrusively, pretending to be reading a textbook and taking notes. For each session, the observer recorded the duration of stay and amount of beverage consumption for three lone individuals and three people in groups. Across the three settings it was found that facilitation effects were more apparent in length of stay than in beverage consumption, with most patrons drinking only one cup of coffee.

Though carried out with a specific purpose in mind, the coffee-house observations were not much different than the sort of informal people watching that regularly occurs in public places where people eat, drink, meet with friends, and so on. The observations presented no risks to the subjects, did not infringe upon their rights, and the data were recorded anonymously and in aggregate form. In such settings, people come to expect that they may at times be subject to the roving eyes of others, so the research observations can be said to be within the range of experiences of everyday life. Under such conditions, informed consent is unnecessary and probably would create more problems for the researcher than it would resolve. By obtaining the informed consent of patrons in the coffee-houses, the advantage of observing spontaneous and natural behavior would be lost. After all, the effect already had been observed in the laboratory, so nothing would have been gained by doing the study in the field with informed subjects. At a more practical level, subjects might resent being bothered by a researcher attempting to explain a study that has little consequence for them, and the informed consent procedure would be time consuming and logistically difficult to carry out effectively.

For similar reasons, debriefing subjects once the observations have been made would not be recommended in this type of situation. There is not much a researcher can do if a subject indicates his or her displeasure at being secretly observed, given that the observation already had been carried out. Additionally, the debriefing could serve to raise subjects' level of discomfort or paranoia in other public settings and, without a full understanding of the potential scientific value of the study, could serve to lower their general opinion of scientists who "don't have anything better to do than to watch me drink a cup of coffee!"

In other situations, simple observations may be viewed as more invasive and, as a result, more problematic from an ethical point of view. It is one thing to observe common non-verbal behavior (e.g. gestures, postures, seating arrangements) in public settings, but another to listen in to the conversations of strangers. Most of us, of course, frequently engage in "people listening" in addition to simply watching people; however, unlike researchers, casual listeners are not apt to take notes or keep careful records of that which they overhear. Indeed, one can find a number of nonreactive studies of conversations in the behavioral science literature, dating back to H. T. Moore's (1922) early investigation of sex differences in conversation.

Moore was convinced that there was a clear mental differentiation between men and women, and so he sought to obtain evidence of this by analyzing the content of their "easy conversations." For several weeks Moore walked along a 22-block area in midtown Manhattan during the early evening hours and jotted down every bit of audible conversation he could overhear. He eventually recorded 174 conversation fragments, which were coded in terms of the sex of the speaker and whether the participants in the conversation were of the same or mixed sex. Moore reported some interesting gender differences, such as the finding that female conversations included many more references to the opposite sex than did male conversations, and his study triggered a number of other hidden-observer language studies (see Box 5.3).

In evaluating the propriety of nonreactive conversation research in terms of invasions of subjects' privacy and other ethical concerns, one must bear in mind that each of the cases described in Box 5.3

Box 5.3

Conversation-sampling research in the behavioral sciences

Behavioral scientists have long recognized the potential for the understanding of language and individual differences that comes from analyzing samples of naturally occurring speech. Around the turn of the present century the French sociologist Gabriel de Tarde (1901) recommended that conversation analysis could serve as a useful technique for studying cross-cultural and social-class differences, and since H. T. Moore's non-reactive conversation study two decades later (see p. 136) researchers have been intrigued by the possibilities of this form of naturalistic research.

Moore's observations were restricted to a small area of New York City and thus the conversation fragments he obtained can be questioned in terms of the representativeness of the subjects and their conversations. Other researchers have attempted to overcome such sampling rigidity by obtaining samples of conversation from a wider variety of places and situations. For example, Landis and Burtt (1924) unobtrusively sampled conversations in a broad range of settings, including railroad stations, campuses, parties, department stores, restaurants, theater and hotel lobbies, and streets in both residential and commercial areas. Cameron (1969) had his students listen in on the conversations of more than 3,000 persons in different settings in order to investigate their usage of profanity. Among those most likely to use profane words were factory and construction workers and college students; secretaries tended to swear less often than people in other professions. Carlson et al. (1936) studied sex differences in lobby conversations overheard during the intermissions of Minnesota symphony concerts. The fact that Carlson et al.'s subjects were observed in a setting in which they nearly had to shout in order to be heard reduced the potential for whispered speech (which may contain significantly different speech content).

Several conversation-sampling studies have been restricted to university campuses, with trained students serving as unobtrusive observers. Stoke and West (1931) studied undergraduate college students' random "bull sessions" held at night in campus residence halls. In a more recent analysis of sex differences in gossip, Levin and Arluke (1985) had students at Northeastern University eavesdrop on the conversations of other students in the campus student union. For each instance of overheard gossip (defined as small talk about other persons), the student observers recorded the gender of the gossipers, and the tone (positive or negative) and focus (i.e. who and what the gossip was about) of the gossip fragments. Levin and Arluke's analysis revealed that the two genders were equally positive and negative in tone, with females devoting only a slighter higher proportion of their conversations to gossip about a third party. The most interesting finding, however, was in content differences: female students were most likely to gossip about close friends, relatives, and their coursework (teachers and classmates), whereas male gossips were most likely to discuss celebrities and distant acquaintances.

As is typically the case in nonreactive observations of public behavior, Levin and Arluke's gossip study was carried out as a systematic observation. The student-observers were armed with a coding sheet that simply required them to check appropriate categories for describing the nature of the overheard gossip fragments. The observers did not record verbatim the actual content of what was said and the identities of the gossipers never were revealed. One way to think about whether such research is appropriate from an ethical point of view is to consider how you might have felt had you been included as a subject in the study. Would you mind having others listen in to your public conversations for research purposes?

took place in public settings in which the subjects were aware that what they were saying could have been overheard by strangers in relatively close proximity. We might assume that when people have something very private and personal to discuss they either choose a more isolated setting for their conversation or else lower their voices so that they cannot be overheard. Thus, most nonreactive observation studies of this type can be justified on ethical grounds by the public (and anonymous) nature of the observations, and by the corresponding fact that the risks posed (in this case, having information of a private nature revealed to others) are no greater than those one encounters in daily experience.

By contrast, consider the early case involving surreptitious observations carried out by Henle and Hubble (1938). According to the researchers, "special precautions" were taken to keep the student subjects unaware that their conversations were being recorded verbatim for a research study. As a result, the investigators hid under beds in dormitory rooms, eavesdropped on conversations in washrooms and lounges, and listened to telephone conversations. The researchers also pointed out, without further explanation, that the "unwitting subjects were pursued in the streets, in department stores, and in the home" (p. 230). Without additional details about how Henle and Hubble protected the rights of their participants, it would be difficult to regard the research as ethically acceptable on any of the grounds discussed above. Although the results of the study were presented in statistical summaries without reference to specific individuals, the subjects' privacy (in terms of the observation settings and the potential nature of the information disclosed) was clearly intruded upon. In lieu of obtaining informed consent in such studies, researchers probably should at least obtain the permission of participants to use any sensitive data after the fact (Diener and Crandall, 1978).

In many circumstances, the simple observation approach can serve as a more ethically acceptable alternative to surreptitious observations carried out in private settings. The challenge is in finding the appropriate public setting for testing one's hypothesis. As an example, consider the researcher who is interested in investigating the relationship between physical attractiveness and the amount of time people spend grooming themselves. The researcher might predict that one's own physical attractiveness is self-reinforcing; that is, people may respond positively to their own physical attractiveness as well as to that of others. There are a number of ways to go about testing this hypothesis, including simply asking people who vary in degrees of attractiveness how much time they spend in front of the mirror on an average morning. Of course, this approach is reactive in nature and may lead to biased responses in addition to potentially embarrassing the subjects. As an alternative, the researcher might choose to set up a hidden video camera and one-way glass mirror in campus restrooms, thus enabling the observation of unsuspecting subjects. Observers could then record the approximate attractiveness level of each subject (say, on a ten-point scale) and the amount of time each spends gazing at his or her reflection in the restroom mirror. While this approach would eliminate the problem of observational reactivity, the researcher would now be faced with the ethical problem of having to justify the invasion of subjects' privacy. In essence, the issues would be very similar to those discussed in reference to Middlemist et al.'s (1976) micturition study.

As a solution to the problems raised by these two research approaches, McDonald and Eilenfield (1980) ingeniously exploited an ideal public setting in order to test the attractiveness and mirror-gazing hypothesis. They observed university students as they walked past a mirrored wall on the outside of one of the campus buildings. Hidden observers rated each randomly selected passerby as being either low, medium, or high in physical attractiveness and timed (in seconds) how long each subject spent gazing toward his or her reflection. The predicted positive correlation between attractiveness judgments and gazing time was obtained for both male and female subjects. Informed consent and debriefing procedures were not employed by the researchers, a decision that appears to be justified on the basis of the innocuous and public nature of the observations.

While it may not be possible to overcome the ethical issues raised in the context of most simple observations of private behavior, such studies may be carried out if the researcher judges that certain conditions have been met. Obtaining informed consent from potential participants is the most obvious and effective safeguard when behavior to be observed is private in nature, although, as described, it is not possible to obtain informed consent and preserve the nonreactive nature of the observation. The researcher should also consider whether there is an alternative research approach that can be implemented in order to obtain the desired data. As McDonald and Eilenfield's (1980) mirror-gazing study illustrates, there may be a more acceptable setting in which to carry out potentially sensitive observations than is first apparent. If an alternative approach is unavailable, researchers should take special care to protect the anonymity of the private information obtained from participants. It also is necessary to consider whether there are legal restrictions barring observation in the intended research setting. Investigators should attempt to establish how private the setting actually is and the degree of sensitivity of the information desired, perhaps by asking individuals from the subject population to consider whether they would view the observations as an invasion of their privacy. When in doubt, it would be wise for the researcher to consult the opinions of colleagues or else to postpone the study until some of the ethical issues can be more effectively resolved.

Nonreactive measures

As an alternative to the direct observation of naturally occurring behavior, researchers may choose to obtain data through the use of a variety of nonreactive (or unobtrusive) measures. These measures, along with some of the issues inherent in their use, are fully described in Webb et al.'s (1981) classic compendium, *Nonreactive Measures in the Social Sciences*. Although rarely adequate as a sole means of testing hypotheses, nonreactive measures are especially suited for the indirect study of behavior in naturalistic settings. While some ethical problems may be encountered in their application, most nonreactive measures have little or no impact on the research participants and thus are permissible in a majority of situations.

Among the potential sources of nonreactive measures of behavior are physical traces and archival records. Physical trace measures involve the examination of physical evidence or traces of past behavior that have accumulated or built up in the environment (so-called "accretion" measures) or that have selectively deteriorated or worn away (so-called "erosion" measures). Some examples of accretion measures that are revelatory about behavior or a society's lifestyles and values include:

1 the accumulation of dust on library books to estimate the extent of their use;
2 the prevalence of politically oriented bumper stickers on automobiles as an indication of interest in an approaching election;
3 the analysis of graffiti to investigate sex differences, social concerns, and lifestyle differences in communities differing along socio-economic lines;
4 length of discarded cigarette butts to gauge concerns about the health risks of smoking among current smokers (with longer butts signifying greater concern);
5 sifting through residents' garbage cans for empty bottles of alcoholic beverages to assess level of home consumption in a "dry" town.

Examples of erosion measures include:

1 the selective wearing away of floor tiles to determine the paths most frequently taken by shoppers in supermarkets, the popularity of museum exhibits, etc.;
2 the wear on library books (especially at the corners where pages are turned) to assess their popularity (this measure can be combined with check-out rates to determine whether borrowed books actually are read);
3 estimating the skills and habits of drivers by noting the prevalence of dents, broken glass, and missing ornaments on their automobiles.

As may be evident from some of these examples, subject anonymity often can be maintained because it is not known who produced the physical traces (such as graffiti and cigarette butts) in the first place. It is the remnants of behavior that are observed rather than the behavior that created the physical traces. However, for some nonreactive measures anonymity cannot be guaranteed. For example, when researchers rummage through the garbage bins in a neighborhood, it usually is clear which household has produced what garbage. Even when garbage is collected at a local dumping site, certain contents may contain identifying information (e.g. magazines with address labels still attached, discarded envelopes and letters, credit card receipts, and bank statements). Because the contents of garbage may contain sensitive information that the previous owner may not want (or expect to have) revealed to others, privacy concerns need to be considered prior to the carrying out of this form of nonreactive measurement (see Box 5.4).

Archival analysis represents another potentially rich source of data. This type of nonreactive research makes use of data obtained from existing records which were produced for purposes other than behavioral research. Archives may consist of the ongoing, continuous records of a society routinely collected and maintained by government agencies, such as actuarial records (e.g. birth, marriage, death records), political and judicial records (including the voting records and speeches of legislators), other government records (e.g. crime reports, automobile ownership records), and the mass media (e.g. news reports, advertising, editorials). Other archival records are more discontinuous (i.e. episodic) and private than running, public records. Some examples of episodic records include sales records (e.g. casualty insurance sales as an indicator of the effects of disasters, sale prices of autographs, sales levels of consumer goods), industrial and institutional records (e.g. job absenteeism and tardiness, accident reports, consumer complaint letters), and various other written documents (such as suicide notes, rumors reported by rumor control centers, journals and diaries).

Research conducted by Phillips (1977) provides a fascinating example of how archival records can be exploited for behavioral research purposes. Phillips reasoned that many driving fatalities may be the result of motorists using their cars to commit suicide, and hypothesized that motor vehicle fatalities, like suicides in general, would increase after a well-publicized suicide story. In order to test this hypothesis, he used running records of automobile accident fatalities kept by the California Highway Patrol and a measure of suicide publicity obtained from California newspapers. The publicity measure was based on the daily circulation of each newspaper and the number of days the suicide story was carried. It was found that the publicity measure indeed correlated significantly with changes in motor vehicle fatalities after each story, with the fatalities increasing significantly during the few days after a well-publicized suicide story.

Box 5.4

Ethics and garbage research

In recent years, researchers have found that an analysis of the contents of garbage can be useful for shedding light on certain aspects of a society, including its citizens' habits, interests, and values. An article on "urban archeology" in a 1980 issue of *The Christian Science Monitor* reported that there were about 12 federally funded garbage projects in the United States at the time. One of the longest-running of these investigations is the Garbage Project, directed by William Rathje, a University of Arizona archeologist who believes that the study of garbage actually represents the study of "contemporary material culture" (Webb et al., 1981, p. 15). When Rathje's team of researchers compare interview data with the actual contents of garbage bags they often find that the verbal reports do not match the garbage. For example, when interviewed, respondents claimed to have healthier eating habits than their garbage reveals. Subjects reported that they consume higher quantities of high-fiber cereals, vegetable soups, and skim milk, and lower quantities of alcoholic beverages, breakfast pastries, and desserts, and high-fat foods than is apparent from their discarded food containers.

Despite the informative nature of such research, the garbage studies raise certain ethical problems, particularly in terms of the risk of invading the privacy of the previous owners, the possibility of discovering incriminating material, and problems in maintaining the confidentiality of data. However, according to Webb et al. (1981) these ethical problems may not be as severe as they first appear. Legally, it seems that once people put things in their garbage cans, the contents are considered abandoned and the previous owners no longer can claim any legal interest in them. Prior to collecting garbage for analysis, the Garbage Project researchers have taken careful steps to explain the research and obtain community approval. For example, residents in the neighborhoods involved have been asked to sign consent forms stating that at some future time their garbage might possibly be collected for study. Later, households are selected at random and the trash is bagged by collectors and taken to an inspection site.

Rathje's student workers also must sign consent forms agreeing in part not to engage in unnecessary inspection or identifying of material. When the studies require that residents be interviewed, garbage collection data are linked with interview responses through the use of codes, which are destroyed once the connection has been made. The Garbage Project has received wide publicity in the geographical region in which the studies have been conducted and so most residents recognize the risk of discarding any incriminating material.

This pattern of fluctuations was not obtained for control periods that were free from suicide stories.

With regard to the ethicality of archival research, at the outset it is clear that the more public the records, the fewer problems will be encountered. For example, in order to determine if there are predictable biases in trial outcomes and prison sentences based on the ethnic status of the accused, a sociologist may decide to consult courtroom records for the necessary data. Such a study would be permissible without having to obtain the informed consent of the individuals involved because the researcher is relying on information recorded in public documents (Diener and Crandall, 1978). In fact, the research use of most archival records, unless they involve private documents, should raise very little concern from an ethical point of view.

The potential problems with archives have to do with issues of confidentiality when the researcher gains access to private records or when various data sets can be linked. Webb et al. (1981) have described several examples of the misuse of data files. For instance, certain records that may be accessible to particularly enterprising researchers, such as copies of driver's licenses and automobile ownership information, may include a wide range of additional personal data, including height, weight, marital status, whether the person is physically handicapped or is vision or hearing impaired, whether the person had been confined to a mental institution or convicted of felonies, and so on. Hospital records are extremely difficult to gain access to, but typically contain much confidential information about the patient. When data sets are very small, confidentiality becomes difficult, if not impossible, to preserve.

A variety of potential methods for protecting the confidentiality of archival data are available to researchers. Although they are too numerous to summarize here, one example provided by Webb et al. (1981) is sufficient to illustrate the kinds of approaches that can be used. The example pertains to a case involving an investigator who is interested in determining the extent of criminal records among persons with venereal disease. In order to determine whether there is a link between the variables, the researcher would have to consult two confidential files. One alternative would be to provide police with five lists, each consisting of the names of ten persons treated for venereal disease. Five more lists, each consisting of the names of ten untreated persons, would be provided as a comparison group. The police then could provide the average number of arrests, along with a variability measure, for persons on each list, which they could recode in a way that only the designation of treated vs. untreated groups is retained. For more complete descriptions of this and other types of confidentiality-preserving techniques, as well as a consideration of their limitations, the interested reader should consult the excellent reviews by Boruch and Cecil (1982) and Campbell et al. (1977).

In sum, by its very nature, nonreactive measurement tends to raise fewer ethical difficulties than the direct observation of unaware subjects. Nevertheless, each case needs to be considered in terms of whether consent must be obtained and confidentiality of data protected. A study involving the prevalence of empty whiskey bottles or unpaid bills in household garbage containers is more problematic than a study of the erosion of floor tiles (Robson, 1993). Under most circumstances, the confidentiality of data obtained from the use of nonreactive measures can be sufficiently preserved. Perhaps the most compelling question that can be asked about nonreactive measures in general is why they have not been more frequently exploited by researchers as alternatives to deceptive research methodologies.

Contrived observation

Unlike the field research approaches thus far described, contrived observation involves the investigator's intervention into the research setting. By systematically varying aspects of the observational setting it is possible to gain greater control over the variables involved than would be possible in simple observational studies. Contrived observations thus provide the researcher with greater facility for precipitating behavior that otherwise might occur infrequently or unpredictably. There are a number of ways to "tamper" with nature or intervene in situations for research purposes. Among the contrived methods available to researchers are structured observations, field experiments, and participant-observation studies.

Structured observation

Observations of this kind typically involve a minimal amount of intervention or control over the situation in which the observations are to be carried out. Rather, researchers set up a situation by arranging conditions or controlling antecedent events in such a way that consequent behavior can be readily observed. As an example, in a study of self-awareness in chimpanzees, Gallup (1979) equipped the animals' cages with mirrors and slightly altered their appearance (e.g. by putting a dab of red coloring on their forehead). He then considered the chimpanzees' touching of the altered spot as an indication of their having recognized a change in their appearance. In another study, Zimbardo (1969) left automobiles along the roadside in different settings and made it appear as if the cars were abandoned (the license plates were removed and the hoods were raised). Hidden observers then watched from a distance and recorded any instances of contact with the cars, including acts of vandalism and the theft of parts.

As with simple observations, so long as the behavior observed is of a public nature and the risks involved are no greater than those encountered in everyday experience, most structured observations raise few, if any, ethical concerns. For example, while it might be said that the abandoned car study may have encouraged unsuspecting individuals to engage in illicit behavior, the investigation was carried out partly as an attempt to understand behaviors that already were occurring with high frequency in similar settings. To the extent that the structured setting reduces participants' usual range of response or exposes them to certain risks, the research will be more ethically problematic. An early laboratory investigation by Landis and Hunt (1939) may have seemed innocuous at the time, but today may be viewed as more troublesome, especially if replicated as a structured field observation.

Landis and Hunt studied movement responses in subjects by exposing them to various stimuli and filming their gestural response patterns. In one case, when a gun was fired the subjects reacted by drawing the shoulders forward, contracting the abdomen, bending the knees, closing the eyes, and widening the mouth. Imagine the potential problems if an investigator chose to expose subjects in natural settings to the unexpected sound of a gunshot in order to study their behavior free from laboratory contaminations (cf. Webb et al., 1981). Given the high fear of crime and the prevalence of random shootings in contemporary American society, the manipulation might be expected to cause an extreme fear or stress response in some subjects.

Field experiments
As another form of contrived observation, field experiments involve the researcher's attempt to evoke behavior by manipulating an experimental variable or by otherwise modifying some aspect of a naturalistic setting. Despite the potential loss of control over extraneous variables when experiments are conducted in the field, researchers typically gain the ability to observe behavior under conditions presumably free of observational reactivity. This is because for most

field experiments the research participants are unaware of the experimental manipulations and the recording of their reactions. Thus, from the researcher's perspective, the main advantage of this approach is the high confidence that the observed behavior reflects natural behavior on the part of the participants.

Numerous field experiments have been carried out by behavioral scientists, on such topics as altruism and helping behavior, social influence, personal space, and community norms (cf. Reynolds, 1979). Two examples of field experiments already have been discussed in this chapter: the Piliavins' bystander intervention research and Middlemist et al.'s micturation study. To help evaluate the ethical implications of this approach, additional examples follow.

As mentioned, a number of field experiments have been conducted in order to investigate the conditions influencing altruism and helping behavior. Overall, these studies have posed very few risks and have placed few demands on the potential participants. For example, Bryan and Test (1967) observed the effects of having either black or white Santas ringing bells next to a charitable organization's donation buckets. All of the passersby (busy shoppers) represented potential subjects, who either donated or did not donate money. Obviously, the researchers could not have expected to obtain the informed consent of each and every person who passed, or tried to debrief them about the study. With the possible exception of momentary guilt for not contributing to charity, there were no risks in this experiment any greater than everyday life, and all of the donations actually went to the charity.

In a somewhat related study, Crusco and Wetzel (1984) investigated the effect of touching on restaurant diners. It was expected that the size of the gratuity left by the diners would be influenced by the non-verbal behavior of the waitress. Subjects in two restaurants were assigned to one of three conditions: fleeting touch (the waitress twice touched the diner's palm for one-half second while returning change), shoulder touch (the waitress placed her hand for up to one and one-half seconds on the diner's shoulder while giving back change), and no touch (no physical contact was made with customers). The researchers speculated that touching customers on the hand would produce a positive affect toward the waitress and hence a larger gratuity than the shoulder touch, which might have been viewed as a sign of dominance, especially among male customers. Contrary to prediction, there was no difference between the two touching conditions; both male and female diners gave a significantly larger tip after being touched than when they were not touched.

Although the restaurant patrons were not debriefed, we can imagine that some would have voiced resentment towards the researchers for attempting to manipulate the size of the gratuity. Of course, one can argue that the research situation was not fundamentally different from that which normally occurs in restaurants when employees attempt to maximize their tips.

In chapter 3 we considered Latané and Darley's laboratory experiments for investigating bystander intervention during emergencies (see Box 3.3). In order to determine whether similar findings could be obtained in natural settings, the two researchers employed confederates to pose as thieves in a liquor store (Latané and Darley, 1970). As the proprietor (who was in on the experiment) went to the back of the store to search for a requested beverage, one of the confederates picked up a case of beer that was on the counter and stated "They won't miss this." The two confederates then left the shop. This "theft" was carried out in full view of either one or more than one real customers. As had been found in the laboratory studies, a bystander was more likely to intervene (by informing the proprietor of the theft) when it was believed he or she was the only witness to the crime than when another bystander also was present.

Staged thefts have been utilized by other researchers in order to study bystander reactions (e.g. Bickman and Rosenbaum, 1977; Moriarty, 1975). In Moriarty's (1975) study, a male confederate approached unsuspecting sunbathers at a New York beach and, depending on the experimental condition, either asked them to watch his things on a nearby blanket for a few minutes or asked for a cigarette light before leaving. A few minutes later, a female confederate approached the man's blanket, grabbed his blaring radio, and ran away. It was hypothesized that subjects who agreed to watch the man's things would feel that intervention was their responsibility and would be more inclined to help than bystanders who were asked for a light. This indeed is what happened: 95 percent of the bystanders in the responsibility condition intervened by stopping the thief or by grabbing back the radio, compared with only 20 percent of the bystanders in the no responsibility condition. In contrast to some of the other examples described here, field experiments involving apparent crimes add a degree of physical risk to the situation, to the confederates as well as the subjects.

In addition to staging emergencies and crimes in natural settings, behavioral scientists have designed other procedures for studying helping behavior that are relatively harmless in nature. One of these approaches is the "wrong number technique," in which a researcher telephones individuals selected at random, pretends that a wrong number has been reached, and then explains that he was trying to reach a garage because his car had become disabled on a highway. Claiming that he has no more change for another call, the researcher asks the participant to call a garage for assistance and gives a phone number. The measure of helping behavior consists of whether the participant makes the call; if so, a confederate answers and promises to take a message for the "garage." The researchers who developed this technique attempted to determine what effect the race of the subject and the "victim" would have on helping rates (Gaertner and Bickman, 1971). One finding was that a smaller percentage of white subjects (evident from vocal characteristics and neighborhood of residence) attempted to help black victims (as manipulated through actual race and vocal characteristics of the caller) than when the victim was white and the subject black.

Another interesting experimental procedure known as the "lost letter technique" has been extensively utilized by field researchers (e.g. Jacoby and Aranoff, 1971; Merritt and Fowler, 1948; Milgram et al., 1965; Montanye et al., 1971). This technique involves leaving stamped, addressed letters, containing innocuous messages, in conspicuous locations (e.g. sidewalks, parking lots, under automobile windshield wipers) so that it appears that the letters had been dropped or otherwise lost by the intended sender (Milgram, 1969). The helping measure in studies employing this technique is the number of letters returned to a post-office box (where they can be retrieved by the researcher) under varying conditions. Researchers have assessed a wide variety of attitudes with the lost letter technique, including attitudes toward handicapped children, rapid transit, the death penalty, busing, and race. In one study, Milgram et al. (1965) planted letters addressed to one of the following: "Friends of the Communist Party," "Friends of the Nazi Party," "Medical Research Associates," and "Mr Walter Carnap." As expected, fewer of the letters were returned for the first two conditions. Merritt and Fowler (1948) obtained lower return rates for dropped letters that apparently contained a coin (actually a lead slug) than for those that seemed to contain no money. While a number of variations of the lost-letter technique have been used by researchers, the deceptions typically are minor and are ones of omission that pose no risk to participants.

The range of methodologies used in experimental studies of altruism and helping behavior in natural settings provides a useful basis for our evaluation of field experiments in general. When one con-

siders the research examples described here, it is apparent that in
most cases only modest demands are placed on the participants,
involving relatively trivial, low-cost behavior (Reynolds, 1979).
Potential participants are asked to mail a lost letter, make a phone
call to a mechanic, help a subway passenger to his feet, leave a
gratuity, make a donation to charity, report a minor theft, and so on
– behavior that rarely exceeds the range of normal public activities.
The likelihood of subjects experiencing a high level of physical or
psychological stress, embarrassment, or long-term effects is low.
Further, the research procedures typically are carried out in a way
that ensures anonymity and the confidentiality of response; as a
result, the experiments are unlikely to involve intrusions of privacy.
Given these considerations, it is not surprising that a majority of
adults surveyed in public places about the acceptability of eight nonreactive field studies did not react negatively to the studies (Wilson
and Donnerstein, 1976). Among the studies viewed as ethical and
non-invasive of privacy were four involving procedures that have
been described here – the bloody victim manipulation, the abandoned automobile procedure, and the wrong number and lost-letter
techniques.

Because it typically is impractical to obtain the informed consent
of participants for field experiments, some researchers have made a
special effort to ensure that the persons selected for study do not
experience any harmful effects. For instance, in their wrong number
experiment, Gaertner and Bickman (1971) attempted to reduce
participant guilt for a non-helping response by having the caller,
prior to hanging up, say "Here come the police, maybe they can help
me." As previously mentioned, Middlemist et al. (1976, 1977)
conducted a pilot study prior to the decision to go ahead with their
surreptitious observations in restrooms. For the pilot study, an overt
observer stood with a stopwatch and clipboard and, following the
observation, explained the nature of the study to the participants.
The researchers based their decision to carry out the hidden observations only after the pilot subjects had given their permission to use
the data and had expressed their lack of concern about the research.
In the Piliavins' bystander intervention studies, the "victim" was
always assisted by a confederate if none of the bystanders came to his
aid before the next subway stop, suggesting that the "victim" was
not seriously injured (Piliavin et al., 1969; Piliavin and Piliavin,
1972). This step may have reduced any lingering after-effects, such as
guilt or upset on the part of the subjects. Actually, Piliavin and
Piliavin's (1972) "bleeding victim" study was quickly terminated

because of problems with subway security and the fact that some of the reactions were dramatic and potentially dangerous for passengers.

Despite the low risk of harm in field experiments, it is apparent that the direct potential benefits for participants also appear to be minimal, with the exception of the temporary satisfaction that comes from interacting with others or, as in the case of the helping experiments, engaging in altruistic behavior. Rather, the benefits can more generally be described in terms of scientific interest and the potential practical significance of knowledge gained.

One additional concern regarding field experiments that was alluded to in our consideration of the Piliavins' research (see pp. 118–19) has to do with the possibility that surreptitious procedures in natural settings can raise the suspiciousness of ordinary citizens when the events occur naturally. As a result, the tendency to provide help or assistance to others, to obey the orders of legitimate authorities, and the like, might be reduced. However, the likelihood of such outcomes probably is small. The number of surreptitious field experiments that have been carried out has not been unduly large, nor have they been particularly well publicized, at least to the extent that they would be expected to have a major effect on ordinary citizens' social reality (Reynolds, 1979). Nevertheless, the author knows of a case involving a psychology professor who one day noticed a US$20 bill lying on the floor in the lobby of the psychology building. According to the psychologist, he hesitated for some time prior to pocketing the money, not knowing whether this was his lucky day or whether he instead had unwittingly become a subject in a psychology experiment. Such reactions may not be atypical among behavioral scientists and perhaps a growing number of non-scientists!

Entrapment studies

In contrast to the relatively innocuous field studies previously described, entrapment research is apt to raise some serious ethical concerns. Entrapment experiments typically are conducted to investigate moral character by providing opportunities for subjects to engage in dishonest behavior or perform otherwise reprehensible acts. A few examples should suffice to clarify the basic nature of this type of research approach. In one application, Freeman and Ataov (1960) contrived a situation in which students were given a chance to cheat on a classroom test by grading their own examination scoring sheets (which previously had been collected and then returned under some pretense). The researchers were interested in determining whether more changed answers would be apparent for different kinds of question formats. In another study, an experimenter pretending to be a foreigner with very limited English and a poor knowledge of the taxi system tested the honesty of New York City cab drivers by requesting a ride to a nearby location and then overpaying the fare (Brill, 1976). As it turned out, the cab drivers were not as apt to cheat their passenger as one may have expected, with two drivers refusing to take the experimenter to the requested address because it was so close. A few other drivers did take some advantage of the opportunity by cheating the experimenter out of a small amount of money.

In one intriguing entrapment experiment stimulated by the well-known 1972 Watergate break-in, West et al. (1975) attempted to induce participants to agree to take part in a burglary that paralleled the real break-in. The study was intended to test a hypothesis derived from a theory pertaining to the determinants of causal attribution. Consistent with the so-called "actor/observer effect," the researchers noted that participants in the actual Watergate burglary explained their behavior in terms of environmental influences external to themselves, whereas observers in the mass media attributed responsibility for the crime to the personal character of those involved.

In order to test the actor/observer hypothesis under more controlled circumstances, the researchers set up a situation in which criminology students were offered inducements to help take part in the theft of records from a local business firm. The experiment was carried out by having a private detective present the students with an elaborate plan for the burglary. For their part, subjects were offered inducements which included a US$2,000 payment or the chance to perform a patriotic service for the federal government. Other subjects acted as observers in the study; that is, they were given descriptions of the crime and asked to explain why the burglars took part in it. From our perspective, the interesting finding was that a substantial number of the students agreed to take part in the crime (which of course was never carried out).

One of the more serious risks posed by entrapment studies is the possibility that they may change the morals of participants who have been enticed into engaging in disreputable behavior. By encouraging dishonest behavior and creating opportunities for its convenient occurrence, subjects are placed in situations that differ from their everyday life experiences. Thus, a student who finds it relatively easy to cheat on an examination during an experiment may be tempted to try again on future occasions, especially if he or she is not informed about the experiment. Another possibility is that a subject who has either agreed to (or actually performed) a dishonest act such as cheating or a Watergate-like burglary as a result of entrapment might experience subsequent guilt or a loss of self-esteem. Anxiety about the behavior being found out by the experimenter or others (police, college administrators, etc.) is another potential outcome.

Although for a majority of entrapment studies the infractions are so minor or common that serious negative effects are unlikely, the closer the contrived situations are to real life, the more difficult the issues become. Providing students with easy opportunities to commit infractions of the rules in a computer simulation exercise or video game produces fewer risks than enticing them to cheat on an actual course exam. Freeman and Ataov (1960) carried out their cheating study in such a way so as to ensure that students' grades were not influenced by their cheating. However, in such studies there is always the possibility that the results can have lingering effects. A professor who learns that his or her students cheated during an experiment or agreed to commit a burglary may become unduly suspicious of their honest academic efforts, and so on. Thus, it is essential that researchers who use this experimental procedure do everything possible to protect the confidentiality of data and take steps to eliminate the likelihood that individual subjects can be identified.

Webb et al. (1981) have suggested that entrapment studies may be less troublesome when the honesty of people in positions of public trust are studied, such as taxi drivers, physicians, and automobile mechanics. Because these individuals occupy roles that have some bearing on public welfare, the potential benefits to society may outweigh the risks involved. Even so, after considering the various issues involved, Webb et al. (p. 270) reasonably advise that researchers should be "cautious and sparing in their use of procedures which involve entrapment" because of the potential ethical jeopardy in which participants are placed.

Participant-observation studies

For many aspects of group behavior that do not lend themselves to ready observation, such as delinquency in street gangs, homosexual practices, and the customs of primitive cultures, researchers may elect to gain access by becoming a party to the ongoing activity. Participant observation, the last field approach to be considered here, involves a method of observation in which a group or community is studied from within by a researcher who makes systematic notes about what is observed (Rosenthal and Rosnow, 1991). In so doing, it is often possible to grasp the point of view of the participants and to become privy to certain insights about the group (such as its goals, attitudes, informal conversations, etc.) that otherwise may not have been available.

Participant observations vary in terms of the researcher's level of involvement in the ongoing activities of the group. The observer may be present without interacting with group members (so-called "passive observation") or may actually engage in the same activities and interact with others as an ordinary participant (so-called "active observation"). Additionally, the extent to which those being studied are aware of the participant-observer can be seen as falling along a disguised – undisguised continuum (Diener and Crandall, 1978). Studies which fall at the undisguised end of the continuum are those in which the researcher openly participates in the daily lives of those being studied; his or her role is something that group members are aware of. At the disguised end of the continuum are studies in which the individuals who are being observed are unaware of the observer's role or that their behavior is being investigated.

Sociologists and cultural anthropologists are apt to use undisguised participant observations when seeking to describe cultures, an approach also called *ethnographic research*. Because the language of the culture being observed typically is not the native language of the researcher, bilingual assistants may be required in order to frame the interview questions in the indigenous language of the culture. One important advantage of the undisguised approach is that after a brief period of adjustment, participants tend to grow used to the presence of the researcher and behave in a natural manner, unaffected by their awareness of the observation.

One example of undisguised participant-observation is apparent in the work of John Haviland (1977), an anthropologist who lived for ten years in Zinacantan, a small Mexican village. Haviland kept meticulous ethnographic records, including tape recordings of conversations, which he analyzed in order to determine the functions of gossip in the Zinacantan community. While the participants were aware of his anthropological role, they were not informed that one of Haviland's main interests was their gossip. By living and working within the community as an active participant, Haviland eventually was treated as just another villager and so he was able to record relatively natural conversations.

Despite the apparently benign nature of Haviland's research, the fact that he lived for an extended period of time among the Zinacantan people suggests that their behavior was under continuous, long-term scrutiny. As a result, it is clear that issues of privacy are relevant even when a participant observation is undisguised. For example, the observed individuals can become so comfortable with the researcher that they unintentionally reveal opinions or information that they would have preferred not to have recorded. This is particularly likely to happen in informal contexts, where the researcher may be viewed as having stepped outside of the role of scientist (Davis, 1961). It also is possible, because of the reactive nature of this approach, that the researcher's presence will affect the behavior of individual group members, thereby influencing not only their interactions with the researcher but with each other as well. Social relationships could be affected and, ultimately, group cohesion could be seriously weakened. In order to overcome such shortcomings, researchers may prefer to disguise their observations.

One example of a disguised participant observation that we previously considered is Laud Humphreys's (1970) "tearoom trade" investigation. It may be recalled that Humphreys frequented public restrooms as a lookout (or "watch queen") in order to observe homosexual encounters and later disguised himself in order to interview participants in their homes. In another study, social psychologist David Rosenhan (1973) utilized the disguised participant-observation approach to investigate the basis of psychiatric diagnosis and the stigmatizing effects of labeling people as "mentally ill." Eight volunteers (including psychologists, a housewife, a pediatrician, and Rosenhan himself) contacted different mental hospitals, misrepresented their identities, and sought admission by claiming to be hearing voices. Once admitted as patients, the researchers stopped complaining of any symptoms and attempted to act normal. Of interest to the observers were patient–staff interactions and the length of time it took before their release from the hospitals. During their stay, the researchers took detailed notes of their interactions with the hospital staff and other patients. Interestingly, the researchers never were recognized as sane by the hospital staff, but eventually were discharged (after a period ranging from 7 to 52 days) as "schizophrenics" whose illness was said to be "in remission." Rosenhan concluded from the results that diagnostic labels tend to stick, regardless of one's subsequent behavior. It also was noted that the staff frequently seemed to avoid interacting with patients, perhaps accounting for the feelings of "depersonalization" and powerlessness experienced by the researchers and often reported by institutionalized mental patients.

As these examples suggest, disguised participant observation is an effective approach for gaining access to phenomena otherwise unavailable to researchers, and has the advantage of minimizing the potentially disruptive effects of the observations on the individuals being studied. The fact that the observations are disguised, however, raises certain issues that are less relevant when the observer is known to the participants. For example, there is the possibility that embarrassment, anger, or other negative reactions may result if participants learn of the deception during the course of the study or if confidential information is made public. Because of the covert nature of the observations, possible invasions of privacy also must be considered. While mentioned in reference to undisguised studies, there is a greater likelihood that the individuals who are unknowingly observed will reveal information about themselves or behave in a way that they would rather not have made known to a researcher.

If the criterion of publicness of the behavior is applied to considerations of privacy infringement, it is clear that while most disguised observations take place in pubic settings, there may be a degree of privateness involved (e.g. when observing psychiatric patients). An important related issue has to do with the extent to which subjects' anonymity is protected. It appears that most participant observers are inclined to protect the identity of their participants. Humphreys, for example, kept all identifying names of his participants in locked storage and destroyed them after the completion of the interviews so that their identities could not be traced. In fact, while there are those who have argued that disguised participant observations are unethical because of the misrepresentation involved (Erikson, 1967; Jorgenson, 1971), there is substantial evidence that behavioral scientists are aware of the ethical problems and typically take steps to minimize any negative effects (Reynolds, 1979). The potential benefits for the groups involved (e.g. greater sensitivity to the problems experienced by homosexuals or hospitalized mental patients) can be considerable.

Summary

It is difficult to arrive at a general conclusion regarding the ethics of field research because of the wide variety of procedures that may be employed by researchers. Some of these procedures involve the simple observation of ongoing behavior in natural settings, while other approaches rely on some form of contrived, covert observation. As a result, the ethicality of each study should be evaluated on an individual basis in light of methodological requirements and participant rights. Because of the tendency for field research techniques to be conducted without the awareness of the participants, these approaches have the advantage of causing a minimum of disruption to the behaviors under scrutiny and thereby provide an excellent means for investigating ongoing events and activities. However, because of the covert nature of some of these methods, their potential for infringements of participants' rights to privacy and informed consent must be fully considered prior to undertaking a study.

In a sense, participants in disguised studies, such as simple observations and participant observations, can be seen as indirectly providing their consent to observation when they freely choose to engage in an activity in the presence of another person, especially when the setting is a public one. This point is less applicable to other approaches, such as entrapment studies, in which the contrived situation may vary significantly from the course of participants' everyday life experiences. Regardless of the specific approach, the potential risks posed by field research can be minimized through the use of strategies designed to protect subject anonymity and confidentiality of data.

One problem with many of the research techniques considered to this point is that while they pose few direct negative effects to participants, they may not have much to offer them in the way of direct positive effects either. In chapter 6, we turn our attention to a consideration of issues in research which is specifically intended to have an impact on society.

6

Ethical Issues in the Conduct of Human Subject Research III: Applied Research

The true and lawful goal of the sciences is none other than this: that human life be enriched with new discoveries and powers.
Francis Bacon, *Novum Organum*

There's nothing as practical as a good theory.
Kurt Lewin, Problems of research in social psychology

The application of scientific knowledge to practical issues has long been an important goal of behavioral scientists. The seventeenth-century English philosopher and essayist Francis Bacon was one of the first great thinkers to recognize the value of empirical research as a tool for improving humankind. Similarly, one of the founders of social psychology, Kurt Lewin, firmly believed that theories developed through systematic research are often useful in a practical sense. In the much quoted statement that appears above, Lewin implied that once we have obtained scientific understanding of some aspect of behavior, it should be possible to put that knowledge to practical use. In recent decades, the trend toward application in the behavioral sciences has steadily grown, bringing to the fore a set of special ethical problems inherent in what might be called "real-world" research.

In this chapter we begin with a brief discussion of the nature and goals of applied research in the behavioral sciences, followed by a description of two illustrative cases of applied studies that have raised ethical issues. We then focus on two areas involving behavioral science applications – social intervention research and marketing research – in order to illustrate some of the difficult ethical dilemmas that can emerge in applied contexts, along with some suggestions for coping with them.

Applied Research in Perspective

Much of the research that is carried out by behavioral scientists is oriented toward the single goal of understanding a phenomenon, without concern for the relevance or direct usage of the knowledge gained. This is the essence of basic (or "pure") research, which typically is performed in order to obtain knowledge merely for knowledge's sake through the testing of theoretically important hypotheses. For example, researchers might be interested in studying purely scientific questions pertaining to the processes underlying the formation of impressions, the development of group norms, the capacity of short-term memory, and the like.

Applied research, in contrast, is oriented toward the attainment of information that will prove relevant to some practical problem (see Box 6.1). That is, the goal of applied research is to gain knowledge in order to modify or improve the present situation (Shaughnessy and Zechmeister, 1990). While applied researchers also are interested in discovering the reasons for a phenomenon, their primary goal is to practically apply their findings to "real-life" problems or situations. For example, researchers may carry out studies in order to develop an effective program to reduce racial tensions in a community, to design an advertising campaign for the use of condoms in an attempt to control the spread of AIDS, and so on. In essence, while basic researchers are more interested in comparing what is with what should be in order to determine if a particular theory is adequate, applied researchers attempt to understand the nature of a problem in order to do something to resolve it (Bickman, 1981; Forsyth and Strong, 1986).

The basic and applied facets of scientific inquiry add to the research diversity within each behavioral science discipline. Behavioral scientists examine general questions relevant to their areas of interest, and also seek the solutions to practical problems. For instance, Milgram hoped to obtain the answers to basic questions about the nature of obedience by conducting a series of laboratory investigations to find out who obeys, what social factors influence the level of obedience, and what are the accompanying emotional processes that maintain obedience. His initial interest in examining these questions was stimulated by the atrocities committed during World War 2 by the Nazis at the direction of malevolent authority.

In addition to shedding light on the psychological variables related to obedience, Milgram's research also yielded insight into socially relevant concerns, suggesting ways that excessive obedience might be

Box 6.1

Kurt Lewin and applied research

As an early proponent of the application of behavioral science research to practical problems, Kurt Lewin advanced the idea that theoretical progress and the understanding of social problems are interdependent. Consistent with this notion, Lewin proposed an approach known as "action research" that centers on studying things by changing them and observing the effects of the change. In his view, such an approach could be applied to the analysis of existing social problems and could assist in solving them.

An example of an applied research program carried out by Lewin during World War 2 is illustrative of this approach. In a series of experiments, Lewin (1947) set out to determine the most effective persuasive techniques for convincing women to contribute to the war effort by changing their families' dietary habits. The goal was to influence the women to change their meat consumption patterns to less desirable, but cheaper and still nutritious meats; to buy more milk in order to protect the health of family members; and to safeguard the well-being of their newborn babies by feeding them cod-liver oil and orange juice.

Lewin compared the effectiveness of two kinds of persuasive appeals by randomly assigning housewives to an experimental condition involving either a lecture or group discussion on the recommended changes. In both cases, the same information about the benefits of the changes was provided. The results of Lewin's research revealed that actively discussing ways to achieve good nutrition resulted in greater changes toward healthier eating habits than passively listening to lectures. For example, 32 percent of those women who participated in a group discussion later reported using the recommended, unfamiliar meats, compared with only 3 percent of the women who had heard a lecture.

Lewin explained the findings by suggesting that group processes had come into play to reinforce the desired normative behavior for those individuals who had participated in the discussions. To some extent, contemporary support groups, such as Alcoholics Anonymous and Weight Watchers, can be seen as part of the legacy of Lewin's wartime research (Brehm and Kassin, 1993).

prevented. By implying that people who obey malevolent authorities are not necessarily evil and sadistic individuals, his research was useful in revealing the importance of increasing feelings of personal responsibility and recognizing the power of social pressure in combatting mindless obedience (Forsyth, 1987). Similarly, some industrial-organizational psychologists are engaged primarily in theoretically based research programs to better understand the factors that are associated with work motivation, job performance, and satisfaction in the workplace – the results of which ultimately might be used to solve practical problems within business and organizational settings.

Behavioral scientists may be employed part or full time as applied researchers in industrial settings, market research firms, advertising agencies, hospitals and psychiatric treatment facilities, and government agencies to evaluate the effectiveness of current policies and programs, or to design new ones. For investigators who conduct research in these settings, attention typically is focused on the following kinds of practical questions. What factors help workers resist the adverse effects of job-related stress? How can work be redesigned to maximize efficiency and productivity, without exploiting employees? What variables influence workers' satisfaction with their jobs? Do non-job-relevant personal characteristics such as an applicant's appearance, color, or religion influence the outcomes of job interviews? Which persuasive techniques are likely to influence consumers' interest in a new product? What kinds of argument influence the decisions of jurors? Are eyewitnesses accurate in their recall of events surrounding a crime? Do special education programs, such as Head Start (a compensatory pre-school program for poor children in the United States), lead to an increase in students' skills (Baron and Byrne, 1994; Forsyth, 1987)?

Another approach to applied research in which investigations are carried out in naturalistic settings is called *social experimentation* (Riecken, 1975). This involves the application of experimental methods to the analysis of social problems and to the development and assessment of interventions to reduce the problems. For example, during the 1970s, the United States government experimented with a program designed to assist individuals following their release from prison (cf. Berk et al., 1987). It was hoped that the program could help ex-prisoners overcome adjustment problems linked to limited resources and difficulties in finding steady employment. Such problems had been identified as contributing to a return to crime among ex-convicts. For the government program, about two thousand newly released prisoners in Texas and Georgia were randomly assigned to a condition in which they either received or did not receive financial assistance. As is sometimes the case with social experiments, the outcomes of the prison release program were not entirely beneficial. The program resulted in some identifiable positive results, but also turned out to be responsible for an increase in crime among some of the financially assisted ex-prisoners, a consequence apparently due to the freedom they were given from having to work.

Social experimentation frequently involves quasi-experimental research designs in which treatments are assigned to subjects on a non-random basis. In essence, quasi-experiments resemble true experiments in the sense that they both involve some type of intervention or treatment and outcome measures. Quasi-experimental designs, however, lack the degree of control found in true experiments because there is no random assignment to create comparisons from which treatment-based changes are inferred (Cook and Campbell, 1979). If a researcher wanted to evaluate the effects of low-cost public housing, for instance, it would be difficult or impossible, for administrative and ethical reasons, to randomly assign families to different housing conditions (Rosenthal and Rosnow, 1991). As a result, the researcher might instead use a non-equivalent-groups design by comparing families already residing in low-cost housing with families residing in other types of housing.

In one example of an informative quasi-experimental study, researchers investigated the effectiveness of the British breathalyzer crackdown of October 1967, a program initiated as an attempt to reduce drunken driving and, consequently, serious traffic accidents (Ross et al., 1970). Through a comparison of driving casualties (fatalities plus serious injuries) before and after the breathalyzer crackdown, a temporary drop in the accident rate following the onset of the program was revealed. In order to rule out alternative interpretations of the data, such as weather changes, the introduction of safer cars, a police crackdown on speeding, and so forth, the researchers considered accident rates during the hours British pubs are open (particularly during weekend nights) and during commuting hours when pubs are closed. This analysis revealed a marked drop in the accident rates during weekends and a less notable decline during non-drinking weekday hours. A follow-up analysis by Ross (1973) served to rule out other possible explanations for the apparent breathalyzer effects and also helped to explain why the effects were short lived: the British courts increasingly failed to punish drinkers who had been detected by the breathalyzer. Taken together, applied research on the British breathalyzer crackdown led to the highly useful conclusion that such a program will help reduce serious traffic accidents when used to discourage drunken driving, but only so long as the courts enforce the drinking and driving laws (Cook and Campbell, 1979; Ross, 1973). A similar research approach was utilized to study the influence of the 1900 revision of German divorce laws (Glass et al., 1971). Despite some potential technical problems, social experiments have been used effectively in a number of countries for the development and evaluation of a wide range of social programs and policies.

Ethics and the basic versus applied research distinction

That behavioral scientists may choose to direct their scientific activity from a purely theoretical or applied orientation has certain implications in terms of ethical responsibilities and the role of values in the research process. While sharing certain fundamental principles of research, the main distinction between these two approaches is that basic science remains unchallenged by practical problems while applied research is essentially atheoretical in nature (Pepitone, 1981). Basic researchers traditionally have viewed their role as an active one in the discovery of truth but a passive one in determining the societal use of their findings. Underlying this position of scientific non-responsibility is the assumption that although research findings can be used for good or bad ends, knowledge in and of itself is ethically neutral. In this light, basic researchers generally contend that their work is value free and morally neutral – as implied by the labels "basic" and "pure" – because their goal is the impersonal pursuit of scientific knowledge for its own sake.

These views pertaining to the basic science tradition have been criticized on various grounds by those who disagree that basic research is value free and morally neutral. For example, some critics consider it immoral *not* to use the knowledge derived from basic research to attempt to reduce real-life behavioral and social problems (e.g. Baumrin, 1970). Others argue that basic research often entails the use of unethical procedures for obtaining knowledge (as when human subjects are mistreated during the course of a theoretical study) and point out the potential destructiveness of some knowledge for personal and social life, such as the undermining of character and social customs (e.g. Smith, 1978). Indeed, it is possible to find examples of past abuses in applications of "pure knowledge," such as research leading to development of the atomic bomb.

Turning to the applied side of the behavioral sciences, it may be argued that research ethics become increasingly important as the results of investigations acquire public policy and other societal implications outside the research professions. For reasons expressed above, this does not mean that responsibility for knowledge produced can be avoided by restricting one's scientific activity to a basic science approach. Conversely, the view that only applied research is ethical because of its assumed potential for social benefits is also inadequate when we consider that some of the greatest breakthroughs in science have come about through basic theoretical research. In fact, moral distinctions tend to blur when the dichotomy between basic and applied science is more closely evaluated, perhaps because the two types of research are not as different as they first may appear. Applied research often leads to theoretical understanding, and theoretical advances permit practical applications. That is, theory does not arise in a social vacuum apart from the concrete events that gave impetus to it, and yet theories of behavior must withstand tests of practical application in order to become established within the scientific community (Georgoudi and Rosnow, 1985; Sarason, 1981).

A shift away from a purely theoretical approach has occurred at various stages in the development of certain behavioral science disciplines, such as psychology and political science. However, each attempt at relevance has brought with it a corresponding set of subtle and complex ethical issues (cf. Isambert, 1987). It has been suggested that some of these issues may have arisen as a result of researchers having failed to consider the ethical implications of using many of their established laboratory and field procedures in applied settings. Because so much experimental research in psychology, for example, has been based on the manipulation and control of variables, psychologists who attempt to solve practical problems may similarly try to manipulate and control variables in the natural world, perhaps inappropriately (Argyris, 1975). If so, it is easy to understand why many of the ethical dilemmas encountered in theoretical research (such as those involving informed consent, debriefing, and confidentiality) also have appeared in applied settings.

Illustrative examples of ethics and applied research

In contrast to many theoretical investigations in controlled settings, applied studies often deal with significant aspects of social, psychological, political, and economic functioning, and require considerable resources and staff to be conducted adequately. As with other types of research, informed consent, the protection of subjects' privacy rights, and confidentiality constitute the primary responsibilities of applied researchers in maintaining respect for their research participants. In the event that the participants in applied studies are inadvertently harmed as a result of the research, compensatory measures, including health and medical care, and economic, social, and other types of reparation must be available.

One of the most serious issues to be considered in applied contexts has to do with the possibility that an investigation will have unintended adverse effects. Some applied studies in fact have resulted in unanticipated consequences for research participants or society in general, such as the previously described program to assist ex-prisoners and the Cambridge – Somerville Youth Study (discussed below). These studies serve as stark reminders of the potentially damaging effects of applied research in general. Among the additional ethical issues inherent in applied research are those involving the methodological requirements of random selection and the assignment of participants to experimental conditions. In this regard, the use of volunteer subject groups and the appropriateness of including untreated control groups in social intervention studies have posed ethical problems for researchers. For researchers employed in applied settings, other issues are likely to emerge when the multiple roles and responsibilities inherent in their work give rise to ambiguous and conflicting expectations. A behavioral scientist employed in an organizational setting, for example, might disapprove of the goals and strategies of the organization or question the fairness and justification of certain management decisions (Walton, 1978). Likewise, a researcher might be asked to implement an intervention which has consequences that conflict with his or her own values. These types of conflict can present troublesome ethical dilemmas for the researcher.

Prior to considering these issues in greater detail as they apply to the areas of preventive intervention and marketing research, two brief cases of applied research are presented: (1) the New Jersey negative income tax study and (2) the quality-of-work-life experiment at the Rushton Mining Company. These examples are illustrative of some of the ethical problems that can arise as a result of the applied research process.

The New Jersey negative income tax experiment
This social experiment was designed to assess the impact of guaranteeing poverty-stricken American families a certain minimum income (Kershaw and Fair, 1976; Kershaw and Small, 1972; Pechman and Timpane, 1975). Despite its positive goals, the study ultimately led to considerable public and governmental concern about privacy issues. The project involved the support of low-income families with monetary sums (i.e. "negative taxes") representing various percentages of the poverty level, ranging from US$1,000 to US$2,000 per year. In order to avoid the disincentive to work, the amount provided was reduced by a graduated percentage of the base level if the family's income rose above the level of the payments.

Because the study could not be distributed randomly on a nationwide basis, the decision was made to restrict it to an area including the state of New Jersey and nearby Pennsylvania cities; however, similar studies also were carried out in other states (see Haveman and Watts, 1976). In the New Jersey study about 1,300 families were randomly divided into seven experimental groups, which differed in terms of income guarantee levels and tax rates assessed on earned income. A no-payment control group also was included, consisting of families who were free to use the regular welfare services available to poor families. (The control families did receive US$10 per month from the researchers to compensate for time spent completing monthly questionnaires.)

Participants in the study were interviewed before, after, and every three months during the experiment, and were asked to complete detailed questionnaires consisting of a number of items about their income and work-related activities. Among the most important findings of the experiment was that the subsidy did not appear to have demoralizing effects on the participants; that is, the families did not stop working when they received a guaranteed minimum income. Nonetheless, the experiment was not without its pitfalls. Although the families had been promised confidentiality, local law enforcement agents viewed the project as an opportunity to check on welfare cheating and attempted to obtain the confidential information in order to expose tax fraud. Individual records containing participant identification and limited information eventually were demanded by a grand jury and a congressional investigating committee. Having promised in good faith to protect respondents' data, these actions placed the researchers in a difficult position, uncertain as to whether that promise could be kept if a court issued a subpoena. Eventually, the officials were convinced to discontinue their demands for relevant data, but not before the release of 14 records at one site and the identification of 18 families participating in the experiment at a second site.

Despite the lack of serious consequences, beyond the decision of some participants to discontinue their involvement in the project, the negative income tax experiment illustrates the problems that can arise when external demands are made for the disclosure of information elicited under a promise of confidentiality. Similar conflicts between research requirements and legal standards are prevalent in applied research on other sensitive topics, such as drug use and criminal behavior. One additional consideration regarding the negative tax experiment has to do with the dilemma that occurs when a beneficial experimental program is terminated. In that study, the income payments eventually were cut off after the participants had become accustomed to receiving them for three years. The abrupt reduction of family income may have resulted in adverse effects for some of the individuals involved.

The Rushton Coal Mine experiment

Whereas the negative income tax study mainly involved privacy and confidentiality issues, the Rushton field experiment raised issues concerning perceptions of unfairness relative to the establishment and treatment of experimental groups. Developed on the basis of earlier research carried out in the United Kingdom (Trist et al., 1963), the 1973 Rushton project was designed to improve the quality of work life at the Pennsylvania mining company (Blumberg, 1980; Susman, 1976). Specifically, the researchers had hoped to improve employee skills, safety, and level of job satisfaction while at the same time maximizing the company's productivity and earnings (Blumberg and Pringle, 1983).

Following an extended period of preparation, an experimental group of volunteers was organized which was to have full responsibility for the production in one section of the mining operations. The volunteers were encouraged to step out of their traditional roles and were extensively trained in effective mining and safety practices before they were left to coordinate their own work activities. Relative to the other workers in the rest of the mine, the volunteer group had the distinct advantages of being paid at the top rate as the highest-skilled job classification in the section and had greater control over the nature of their work. Within a short period of time they became enthusiastic proponents of their newly acquired approach to work.

At this point, the reader perhaps can imagine what happened next at Rushton. The other workers (i.e. the untreated control group), angered by the "haughty" attitude of the volunteers and suspicious of their receiving additional benefits from the company, developed a strong resentment towards the company and the volunteers. Control subjects complained that the volunteers did not deserve a higher salary than more senior employees in the company. Additionally, rumors began to circulate throughout the mine that the volunteers were "riding the gravy train," that the company was "spoon-feeding" them, and that the researchers were "pinko" college people who intended to "bust the union" (Blumberg and Pringle, 1983). With these signs of impending conflict in the air, a decision was made to abruptly terminate the experiment.

Unlike the other cases that we thus far have considered, the Rushton experiment is unique in the sense that the ethical problems did not emerge as a result of the use of deception, invasion of privacy, or related procedures, but because some participants perceived that they were at a disadvantage relative to others in the experiment. That is, a sizable number of non-volunteers believed that it was unfair not to receive the same benefits enjoyed by the experimental work group (Rosenthal and Rosnow, 1991).

The issues raised in the Rushton case are reflected in related situations, such as in biomedical research that includes placebo-control groups (i.e. comparison conditions consisting of subjects who are administered substances without pharmacological effects as "drugs"). In such studies there are potential ethical costs in not providing treatment to some subjects, particularly if the experimental treatment is shown to be effective. As a result, it may be more ethically appropriate to provide a "common" or "usual" treatment to control subjects. In another example, Broome (1984) described the ethical issue of fairness in the selection of people for chronic hemo-dialysis, a potentially life-saving medical procedure for individuals whose kidneys have failed. This procedure happens to be quite expensive and facilities are not available in many communities for treating everyone who could benefit from it. While the inventor of the procedure proposed such selection criteria as age (under 40 years old), health (free from cardiovascular disease), lifestyle (churchgoer and financial supporter of the community), and marital status (married), Broome suggested that the most ethical selection approach would be based on randomness.The ethical dilemma in this situation involves questions of which procedure is more "ethical" – randomness or selecting on the basis of who is most likely to survive (Rosenthal and Rosnow, 1991).

The negative tax experiment and the Rushton mining project suggest that action research can frequently have its own problems, quite apart from those encountered in more theoretically oriented studies. Some of these problems are described in greater detail below as we turn our attention to two quite different areas of applied research activities.

Ethical Issues in Prevention Research

As a form of social intervention, prevention research involves the experimental or quasi-experimental investigation of natural behavior in community settings. The overriding goal of such research is the prevention of psychological and related problems, not their treatment once they have become manifest. Unlike most theoretical research programs, the participants in preventive research projects, who often come from low-income and minority populations, stand to benefit directly from their involvement in the research. In essence, prevention research is intended to discover ways to offset human suffering by intervening prior to the appearance of a maladjustment problem.

Prevention studies tend to involve interventions that are educational in nature, oriented toward teaching people different ways to think, behave, and interact in order to protect them from depression (e.g. Muñoz et al., 1982), difficulties in interpersonal relationships (e.g. Spivak et al., 1976; Spivack and Shure, 1974), behavior putting them at risk for AIDS (Rotheram-Borus, 1991), drug abuse (e.g. Polich et al., 1984), and other potential adjustment problems. For example, pre-schoolers from disadvantaged homes, recently divorced people, individuals functioning under high-stress conditions, adolescent runaways, and parents fitting patterns indicating that they may become abusive towards their children might be seen as "at risk" populations. Once such groups are identified, they can be recruited for appropriate intervention programs, including educational, psychotherapeutic, and coping-skills training.

Values play an important role in preventive intervention studies because the ultimate goal is to prevent mental health patterns which the researcher believes are potentially damaging to others (Mulvey

Table 6.1 Ethical issues in prevention research

1 Privacy and confidentiality, especially in longitudinal prevention studies
2 Informed consent and coercion of participants in studies that may significantly affect their life experiences
3 Unanticipated adverse effects resulting from interventions
4 Effects of labeling research participants as "at risk"
5 Creation of dependencies on research-related resources
6 The use of untreated control groups
7 Selection of individuals for participation in preventive programs (see chapter 7)

and Phelps, 1988). As a result, ideologies often enter into decisions as to what should be prevented (Keith-Spiegel and Koocher, 1985). Despite the humanistic intentions of prevention research programs, several complex ethical issues can arise from this research activity (see Table 6.1).

In measuring the success of prevention efforts, it often is necessary to collect data concerning subjects' economic and social status, home and family relationships, and mental and physical well-being longitudinally over extended periods of observation (sometimes lasting for several months or years). Data often must be stored for long periods of time and may become accessible to a variety of persons and interest groups. Thus, ethical problems relevant to invasion of subjects' privacy and the extent to which confidentiality can be guaranteed for future use of data may become particularly salient in ongoing prevention studies (Blanck et al., 1992; Sieber and Sorensen, 1991). To illustrate, one planned investigation involved a private consulting group's attempts to develop a drug prevention program in a local high school (cf. Boruch and Cecil, 1979). The researchers intended to use a personality inventory in order to predict which of the students were likely to become drug abusers. Once identified, the potential drug abusers then were to be referred to supportive services. However, as a result of a civil suit, *Merriken* v. *Cressman*, the court ordered the project to be halted, partly on the grounds that records on the students' propensity to use drugs would not be kept confidential. Despite the researchers' good intentions to the contrary, it was concluded that record dissemination was virtually uncontrollable.

Like privacy and confidentiality, informed consent becomes especially important in preventive intervention programs, primarily because individuals may be strongly influenced by the research, much more so than is typical in purely theoretical investigations. When subjects may be permanently changed by an experiment, the importance of voluntary participation is magnified. Special care must be taken to inform potential subjects about their rights and obligations, the procedures they will undergo, potential risks and anticipated benefits, and their freedom to withdraw at any time. When studying minority group members, children, and non-English-speaking subjects, informed consent may be especially problematic. For these persons, there are likely to be limited possibilities of articulation and understanding to adequately judge the purpose, risks, and other aspects of the investigation. These limitations further threaten the adequacy of the informed consent process and increase the danger of coercing relatively powerless individuals into prevention programs. (Issues relevant to the use of special populations for research purposes are discussed in detail in chapter 7.)

Unanticipated consequences of prevention research

As previously suggested, there are a number of unforeseen ways in which new research innovations can have harmful effects on individuals, and investigators need to be aware of potential unintended consequences prior to carrying out a prevention study. That preventive research can be damaging to the individuals they are designed to benefit is a danger that can be easily overlooked or minimized (Lorion, 1983). This was apparent in the aforementioned program that led to an increase in crime among financially assisted ex-prisoners. Another illustration was provided by Joan McCord (1978) in her 30-year follow up of the Cambridge–Somerville Youth Study (Powers and Witmer, 1951), a prevention program aimed at delinquent youths. The origins of the Cambridge–Somerville project can be traced to 1939 when Richard Clark Cabot, a social philosopher and physician, began an experimental program intended to prevent delinquency among adolescent males in Boston. Cabot's research involved 506 boys (aged 5 to 13), half of whom had been judged as "difficult" or delinquency prone, while the others were considered as "average." An equal number of boys from each group were randomly assigned to a prevention-oriented counseling program, where they received tutoring, medical assistance, and friendly advice for approximately five years. The remainder of the subjects were assigned to a no-treatment control group.

More than 30 years later, McCord and her research team conducted an evaluation of the Cambridge–Somerville project in order to assess its long-term effects. Through the use of official records and personal contacts, the researchers were able to obtain information on 95 percent of the original participants. Although it was found that the subjective evaluations of the program were generally favorable among those who experienced it, the objective criteria revealed a quite different, disturbing picture. None of the matched comparisons between the treatment and control groups showed that Cabot's prevention program had improved the lives of treated subjects; rather, the only significant differences favored those who had *not* experienced the intervention. Subjects who had participated in the delinquency prevention program were more likely than non-treated subjects to have become alcoholics, suffered from more stress-related and mental illnesses, died at a younger age, tended to be employed in lower-prestige occupations, and were more likely to commit second crimes.

In short, McCord's findings revealed not only that the Cambridge–Somerville project failed to achieve one of its basic goals in preventing treated subjects from committing crimes, but also apparently produced negative side-effects. Unfortunately, such findings in prevention research are hardly unique. In another community-based intervention program, delinquents and adolescents identified as potential delinquents were matched with adult "buddies" from their neighborhoods (Fo and O'Donnell, 1975). The program apparently helped those with criminal records, but those participants without records of criminal offenses from the year preceding the program committed *more* major offenses in the following year than untreated controls. Other examples of prevention programs that have resulted in unintended negative consequences have been described by Gersten et al. (1979) and Lorion (1984).

Preventive interventions can result in the kinds of unanticipated consequences described above in a number of subtle ways. For example, McCord suggested that treated subjects in the Cambridge–Somerville project may have become overly dependent on agency assistance, and that these individuals then experienced resentment when the outside assistance no longer was forthcoming. McCord also conjectured that the treatment program may have increased the likelihood that participants perceived themselves as being in need of help in order to justify receiving the program's services. Fo and O'Donnell (1975) reasoned that treated youngsters in their study with no criminal records may have formed relationships with offenders as a result of their participation in the buddy program, which consequently raised the probability of their committing crimes as well. Thus, exposure to certain uncontrollable features of preventive treatments can produce self-defeating psychological and behavioral effects in research participants.

Some additional explanations for the occurrence of unintended effects in prevention research have been suggested by Muñoz (1983). According to Muñoz, intervention strategies that are targeted to alleviate the incidence of conditions known to place people at high risk for psychological problems might instead serve to increase their occurrence. For instance, a program found to be successful in preventing the negative consequences of divorce, such as a reduced ability to function or traumatic effects on children, could make it easier for people to consider divorce. A similar process could operate to cause a corresponding increase in drug and alcohol abuse or premarital sexual relations as individuals become aware of programs capable of preventing the adverse consequences of those behaviors, such as addiction and sexually communicable diseases, respectively.

Another issue raised by Muñoz has to do with the emphasis on self-control that is implicit in many intervention approaches; that is, participants in prevention programs often are encouraged to believe that what happens to them is in large part under their own control. Acceptance of this message is intended to promote self-fulfillment strategies that allow the individual to approach life experiences in healthier and more adaptive ways. According to Muñoz, however, this self-control orientation could threaten the traditional power structure in households where a certain family member – such as the husband in Latino families – traditionally dominates. An intervention that successfully conveys the self-control message to treated participants might then motivate them to demand more power in their households in an attempt to increase control over their own lives. Similarly, when Cook (1970) successfully reduced prejudice among people who were unaware that they were serving as research subjects, he inadvertently may have caused them great difficulties when they returned to their family and friends who were still prejudiced (Diener and Crandall, 1978).

These examples suggest that the reduction or elimination of certain maladaptive behavior, beliefs, or thinking processes in at-risk persons could cause subjects to question a way of life or traditional values at the core of shared belief systems in their cultural environments. This possibility represents a difficult dilemma for researchers, one that is tied to the fact that the lives of subjects may be affected in ways that cannot always be anticipated without a sensitivity to the nature of the social contexts from which the subject samples are drawn.

Some critics might suggest that prevention programs should not be carried out because of their potential for damaging side-effects; however, it is possible to identify some more appropriate ethical responses to these issues. McCord (1978), for instance, concluded her Cambridge–Somerville follow up by urging readers not to use the findings as grounds for discontinuing social action programs, but rather as a reminder of the need to conduct pilot studies to assess the potentially damaging effects of preventive treatments before introducing them on a wide-scale basis. Another recommendation for dealing with the problem of unanticipated consequences without reducing research on new treatments is to change the focus of the intervention. For example, rather than directing one's research efforts to the prevention of negative outcomes suffered by divorcees, programs could be devised to assist high-risk couples to achieve more fulfilling marriages or to avoid marriage altogether when there are strong indications that it will fail (Muñoz, 1983). Others have suggested that unintended effects can be avoided if investigators are careful in considering the timing of their interventions. Although certain types of behavior targeted for prevention may be apparent in both risk and non-risk populations, they typically will disappear from non-dysfunctional persons. If a researcher attempts remediation efforts too early, the intervention could cause an interruption of this process and the premature labeling of certain individuals as "at risk" (Gersten et al., 1979; Lorion, 1983).

Labeling effects in prevention research

A common strategy in prevention research is to achieve early identification of persons at risk and to modify their environments in order to reduce that risk. This necessitates that significant persons in the environment typically must be alerted of their at-risk status, such as teachers, parents, employers, and perhaps peers. This early identification and labeling of individuals with apparent mental health or behavioral difficulties represents another sensitive issue for prevention researchers. The "at risk" label can result in a stigma that serves to raise the probability that the initial diagnosis will be confirmed (cf. Mulvey and Phelps, 1988; Rosenthal, 1994).

Research on the effects of labeling children in the classroom has demonstrated that teachers' expectations can operate as self-fulfilling prophesies of their students' intellectual growth (cf. Cooper and Hedges, 1994; Rosenthal, 1991; Rosenthal and Jacobson, 1968). Similarly, treatment by parents, teachers, and peers can produce powerful effects on children who are labeled as high-risk delinquents, drug abusers, and the like. The children could become victims of self-fulfilling prophesies in rating and grading, stereotyping, scapegoating, destructive teasing, name calling, and so on. These potential negative effects were taken into account by the courts in the *Merriken* v. *Cressman* case mentioned previously. The agency responsible for developing the prevention program had not anticipated the possible scapegoating and labeling effects of the process used to predict potential drug abusers. Exacerbating the problem was the fact that evidence for the validity of the personality inventory used to identify "at risk" high schoolers was weak (Boruch and Cecil, 1979).

Designing interventions that minimize or avoid the labeling and stigmatizing process represents one way that the potentially negative unintended effects of preventive research can be counteracted. Towards that end, Jason and Bogat (1983) recommend that an entire classroom of youngsters could be exposed to a treatment, with none identified as the specific target of the project. At the conclusion of the intervention the researchers could ascertain whether those youngsters who initially were lagging in basic skills had progressed in performance. Thus, the success of the intervention would be determined without resulting in potentially damaging labeling effects. Other recommended solutions to the labeling problem, such as regional "broadcast" designs and regression-discontinuity designs, are beyond the scope of this book. Discussions of these strategies can be found in Boruch and Cecil (1979), Cook and Campbell (1979), Kimmel (1988a), and Trochim (1982).

Untreated control groups

When preventive intervention studies are experimental in nature, ethical issues are likely to emerge regarding the use of untreated control groups and other comparison groups. These issues tend to revolve around questions of fairness in terms of who from among a population of individuals is to receive an experimental treatment when all would stand to benefit from it. Is it ethical to designate some needy persons to an untreated control condition for the purpose of comparing the effectiveness of the treatment under investigation? To a great extent, decisions on such questions must be made on the basis of the relative value of the treatment and current alternatives, as well as the immediate needs of participants.

Applied researchers tend to be in agreement that the random

assignment of persons to untreated control groups is an ethical practice, so long as the benefits of the treatment are discovered *after* the results are analyzed (Campbell and Cecil, 1982; Schuler, 1982). However, when some direct loss to untreated control subjects can be foreseen prior to carrying out the study, this approach is more difficult to defend, especially when failing to receive a benefit is experienced as a loss by those persons affected. A methodological problem also may arise as a consequence of the random assignment if untreated control subjects become disappointed about their relative deprivation. Such subjects may drop out of the experiment at a greater rate than treated subjects, thereby making it more difficult to assess the true effect of the treatment.

Decisions about the use of untreated control groups can be based largely on a determination of whether there is good reason to believe (based on evidence obtained from previous research) that the effects of a preventive technique will be highly effective, weak, or unknown. It would be difficult to justify the use of no-treatment groups if, in the consensus of experts, a preponderance of prior evidence suggested that the effects of a prevention technique would be strong. In this case, the experimental treatment should be compared not with a no-treatment control group, but with the best treatment available at the time. This approach is potentially more valuable from a methodological standpoint as well. Rather than making a comparison with the untreated control group and perhaps learning that the intervention is better than nothing, one can determine whether the intervention is better than the current treatment of choice.

The use of untreated controls is likely to be viewed as ethically appropriate, even though the preventive intervention is expected to be strong, in research situations where the beneficial treatment could not reasonably be offered to everyone in need because of limited resources. This point has been used to justify the inclusion of untreated, randomly assigned control subjects in Head Start programs for deprived pre-schoolers (Diener and Crandall, 1978). Although placed at a disadvantage, the control subjects would not be unlike thousands of other disadvantaged children around the nation whom the investigators could not possibly have been expected to help. In fact, one can argue that perceptions that one has been deprived of a potentially beneficial intervention is but a "focal deprivation," made salient by the existence of the experimental group (Kimmel, 1988a). Except perceptually, control group participants would not be less deprived if the experiment were canceled and no one received the treatment.

Another case in which untreated control groups would be acceptable is that in which participants can experience a successful treatment once the research is completed, or as soon as the effectiveness of the treatment becomes evident, as long as they do not suffer substantially in the meantime. Researchers should recognize their ethical obligation to assist untreated control subjects to whatever extent possible upon completion of an investigation. When a researcher lacks evidence to substantiate the effects of an untested intervention, having subjects serve in an untreated control group would be justified because it is not known whether they would be in a comparatively deficient condition. For situations in which the effects of a preventive intervention are unknown or doubtful, the key issue is whether treated subjects will be harmed or placed at an unfair disadvantage to persons receiving the current accepted treatments.

Summary

As should be apparent from the foregoing discussion, the ethical issues involved in areas of applied science in which research participants may benefit directly are numerous and complex. The issues considered above are representative, but they do not comprise all of the potential problems involved in intervention research. For example, there are additional ethical considerations relating to the use of voluntary participants in prevention programs (see chapter 7). Next, we turn our attention to another area of applied research in order to focus on some of the ethical problems that emerge when behavioral scientists carry out research as consultants or as employees within business and organizational settings.

Ethical Issues in Marketing Research

The primary purpose of marketing research is to improve marketing-related decisions by providing information that can be used to identify and define marketing opportunities and problems. Marketing research may also be carried out in order to generate and evaluate marketing actions and to monitor current marketing performance (Aaker and Day, 1990). Consumer goods manufacturers often rely on researchers in order to develop new products or new advertising campaigns for products already available in the marketplace. Even radio stations have relied on marketing research in order to determine how to best improve their standing in a highly competitive business area (see Box 6.2).

Marketing researchers, like their counterparts in related disciplines, have been sensitive to ethics for some time. As early as 1962, the American Marketing Association (AMA) adopted a *Marketing Research Code of Ethics*, which was revised in 1972 (see Box 6.3). The decision to draft a code of ethics came about largely as a result of an apparent rise in research abuses among large and respected business organizations, such as the use of marketing research approaches in order to disguise sales purposes (Twedt, 1963). It was feared that without a code of ethics, such abuses ultimately would lead to a loss of consumer cooperation with researchers and to substantial deterioration of marketing research in general.

The AMA research code, which supplements a more general code of ethics for association members, is somewhat limited in that it

---

Box 6.2

Marketing and radio research

Marketing research has been utilized by radio stations in order to make better decisions on what will draw more listeners and advertisers. In a rudimentary sense, some radio stations carry out their own "research" by keeping track of the age and gender of callers who request certain songs to be played in order to determine which songs are popular with which listener groups. In addition, marketing research consultants may be hired to determine more systematically the characteristics of a station's current or potential audiences, such as their ages, income level, buying habits, political beliefs, likes and dislikes, and leisure activities. This information then is evaluated to help determine how to program a station, from the music it plays to the vocal patterns of its disc jockeys.

Among the research techniques employed by radio researchers are elaborate mail surveys to determine the characteristics of listeners and non-listeners, telephone interviews in which respondents' reactions to different songs are obtained, and in-depth face-to-face interviews and discussions with selected audiences (so-called "focus groups"). Researchers also may interview buyers of CDs and tapes or ask them to complete questionnaires detailing why they made their recent purchases. More controlled, laboratory experiments also have been carried out in order to test the effectiveness of various advertising appeals. As a more surreptitious approach, some radio stations have even sent researchers out on the road as hitchhikers to obtain more unguarded comments.

Box 6.3

Marketing research code of ethics

The American Marketing Association, in furtherance of its central objective of the advancement of science in marketing and in recognition of its obligation to the public, has established these principles of ethical practice of marketing research for the guidance of its members.

Adherence to this code will assure the users of marketing research that the research was done in accordance with acceptable ethical practices. Those engaged in research will find in this code an affirmation of sound and honest basic principles which have developed over the years as the profession has grown.

For research users, practitioners, and interviewers

1 No individual or organization will undertake any activity which is directly or indirectly represented to be marketing research, but which has as its real purpose the attempted sale of merchandise or services to some or all of the respondents interviewed in the course of the research.
2 If a respondent has been led to believe, directly or indirectly, that he is participating in a marketing research survey and that his anonymity will be protected, his name shall not be made known to anyone outside the research organization or research department, or used for other than research purposes.

For research practitioners

1 There will be no intentional or deliberate misrepresentation of research methods or results. An adequate description of methods employed will be made available upon request to the sponsor of the research. Evidence that field work has been completed according to specifications will, upon request, be made available to buyers of research.
2 The identity of the survey sponsor and/or the ultimate client for whom a survey is being done will be held in confidence at all times, unless this identity is to be revealed as part of the research design. Research information shall be held in confidence by the research organization or department and not used for personal gain or made available to any outside party unless the client specifically authorizes such release.
3 A research organization shall not undertake marketing studies for competitive clients when such studies would jeopardize the confidential nature of client–agency relationships.

For users of marketing research

1 A user of research shall not knowingly disseminate conclusions from a given research project or service that are inconsistent with or not warranted by the data.
2 To the extent that there is involved in a research project a unique design involving techniques, approaches, or concepts not commonly available to research practitioners, the prospective user of research shall not solicit such a design from one practitioner and deliver it to another for execution without the approval of the design originator.

For field interviewers

1 Research assignments and materials received, as well as information obtained from respondents, shall be held in confidence by the interviewer and revealed to no one except the research organization conducting the marketing study.
2 No information gained through a marketing research activity shall be used, directly or indirectly, for the personal gain or advantage of the interviewer.
3 Interviews shall be conducted in strict accordance with specifications and instructions received.
4 An interviewer shall not carry out two or more interviewing assignments simultaneously unless authorized by all contractors or employees concerned.

(This code was adopted by the American Marketing Association in 1962 and includes 1972 revisions.)

focuses mostly on the researcher and user (or client) relationship and emphasizes problems inherent in survey research, which was the dominant marketing research paradigm at the time the code was drafted. Codes of ethics have been developed by other marketing organizations, and corporate codes of conduct have been implemented by a number of companies (Ethics Resource Center, 1979). The Society for Industrial and Organizational Psychology (SIOP), the major national organization for the profession of industrial and organizational psychology in the United States, has initiated steps to update its current casebook of ethical issues (Lowman, 1985) and there have been discussions about the feasibility of developing an ethics code which directly addresses issues relevant to the professional practice of industrial and organizational psychology (Lowman, 1993; Seberhagen, 1993). In addition to codes of ethics, insightful considerations of the issues periodically have appeared in the literature (e.g. Day, 1975; Hunt et al., 1984; Tybout and Zaltman, 1975), and have helped define the scope of marketing research problems and the responsibilities and obligations of marketing researchers.

Responsibilities and role relationships

When researchers restrict their scientific activities to the investigation of basic research questions, the pertinent ethical issues tend to center on the experimenter–subject relationship. This is reflected in the ethical principles established for human subject research within the behavioral science disciplines, which emphasize the researcher's responsibility to protect the rights of participants (see chapter 2). To be sure, when engaging in basic research the experimenter also bears additional ethical responsibilities to his or her discipline and to science in general in terms of the unbiased pursuit of knowledge. Basic researchers also must anticipate attempts by others to apply the research results at a later point in time, even if that outcome was not an intended goal or objective of the study (Lamberth and Kimmel, 1981). Nevertheless, in the basic research context, the primary ethical responsibilities pertain to the protection of the well-being of the research subjects.

In applied research fields like marketing, however, the ethical issues are compounded by the added relationships into which the researcher often enters and the additional role responsibilities these relationships entail. For example, along with the obligation to treat research participants fairly, one must attempt to fulfill the expectations of the client or research user. The researcher also has a responsibility to protect the well-being of the public when the results are put into action. These additional role relationships are emphasized in the AMA code of ethics (see Box 6.3). Ethical conflicts are more likely to occur in the applied research context as the researcher recognizes that certain duties and responsibilities toward one group are inconsistent with those toward some other group or with his or her own values. For example, a researcher might find that the only way to obtain reliable and accurate data in order to satisfy certain obligations to a client is by deceiving respondents about the true nature of the study.

Consider the hypothetical case in which a marketing researcher who is employed by a pharmaceutical company is asked to carry out a study to determine physicians' perceptions of the quality of the pharmaceutical company's products (cf. Aaker and Day, 1990). Suppose that the company insists that the researcher should contact the physicians under the name of a fictitious company and provide a false purpose for the research in order to obtain more objective responses. Let us assume that this situation represents an ethical dilemma for the researcher because it involves conflicting sets of values (the desire to fulfill the company's research needs and the responsibility to protect the respondent's rights to choose and to be sufficiently informed).

Various solutions to this ethical conflict might be selected, depending largely upon the strength of the researcher's obligations to the different groups involved. The researcher who feels that it is unethical to blatantly deceive subjects could choose to protect the physicians' right to informed consent. In so doing, there is the possibility that unreliable data will be collected and that the interests of the client will have been sacrificed. On the other hand, the researcher could decide to carry out the study as suggested, while taking special care to debrief the physicians about the study's true purpose upon its completion, thereby fulfilling (at least to some extent) the obligation of fairness to the research participants. However, both of these solutions are likely to be viewed as unacceptable compromises. In the first case, the researcher would likely lose a client for failing to comply with the company's desire to remain anonymous to the physicians. The second course of action might be deemed even more unacceptable, given the likelihood that the physicians would react very unfavorably to having been deceived about the study's intent, perhaps to the point of taking legal action against the researcher.

Perhaps the wisest course in this example would be for the researcher to consider whether a more ethical procedure for obtaining the desired information is available, using an alternative research methodology. Further, the ethical dilemma might have been avoided altogether had the researcher and client clarified their roles and expectations when they first established a professional relationship. This is an approach recommended by those who believe that many potential ethical problems in applied contexts can be avoided through an initial process of role definition, clarification, and resolution of conflicting expectations (e.g. Mirvis and Seashore, 1982). In this view, the researcher should clarify his or her professional and scientific obligations to obtain the voluntary informed consent of participants, to protect the confidentiality of their data, and to maintain their anonymity, especially if these are conditions by which the agreement to participate is obtained. Similar clarifications need to be established with research participants. For example, if the researcher has some suspicions that the study's sponsors might request access to certain confidential information for a follow-up study, potential subjects need to be informed of that possibility during the consent procedure.

Ethical dilemmas that involve conflicting role expectations should be anticipated prior to the carrying out of an investigation, especially when entering a relationship with a client whose company priorities may subsequently change, thereby altering the nature of the agreement. Should researchers determine that the values of a client are unacceptable and cannot be changed, it may be best to leave the situation entirely. In such cases, subsequent conflicts among role expectations might be unresolvable and overly frustrating for all parties involved.

Classifying ethical issues in marketing research

Several suggestions for classifying the ethical issues in marketing research have been offered by marketing professionals. Perhaps the most basic scheme was presented by Kinnear and Taylor (1991), who distinguished between two general kinds of judgments that are involved in considerations of ethics as related to marketing research: (1) judgments that certain types of research activities are inappropriate (e.g. the use of marketing research as a trick to sell products), and (2) judgments that certain types of activities are required (e.g. the necessity to inform the client or research sponsor about certain details of the study's methodology, such as how the subject sample was selected). Although a rudimentary framework, Kinnear and Taylor's distinction reflects the typical thrust of ethics codes in attempting to delineate acceptable and unacceptable conduct on the part of researchers toward their subjects and the users of research results.

A second typology of ethical issues in marketing research focuses solely on issues involving the basic rights of research participants. Drawing on codes of ethics from other disciplines (such as psychology and sociology), Tybout and Zaltman (1974) grouped subjects' rights into three general categories: (1) the right to choose (whether or not to participate in a study), which includes the right to be aware that one has a choice, to be given sufficient information in order to be able to decide, and to be given an explicit opportunity to choose; (2) the right to safety, which includes protection of subjects' anonymity, freedom from stress, and freedom from deception as to the nature and objectives of the study; and (3) the right to be informed (after research participation), which includes the debriefing of subjects and dissemination of data to respondents if desired. As is apparent, Tybout and Zaltman's classification of issues is limited to a focus on the relationship between researchers and the participants in their studies, and serves as an important reminder that researchers in applied settings must be sensitive to many of the same human subject issues that pertain to basic research.

During the mid-1980s, Shelby Hunt and his colleagues (1984) used an empirical approach to determine the most prevalent ethical problems confronting marketing research practitioners. The researchers asked 460 randomly selected marketing professionals to identify job situations that posed the most difficult ethical or moral problems for them. The most frequent responses were classified into the following five areas:

1   Research integrity (33 percent), including such issues as deliberately withholding information, falsifying or ignoring pertinent data, altering research results, misusing statistics, compromising the design of a research project, and misinterpreting the results of a study in order to support a predetermined point of view. The fact that this category emerged as the most frequently cited source of ethical problems suggests that marketing researchers perceive significant problems in maintaining the integrity of their research.

2   Treating outside clients fairly (11 percent), including issues involving clients outside one's own company, such as the use of hidden charges passed on to the customer.

3   Research confidentiality (9 percent), including issues involved in the protection of data obtained for research purposes (such as the use of data from one project in a related project for a competitor).

4   Marketing mix social issues (8 percent), including issues pertaining to the decision to conduct research within the context of product or advertising decisions that are contrary to the best interests of society, such as advertising to children and the production of products that are hazardous to health.

5   Treating respondents fairly (6 percent), including issues involving the protection of anonymity and the use of deception in order to conceal the purpose or sponsor of the research. Other sets of

issues that were less frequently cited as problems included inter-viewer dishonesty, misuse of funds, and the unfair treatment of others in one's company.

Hunt et al.'s results are informative in that they help us to recognize the range of actual ethical dilemmas that are likely to confront researchers in applied settings. The findings also suggest that there are two faces to the ethical issues – one dealing with issues that emerge in the conducting of research and the other dealing with considerations relevant to the implied use and application of the research results. Some of the ethical issues that have received particu-lar attention by marketing professionals are listed in Table 6.2. While not an exhaustive listing, these research activities are representative of those most likely to pose ethical dilemmas for the marketing researcher (see Box 6.4).

Several of the areas of ethical concern appearing in Table 6.2 already have been considered in our discussion of basic research activities (such as deception and informed consent, invasion of privacy and confidentiality of data, and debriefing), while others are

Table 6.2 Ethical issues in marketing research

1 Involvement of individuals in the research process without their know-ledge or informed consent
2 Coercion of individuals to participate in research
3 Abuse of research participants' rights during the data collection proc-ess
4 Invasion of the privacy of research participants
5 Threats to research participants' anonymity and confidentiality of data
6 Failure to debrief research participants or not sending a promised report of the study's results
7 Use of inadequate sampling procedures to increase the likelihood of obtaining desired results
8 Misrepresentation of non-random subject samples (such as volunteers) as representative
9 Misrepresentation of research results or dissemination of conclusions that are inconsistent with or not warranted by the data
10 Use of survey techniques for selling purposes
11 Use of research strategies to obtain the names and addresses of prospects for direct marketing
12 Pressure to design research to support a particular marketing decision or to enhance a legal position

Source: Adapted from Day (1975).

Box 6.4

Ethical dilemmas in marketing research

The following hypothetical cases illustrate the kinds of ethical dilemmas that can arise in marketing research. There are no uniquely correct resolutions to these dilemmas and, in fact, several courses of action might be taken in each case. These cases were adapted from a set of vignettes developed by Charles Weinberg, Professor of Marketing at the University of British Columbia (cf. Aaker and Day, 1990, pp. 56–7).

1 A researcher employed by a marketing research firm has conducted an attitude study for a client, the results of which suggest that the product is not being marketed properly. However, when the researcher presents the findings to the client's product management team they request that the data be omitted from the formal report of the attitude study, which will be widely distributed. The researcher is told that the verbal presentation is adequate enough for their needs.

2 A study director for a research company has developed a questionnaire to be used for a project that is being undertaken for a regular client of the company. After sending the questionnaire to the client for final approval, however, it is returned drastically modified, having been rewritten by the client. Leading questions and biased scales have been introduced, and an accompanying letter states that the questionnaire must be sent out as revised. The study director does not believe that valid information can be obtained by using the revised instrument.

3 The market research director for a large chemical company has obtained research evidence suggesting that many customers of the company are misusing one of its principal products. Although the misuse does not pose any danger to customers, it is costing them money because too much of the product is being used at one time. When the director is shown the newly developed advertising campaign for the product, she recognizes that the ads ignore the misuse problem and actually seem to encourage it.

4 The professor of a marketing research course has assigned a class project for which each student is to conduct personal interviews with executives of high technology companies concerning their future plans. The professor has stated that all of the information relevant to

These vignettes developed by Charles Weinberg first appeared in D. A. Aaker and G. S. Day (1990). *Marketing Research*, 4th edition, New York, John Wiley and Sons, Inc. Copyright © 1990 John Wiley and Sons, Inc. Reprinted by permission.

the project is confidential and will only be used in the research course. A couple of days after the project has been assigned, a student overhears the professor mentioning to a colleague that the project will be sold to a major technology firm in the industry.

For each of these scenarios, the reader should attempt to identify the ethical dilemma involved and be able to justify his or her decision for resolving the dilemma. The author has used these scenarios for class discussions in his marketing research courses and has sometimes been surprised by students' reactions to them. For example, one group of marketing research students in France suggested that the best approach for resolving the fourth case would be to confront the professor and threaten to "blow the whistle" unless the profits from the sale of the research project were shared with them! While hardly an adequate resolution for the case, this response is consistent with the findings of a Canadian study suggesting that the values of business school students are influenced during their educational training in such a way that the students become more self-oriented (e.g. ambitious) and less society-oriented (e.g. helpful) than their counterparts in non-business programs (Petrof et al., 1982). Other French students, however, have recognized that the fourth vignette illustrates problems involving the unequal distribution of power in research relationships; in this case, the professor has a power advantage over the student researchers, who must satisfy the professor's requirements in order to obtain a good grade. While it might be threatening to individual students to approach the professor (or department chairperson) with their concerns about the use of the project, there often is "safety in numbers." The professor might be more willing to reconsider selling the research when confronted by a group of rebellious students. Another option would be for the students to indicate the potential risk to the executives – the inability to protect the confidentiality of their responses – prior to carrying out the interviews. In any case, the researchers should not promise that the results will be anonymous if it is unlikely that anonymity can be protected. (For additional case examples illustrating ethical dimensions in marketing research, see Kinnear and Taylor, 1991, pp. 126–7.)

described in remaining chapters (such as coercing individuals to participate in research). In order to shed light on some of the ethical issues that are particularly troublesome in marketing research, we next turn our attention to research activities involving the use of marketing surveys.

Ethical issues in the use of marketing surveys

Regardless of the discipline, survey studies need to be carried out in ways designed to minimize the risks to research participants to the greatest extent possible. This requires that survey researchers maintain a sensitivity to general ethical considerations that appear in other research contexts, including respondents' right to be informed and the protection of research participants through the proper treatment and use of information obtained from them during the investigation. When marketing and other applied researchers conduct survey studies, additional care must be taken to avoid potential ethical abuses involving the use of survey findings for decision-making purposes.

Informing marketing research participants

Because respondents to marketing surveys generally are voluntary participants, it is important that researchers adequately inform them about the true nature of what they have volunteered for (Fowler, 1993). Thus, the ethical survey researcher typically is expected to provide potential respondents with such information as the name of the organization carrying out the research (and the interviewer's name, if applicable), the research sponsor who is paying for or supporting the research, a brief description of the purposes of the research and issues or topics the research is designed to cover, an accurate description of the extent to which responses will be protected with respect to confidentiality, and the possible threats to or limits on the right to confidentiality. Respondents also should be assured that their cooperation is voluntary, that no negative consequences will result from their participation, and that they are free to decline to respond to any questions that they choose not to answer (Fowler, 1993). Given these obligations, it is clear that it would have been ethically unacceptable for the marketing researcher in the example mentioned previously to adhere to the pharmaceutical company's demands to misrepresent the survey sponsor and purpose.

Although the informed consent process can be relatively straightforward with respect to marketing research surveys, as we have suggested, researchers might be tempted to use deception in order to conceal the purpose or sponsor of the research. This typically is done in order to reduce the transparency of the survey or to increase response rates (Levin, 1981). A serious ethical problem has emerged in recent years in the marketing industry as a result of the use of survey techniques for selling purposes (Day, 1975; Rothenberg, 1990). That is, respondents are led to believe that they are participating in a marketing research survey, only to find out after answering a series of questions that the actual purpose of the interview was to sell or promote a product or service. This practice is known in the marketing industry as "sugging," an acronym which stands for "selling under the guise of market research." In the words of one woman who claims to have been victimized by this marketing technique three times over a two-year period, "they say they're doing market research, but they're really trying to sell you water treatments" (Rothenberg, 1990, p. D4).

The rationale underlying the use of sugging by marketers is readily apparent, given the fact that the mere mention of certain products and services (such as insurance, magazines, and home improvements) during the early stages of an interview tends to sharply increase the rate of refusal, especially when respondents are contacted by telephone (Day, 1975). It also has been found that prior experience with a false survey increases the tendency of respondents to refuse to participate in subsequent legitimate surveys (Sheets et al., 1974). There is general agreement that this deceptive tactic represents a serious abuse of respondents' rights, and it is prohibited by the AMA's ethics code (though largely unenforceable). Nonetheless, there is some evidence that this practice has been increasing in frequency. According to studies conducted by Walker Research, a St Louis company that tracks trends in the market research industry, the percentage of respondents who said they had been asked to participate in a survey that turned out to be a sales pitch rose from 15 percent in 1982 to 22 percent in 1988 (cf. Rothenberg, 1990). As a result, local authorities in the United States have begun to require licenses and have imposed restrictions on the days and hours when interviewing can be conducted (Day, 1975).

In addition to the ethical problems involved in the use of "surveys" to deceive potential consumers, selling under the guise of research also illustrates a link between methodological and ethical issues in marketing research. This is evident in the widespread fear that the practice increases suspiciousness among potential respondents and ultimately will lead to a reduction in response rates and response quality (see "Methodological issues" below).

Protecting survey respondents
Of all the ethical issues in marketing research listed in Table 6.2, those that are most central to the well-being of survey respondents

have to do with privacy concerns and the manner in which the information obtained from respondents is treated. Specifically, researchers must respect the privacy of respondents, especially when the survey involves highly sensitive topics and when personal information is gathered. For example, some marketing researchers utilize in-depth probing techniques during interviews in order to uncover respondents' personal concerns, desires, fears, and anxieties. In other cases, potentially incriminating information might be sought, such as whether consumers have utilized illegal decoding devices in order to obtain free cable television service. Previous research conducted in both the United States and Sweden revealed that concerns about breaches of confidentiality are particularly evident with regard to questions about income and the trustworthiness of institutions. Moreover, the topic of the survey and the sponsorship were as important as privacy concerns for many individuals (National Academy of Sciences, 1979; National Central Bureau of Statistics, 1977).

For a majority of marketing surveys, respondents' privacy rights can be protected and research cooperation can be obtained under conditions of anonymity and through maintaining the confidentiality of responses (see Table 6.3). Some researchers, for example, attempt to ensure respondent anonymity by utilizing a self-administered questionnaire rather than an interview approach that requires the presence of an interviewer. The self-administered approach has both methodological and ethical advantages, particularly when it can be assumed that respondents are not suspicious about identifying marks on the questionnaire forms. Other researchers have provided written or verbal assurances of confidentiality in the hope that such a procedure will promote more honest disclosures from participants. Research to date on the relationship between such assurances and the subsequent quality of data is somewhat equivocal (Blanck et al., 1992).

Threats to survey respondents' anonymity and confidentiality of data can arise from a number of activities, ranging from pre-coding mail questionnaires with ultraviolet ink to determine the identity of participants to providing subjects' names and addresses to a client after obtaining data under conditions of anonymity. Certain mildly deceptive procedures also represent threats to privacy, including the construction of survey questionnaire items so as to disguise their true intent from respondents or to encourage them to reveal information that they otherwise would not admit (e.g. by requesting year of birth rather than age).

The use of ultraviolet ink on pre-coded mail questionnaires in order to determine the identity of respondents represents a dramatic example of the invasion of privacy in the context of survey research. In a case reported by Boruch and Cecil (1979), a survey contractor engaged in marketing research for the *National Observer* newspaper covertly identified mail questionnaires using invisible ink, despite having included information implying that responses were anonymous. The purpose of the marketing study was innocuous enough: to

Table 6.3 Procedures for protecting survey respondents

According to social psychologist Floyd Fowler, Jr. (1993), the key to protecting the privacy of survey respondents lies in the careful treatment of the information they provide. Fowler has suggested several procedures for minimizing the chances that confidentiality may be breached. These procedures are summarized below.

1 Commitment to confidentiality should be obtained in writing from all individuals who have access to survey data or who participate in the data collection process
2 The linkages between survey responses and subject identifiers should be minimized. In many cases, respondent names need not be obtained during any phase of the research process. When specific identifiers, such as names or addresses, are necessary, they can be physically separated from the interview forms in which the actual survey responses are recorded (e.g. by putting them on separate pieces of paper or coversheets). (For specific procedures, see Boruch and Cecil, 1979, 1982.)
3 Only those persons who are involved in the survey project should have access to completed surveys
4 Identifiers should be removed from completed questionnaires as soon as possible
5 Actual questionnaire responses should not be available to individuals who could identify respondents from their profile of responses (such as supervisors in the case of a survey of employees)
6 Actual data files usually will include an identification number for each respondent. The linkage between these numbers and other respondent identifiers (such as names and addresses) should not be available to data file users
7 Special care should be taken when presenting data for very small categories of respondents who might be identifiable
8 Researchers are responsible for the destruction of completed survey research instruments, or their continuing secure storage, when a project has been completed or when the use of the actual survey questionnaires has ended.

obtain a statistical description of the newspaper subscribers' demographic characteristics. It was hoped that the statistical data obtained from the survey could be used to persuade advertisers to purchase advertising space and to identify the sorts of news stories and advertising most likely to appeal to the newspaper's readers. On the questionnaire sent to subscribers, the researcher promised that responses would be kept confidential and provided additional assurances which likely led most readers to conclude that their anonymity would be protected. For example, no identifying information was requested nor was the questionnaire apparently coded in any way. Eventually, a skeptical recipient exposed the form to ultraviolet light and discovered that an identification number did indeed appear on the questionnaire, resulting in embarrassing publicity about the deception for the survey research firm. The firm admitted the use of the invisible print but claimed that the technique was essential for following up on non-respondents.

The use of secretly pre-coded surveys has been defended by those who have argued that the practice reduces survey costs and response bias, and that the identity of respondents is never disclosed to anyone outside the research organization (Dickson et al. 1977). Regardless of these points, it is difficult to justify the secret coding of questionnaires from either an ethical or methodological perspective. The deceptive practice represents a clear violation of the respondents' right to privacy, and researchers have expressed concern that the use of such techniques will serve to raise the level of public mistrust and skepticism regarding all forms of research (e.g. Skelly, 1977). Moreover, Boruch and Cecil (1979) have convincingly argued that deceptive coding techniques are not necessary in the first place, given the availability of a wide range of other non-deceptive approaches that can accomplish the same goals while assuring the validity of sampling.

One technique that enables researchers to follow up on non-respondents in a way that also reassures respondents that their questionnaire responses will remain anonymous is to attach a separate postcard with a respondent identifier to a questionnaire with no identifier on it. On the postcard a statement addressed to the researcher indicates that the completed questionnaire was mailed at the same time and that the researcher need not send a further reminder urging the respondent to participate in the study. According to Fowler (1993), this procedure apparently does not result in respondents simply returning the postcard alone in order to avoid further reminders; in fact, the number of postcards and questionnaires received tends to be about the same in most studies employing this procedure.

In large-scale survey research, the confidentiality of individual responses can be preserved by application of randomized response methods (Boruch and Cecil, 1982; Fidler and Kleinknecht, 1977; Fox and Tracy, 1980, 1984; Warner, 1965). In a simple variation of this approach, a randomizing device (such as flipping a coin) is used in such a way that interviewers are unable to ascertain which question a respondent is actually answering. As an illustration, this technique can be used to estimate the prevalence of a behavior that is likely to generate a legal risk or embarrassment to a disclosing respondent, such as spousal abuse or cocaine usage. In order to select how to respond to a stigmatizing question (e.g. "Did you beat your wife last month?" "Have you ever used cocaine?") the subject is instructed to flip a coin out of the researcher's sight, and to respond "yes" if it lands heads and to respond truthfully ("yes" or "no") if it lands tails. On a probability basis, it is possible to obtain an estimate of the behavior of interest without compromising the respondent's right to privacy. Knowing that 50 percent of the respondents are expected to get heads to respond "yes," it is then possible to estimate the proportion that actually indicated they had engaged in the sensitive behavior.

In marketing research, the randomized response method might be employed in lifestyle studies involving sensitive and perhaps illegal activities, such as shoplifting, gambling, and the procurement of prostitutes within various consumer populations. It should be noted, however, that recent evidence suggests that if sufficient care is taken to guarantee anonymity, the randomized response technique may not prove more effective than straightforward questioning of respondents in yielding revelations in response to sensitive questions (Linden and Weiss, 1994).

Methodological issues in marketing research

The linkage between methodological and ethical issues is apparent in marketing research in several ways. Perhaps the most obvious ethical concern stems from the recognition that applied researchers sometimes are pressured by organizations to design research in order to support a particular decision or to enhance a legal position. In marketing, recent revelations pertaining to research sponsored by major American tobacco companies clearly illustrate the difficulties that may be faced by researchers working in applied settings. For example, one tobacco company halted the research of two of the company's scientists who had found preliminary evidence in the early 1980s that a substance in cigarettes increased the addictive power of nicotine (Hilts, 1994a). The company also apparently took steps to block efforts to publish that and other work. One company executive who met with the researchers shortly after they were told to stop their studies and to kill their laboratory rats reportedly asked: "Why should we risk a billion-dollar business for some rats' studies?"

In point of fact, it is relatively easy for marketing researchers to use their methodological skills in order to obtain desired outcomes. Studies can be designed to support a particular decision through the adroit choice of subject samples, biased question wording and placement, and the selective coding and analysis of responses (see Box 6.5).

Another methodological concern in marketing research has to do with the impact of informed consent procedures on the outcomes of marketing surveys. Research on the effects of informed consent in survey research primarily has focused on the survey response rate and the quality of subjects' responses. In contrast to the situation in the laboratory setting, the power differential between the survey researcher and the potential respondent is somewhat reversed, with the respondent typically in a powerful position to refuse to participate (Sieber, 1982c).

Overall, studies suggest that the adverse impact of informed consent on survey research is negligible. With the exception of request for a signature to document consent (a procedure that is not always necessary), the elements of informed consent do not appear to have much of an effect on response rates or quality of responses to surveys. In one study, Singer (1978) reported that when more detailed, informative, and truthful information about the content of a survey was provided beforehand to potential respondents (such as a description of the content of sensitive questions), neither the overall response rate nor responses to specific survey questions seemed to be affected. In a follow-up study, Singer and Frankel (1982) varied information about the content and purpose of a telephone interview. The results revealed that the manipulated variables did not significantly affect the overall response rate, response to individual survey items, or quality of response. In fact, the highest response rate and fewest refusals were obtained when respondents were provided with the greatest detail about the interview. Similarly, Loo (1982) reported a high level of cooperation among respondents given an

---

Box 6.5

Biased questioning in survey research

Sources of bias can enter into the questionnaire data collection process in several ways. An awareness of the effects of question biases can be exploited by unscrupulous researchers in order to increase the likelihood of obtaining desired results. As an example, consider the results of a publicized finding from a survey conducted on behalf of the Burger King hamburger chain. It was reported that consumer respondents preferred Burger King's flame-broiled hamburger to McDonald's fried hamburger by a margin of 3 to 1 (Kinnear and Taylor, 1991). The actual question used by Burger King which led to this result was "Do you prefer your hamburgers flame-broiled or fried?"

When an independent researcher rephrased the question by asking consumers "Do you prefer a hamburger that is grilled on a hot stainless steel grill or cooked by passing the raw meat through an open gas flame?," the results, not surprisingly, were reversed (53 percent of those questioned preferred McDonald's cooking process). Further, when the researcher revealed to respondents that flame-broiled hamburgers were kept in a microwave oven prior to serving, the preference for fried burgers increased to 85 percent. These results suggest that the initial Burger King survey pitted the most appealing word it could find to describe its cooking process against the most unappealing word for its competitor's cooking process. In short, it was the question semantics that determined the apparent preference for cooking methods, not the actual product.

In another example of how questionnaire wording can influence the obtained results, a sample of non-working women was asked to respond to one of the following items: "Would you like to have a job, if this were possible?" or "Would you prefer to have a job, or do you prefer just to do your housework?" Of course, the second question makes explicit the implied choice in the first question, a distinction that clearly affected subjects' responses. Whereas only 19 percent of the women said they would not like to have a job in response to the first question, 68 percent said they would not like to have a job in response to the second question (Kinnear and Taylor, 1991).

Another way that bias can enter into the survey research process is through the use of so-called leading questions. Leading (or "loaded") questions are those that provide respondents with cues as to what the answer should be. For example, the question "Do you own a SONY television set?" would be a leading question if it resulted in a higher reported SONY ownership than when the question simply asked "What brand of television set do you own?" When a brand or company name is included in the question, respondents may assume that company has sponsored the study, thereby resulting in a desire to express positive feelings toward the survey sponsor (Kinnear and Taylor, 1991).

Other examples of biased questions are those including words or phrases that are emotionally charged, provide the respondent with a reason for selecting one of the alternatives, or associate a position with a prestigious (or non-prestigious) person or organization. The following questionnaire items reflect these kinds of hidden (or not-so-hidden) biases (Aaker and Day, 1990; Fowler, 1993):

"Do you agree or disagree that the government should stop wasting money on road construction?"
(The researcher's bias is obvious here, and who would ever favor wasting money?)

"Should we increase taxes in order to get more housing and better schools, or should we keep them about the same?"
(Despite most individuals' aversion to raising taxes, this question is biased toward an affirmative answer because only the benefits of raising taxes are specified.)

"Do you agree or disagree that the free trade agreement has destroyed the economy?"

"Do you agree that the Mayor has an annoying, confrontational style?"
(The use of emotionally charged or negative words, such as "destroyed" or "annoying," can have strong overtones that may overshadow the specific content of the question.)

"Do you agree or disagree with the American Dental Association's position that advertising presweetened cereal to children is harmful?"
(In this case the prestigious organization is associated with a particular position.)

opportunity to vent their concerns to respectful and apparently sympathetic investigators.

Despite the apparent lack of significant adverse effects in survey research, researchers should remain cautious of the influence of certain mediating variables that are operative in different consent situations. For example, there is some evidence that respondents may not fully attend to what they are told during the consent procedure (cf. Singer, 1978), and other research has shown that informed consent can reduce participation and response to specific items in surveys (Lueptow et al., 1977). Additionally, the content and purpose of a survey can interact with sensitive response variables. Informed consent might alert respondents in advance about threatening questions, thereby allowing respondents ample time to think up alternative responses or to decide not to respond at all. Such might be the case when welfare clients are forewarned about income questions or when employees in large corporations are asked about their drinking habits (Singer, 1978).

One of the more pressing methodological concerns in marketing research has to do with the increasing difficulty of obtaining random samples for surveys due to declining rates of participation. Many believe this decline in research participation is due in large part to an increase in suspiciousness among respondents after negative encounters with marketing researchers or with salespersons posing as marketing researchers. As mentioned in our discussion of "sugging," when consumers are not told that a company conducting a survey is in a business other than research, there is the potential for this practice to affect perceptions of legitimate research as well. As response rates and response quality continue to decline, the long-term statistical reliability of marketing research may be jeopardized. Further, as it becomes more difficult for researchers to obtain random subject samples, there is the additional possibility that they will be tempted to utilize unethical strategies for inducing people into their samples.

Another factor that adversely affects public willingness to cooperate is the increasingly intrusive nature of the research process. In marketing, excessive interviewing in certain districts, such as those that are especially popular for test markets, may be perceived by consumers as violations of their basic right to privacy (Aaker and Day, 1990). This can lead to resentment and a decrease in willingness to fully cooperate, especially if it is believed that participation in one study will result in further research requests.

Summary

As our illustrations within the areas of preventive intervention and marketing research suggest, applied researchers – like their more theoretically oriented counterparts – must face difficult dilemmas regarding informed consent and privacy issues, although the ways in which they resolve these dilemmas may have more immediate consequences than are apparent in basic research contexts. As with deception, if one were to be absolutist about the principle of privacy in applied research it would be nearly impossible to conduct many meaningful studies because, in a sense, the basic nature of such research – like most behavioral research – is to be intrusive. Ringing a respondent's doorbell or telephone represents an intrusion. After all, much of the information that is sought in studies of human behavior involves human processes that people are trying to protect, defend, or cover up. While respect for the rights of subjects should be held in the highest regard by applied researchers, they have to be balanced by the potential benefits and goals within the applied research context.

Whatever the research situation, it is important for applied researchers to bear in mind that their first obligation is to those persons who cooperate with and participate in the research process. Many of the ethical problems that arise when researchers impose their frames of reference on behavior can be avoided when subjects are treated as partners in the research process. In chapter 7, ethical issues pertinent to the recruitment and selection of research subjects are considered.

7

Ethical Issues in the Recruitment and Selection of Research Subjects

All students are expected to complete at least three *(3) hours of participation in research studies conducted within the university's Psychology Department in addition to the other course requirements. Sign-up forms describing ongoing studies are located on the bulletin board outside the department secretary's office. Students who do not wish to serve as research subjects may instead complete three written reports (each at least five pages in length) on topics relevant to psychology. All topics must be approved by the instructor.*

Research Participation Requirement on Introductory
Psychology Course Syllabus

To the best of the author's recollection, the passage that appears above approximates a course requirement that appeared on the syllabus for his first college course as a student of psychology. Many readers may recall a similar passage from the outlines of courses they had taken in psychology, or perhaps have themselves written such a statement on a syllabus for a course they have taught.

It is quite common to find research participation requirements or opportunities for extra credit on introductory course outlines because the university "subject pool" represents a primary source of behavioral science research participants (Higbee and Wells, 1972; Jung, 1969; Schultz, 1969; Sieber and Saks, 1989). Surveys show that more than 70 percent of the studies conducted in the areas of personality and social psychology and about 90 percent of cognition and perception studies utilized college students as participants (Korn and Bram, 1987; Sears, 1986). One recent survey of 366 psychology departments in the United States revealed that 74 percent had student subject pools, with 93.4 percent recruiting subjects from introductory psychology courses and 35.4 percent recruiting from other lower division courses (Sieber and Saks, 1989). Similar results were obtained from a survey of 50 Canadian educational institutions offering programs in psychology (Lindsay and Holden, 1987). These recent figures represent a continuation of a long-standing tradition in the behavioral sciences, as evidenced by Quinn McNemar's (1946, p. 333) remark, half a century ago, that "The existing science of human behavior is largely the science of the behavior of sophomores." In other disciplines, students may be solicited as volunteers or asked to participate in studies during class time (e.g. by completing questionnaires and the like). The fact that college students comprise an overwhelming majority of behavioral science research participants has prompted some writers to refer to them as psychology's human "fruit flies" (Keith-Spiegel and Koocher, 1985; Rubenstein, 1982).

While there are certain educational (and other) benefits that may be gained from research participation, there are ethical and methodological questions that need to be considered with regard to the extensive use of college students as participants in research investigations. For example, are such individuals, who by nature of their educational status represent a relatively low-power group, unfairly coerced to serve as subjects or is research participation truly voluntary? Do they have freedom of choice to refuse to satisfy a research participation course requirement and are alternative means of completing the coursework offered? Are student participants educationally debriefed, and do the educational and other benefits of participation outweigh the risks for this group? What are the methodological limitations of studying individuals whose behavior in research may not be representative of the general population? Similar kinds of questions are relevant to the use of subjects sampled from other low power or vulnerable groups in society, such as children, the elderly, prisoners, and the institutionalized, or when research participation is required as part of employment or some other professional relationship with the investigator.

Increasing sensitivity to issues having to do with subject pool policies and procedures has been apparent in recent years (McCord, 1991) and the importance of voluntary research participation has been emphasized in professional codes of ethics and governmental guidelines. As a guiding principle for framing this chapter's consideration of issues pertaining to the recruitment and selection of research participants, the earlier APA guideline (principle F) for ensuring freedom from coercion to participate, which the reader may recall from chapter 2, is particularly relevant (APA, 1982, p. 6):

> The investigator respects the individual's freedom to decline to participate in or to withdraw from the research at any time. The obligation to protect this freedom requires careful thought and consideration when the investigator is in a position of authority or influence over the participant.

The application of this principle is not always straightforward and researchers are apt to encounter ethical dilemmas, particularly when they hold power over prospective subjects or otherwise control desired resources.

The University Human Subject Pool

As mentioned, most of the data obtained in behavioral research with human participants has been provided by college students, a majority of whom were recruited through organized subject pools. University subject pools typically are comprised of all those available students in introductory and other lower-level behavioral science courses who are encouraged (and sometimes required) to participate in campus research in order to obtain some sort of incentive. While the incentive may be monetary in nature, it is more likely to take the form of course credit (e.g. Lindsay and Holden, 1987). Research participation also may be required merely as a prerequisite to obtaining a course grade; in such cases, an alternative means of satisfying the research requirement typically is made available to students.

Justification for student research participation

Several arguments have been offered to justify the extensive use of students for research participation. One practical consideration has to do with the basic necessity of having a readily available pool of prospective subjects for the many research studies that are carried out in university settings during a typical academic year. For academic researchers, a student subject pool represents a convenient, inexpensive study population, thereby eliminating the difficult and costly task of having to recruit subjects from outside the university setting. In this regard, undergraduate students stand to benefit society by permitting the carrying out of important research that otherwise might not have been conducted.

Beyond the pragmatic concern for easy attainment of subject samples in research-oriented institutions, an argument can be made that students have an obligation to make a contribution to the discipline that they are studying (Schuler, 1982). By advancing knowledge through participation, students themselves also stand to benefit as that knowledge eventually is conveyed to them during the study of the relevant subject matter in their classes (Diener and Crandall, 1978; King, 1970). More directly, it is commonly agreed that participation in research can provide clear educational benefits to students by demonstrating firsthand what the research process is like. This common educational justification for requiring student participation is based on the assumption that students can gain a better appreciation of the scientific process by experiencing it directly rather than by having the process described solely through class lectures or readings. Universities often require that subject pool participants be educationally debriefed by the researcher or otherwise obtain an explanation of the research purpose and methods.

Despite concerns about coercion and the possibility that students may not receive worthwhile feedback or attend to the feedback which they do receive, surveys generally have shown that students believe that their participation in research was a positive and useful learning experience (Britton, 1979; Christensen, 1988; Davis and Fernald, 1975; King, 1970; Leak, 1981). Of course, it could be argued that such after-the-fact evaluations on the part of subjects may simply be a means for them to cognitively justify having devoted their time to the research enterprise. Whether or not this is the case, subjects inevitably learn something about the research process through research participation, even if they do not retain specifics about the research design or variables investigated in complex studies (Diener and Crandall, 1978). Further, there is evidence that students' reactions to research participation are especially favorable when an organized effort is made to integrate the research experience into the educational program and when participation is specifically used to introduce and illustrate the nature of the scientific process (Davis and Fernald, 1975). As Diener and Crandall (1978) have noted, the research experience is not unlike the laboratory section of traditional science courses, with the additional advantage that students are contributing to science and not merely engaging in a learning exercise.

The fact that subject pool participants often are required to participate in multiple studies can have certain methodological disadvantages, especially when subjects become suspicious of the experimenter's intent in later studies following early exposure to deceptive manipulations (see chapter 4). However, Schuler (1982) has suggested that the routine nature of participation for students can actually be seen as having an ethical advantage as well. In his view, by becoming accustomed to the research situation, the overall experience becomes less stressful for the subjects, thereby reducing the researcher's responsibility to some extent. Rather than leading to a cynical attitude on the part of subjects, routine participation can encourage them to adopt the role of research partner in the scientific process.

Ethical concerns about student subject pools

It is clear that subject pools satisfy their primary purpose in providing "grist for the research mill"; nevertheless, ethical concerns have been voiced about the use of college recruits, particularly with regard to the problem of coercion and related forms of exploitation (McCord, 1991). Specifically, there are fears that alternative means of satisfying course requirements may not be offered or else are excessively time consuming or noxious; that students may receive little if any educational debriefing; and that readily accessible complaint procedures are not available to student research participants (Diener and Crandall, 1978; Keith-Spiegel and Koocher, 1985; Sieber and Saks, 1989). According to one critic of the extensive use of student participants in research, "only a moderate degree of cynicism is required to characterize psychology as a science based on the contributions of a captive population with little power" (Korn, 1988, p. 74).

The results of a comprehensive survey of 366 psychology departments identified as having both graduate programs and human subject pools seem to justify these concerns. Sieber and Saks (1989) reported that a number of the surveyed departments were not in full compliance with professional and federal ethical guidelines in terms of protecting students from coercion to participate in research, offering clear educational benefits to research participants, and providing reasonable alternatives to participation. For example, only 11 percent of the departments appeared to have a subject pool that was truly voluntary; that is, there were no penalties for non-participation, no grades or other incentives for participation, and no alternatives to participation. Only 20.1 percent of the respondents reported that they announced subject pool requirements in the course catalog (contrary to APA guidelines urging departments to do so), although most (92.9 percent) announced the requirement at the first class meeting of the term. Whereas nearly all of the departments provided alternatives to subject pool participation, more than 70 percent offered relatively unattractive alternatives, such as having students write a paper or take a quiz.

When students are provided with undesirable and time-consuming alternatives to a research participation requirement, like having to write two or three short term papers, they essentially are left with little choice at all but to participate in the research. It is the author's experience that very few students actually select extensive written alternatives to research participation, no matter how unpleasant they may view the research requirement as being. In addition to the fact that having to write additional papers is an especially noxious choice for many students, they also may fear that their course grade will somehow be adversely affected if they do not choose the instructor's obviously preferred requirement. Such forces are likely to reduce the voluntary nature of participation, given the professor's power over students, and are apt to have a more serious impact in required courses. Voluntary participation also will be seriously threatened when bonus points can be obtained for volunteering for research in situations where there is strong competition for grades.

The fact that subject recruitment announcements are characteristically vague exacerbates the problems of undue coercion and exploitation of naive research participants. The solicitation of subject pool participants usually involves the posting of sign-up forms in the appropriate academic department on campus. (Such forms also may be distributed during class meetings.) Students are expected to select a convenient appointment time from those listed on the sheet; their signature essentially serves to establish an agreement with the researcher to participate in the study at the designated time. Often, however, the brief summary of the research that appears on the sign-up form is uninformative and at times misleading, so that students may be left unaware of what they are agreeing to. The decision to participate in one study versus another thus may be based more on the availability of a convenient appointment time than the specific details of what the study apparently will involve. As illustrated in Boxes 7.1 and 7.2, subject recruitment announcements may appear rather innocuous, but often belie the true nature of the investigation. For example, the newspaper advertisement used by Philip Zimbardo to solicit participants for his simulated prison experiment (Box 7.1) reflects the passive deception employed throughout the consent phase of that investigation. The announcement for Stanley Milgram's obe-

```
┌─────────────────────────────────────────────────────┐
│                    ┌──────────────┐                   │
│                    │   Box 7.1    │                   │
│                    └──────────────┘                   │
│                                                       │
│   Subject recruitment announcement for the Stanford prison │
│                        experiment                     │
│                                                       │
│   Male college students needed for psychological study of prison life. $15 │
│   per day for 1–2 weeks beginning Aug. 14. For further information and │
│   applications, come to Room 248, Jordan Hall, Stanford U. │
└─────────────────────────────────────────────────────┘
```

dience study (Box 7.2) includes an active deception (the investigation is described as "a scientific study of memory and learning"). It is noteworthy that in these two examples, neither researcher relied on available subject pools for selecting participants; in fact, for the obedience study, Milgram recruited non-students from the general population.

When subjects agree to participate in studies that are only vaguely described on recruitment forms, they may find themselves in difficult positions when they appear for their appointment and learn that the research is not quite what they expected. Faced with the decision to withdraw or to go ahead with participation, most students tend to opt for participation, viewing their initial agreement to serve as subjects as an obligation that cannot be avoided. Once the initial agreement has been made, demand characteristics of the experimental situation are apt to enhance the perceived power of the investigator and importance of the task at hand, thereby increasing the likelihood of the participant's continued compliance (Kelman, 1972; Orne, 1962). These forces underline the ethical responsibility of investigators to maintain a sensitivity to their subjects' needs and to cues which suggest that withdrawal is being contemplated.

In addition to these threats to the voluntary nature of research participation, it has been argued that students typically receive dubious educational benefits from their participation in research (Korn, 1988). In fact, for many behavioral science investigations conducted on university campuses, subject pools tend only to provide direct benefits to the researchers who use them. Investigators may do little to help make research participation a learning experience for subjects, perhaps only providing subjects with a cursory description of their studies on a sheet of paper that is distributed at the end of the experimental session or in the form of an abstract of findings that is mailed to them a few months later. Moreover, some studies are of such minor significance that they hold little educational

Public Announcement

WE WILL PAY YOU $4.00 FOR ONE HOUR OF YOUR TIME

Persons Needed for a Study of Memory

*We will pay five hundred New Haven men to help us complete a scientific study of memory and learning. The study is being done at Yale University.

*Each person who participates will be paid $4.00 (plus 50c carfare) for approximately 1 hour's time. We need you for only one hour: there are no further obligations. You may choose the time you would like to come (evenings, weekdays, or weekends).

*No special training, education, or experience is needed. We want:

Factory workers	Businessmen	Construction workers
City employees	Clerks	Salespeople
Laborers	Professional people	White-collar workers
Barbers	Telephone workers	Others

All persons must be between the ages of 20 and 50. High school and college students cannot be used.

*If you meet these qualifications, fill out the coupon below and mail it now to Professor Stanley Milgram, Department of Psychology, Yale University, New Haven. You will be notified later of the specific time and place of the study. We reserve the right to decline any application.

*You will be paid $4.00 (plus 50c carfare) as soon as you arrive at the laboratory.

- -

TO:
PROF. STANLEY MILGRAM, DEPARTMENT OF PSYCHOLOGY, YALE UNIVERSITY, NEW HAVEN, CONN. I want to take part in this study of memory and learning. I am between the ages of 20 and 50. I will be paid $4.00 (plus 50c carfare) if I participate.

NAME (Please Print). . . . .

ADDRESS . . . .

TELEPHONE NO. . . . . Best time to call you . . . .

AGE . . . . OCCUPATION. SEX . . . . . .
CAN YOU COME:

WEEKDAYS . . . . EVENINGS . . . . . . WEEKENDS. . . . .

Fig. 1. Announcement placed in local newspaper to recruit subjects.

potential for participants. By contrast, certain complex, theoretical investigations may be virtually impossible to explain adequately to introductory level students in the short period of time typically allotted to subject debriefing. We earlier considered the possibility that subjects may give minimal attention to the debriefing information they receive after participating in a study (see p. 85). As a result of these factors, some subjects are apt to view their research experience as a trivial and boring waste of their time, perhaps compounding any initial feelings of resentment at having to participate in the first place.

For those students who choose for whatever reason not to satisfy the participation requirement, opting instead to complete an alternative assignment, the educational merits may be even more dubious. In recent years, course instructors have turned to fairer alternatives to research participation that can provide students with relatively equivalent learning experiences. Thus, instead of requiring extensive term papers, students may be asked to summarize empirical studies appearing in scientific journals, attend departmental lectures or seminars and write about them, and the like. While such activities can offer much to students in the way of educational benefits, they can easily become empty exercises unless they are taken seriously by students and faculty. For example, students might be provided with only minimal feedback from their instructor relative to alternative writing assignments, such as an acknowledgment that they were acceptable (or unacceptable). Instructors rarely have an incentive to grade such reports, which may consist of nothing more than a poor paraphrasing of a journal article or a lecture's key points. In short, although alternative assignments can reduce the likelihood that students are forced into research participation against their will, a poorly implemented program of research alternatives merely adds to concerns about the unfair treatment of student participants.

The ethical use of student subject pools

Several guidelines have been proposed for the ethical use of student subject pools, including the set developed by the American Psychological Association (APA, 1982), summarized in Box 7.3. It should be noted that many of the guidelines that have been suggested relevant to the ethical use of subject pools also apply to the general treatment of semi-volunteer research participants solicited in other ways, such as when students are asked to provide questionnaire data during class time.

The APA guidelines provide a useful starting point for a consideration of the proper administration of subject pools. These guidelines are intended to apply only to the use of a research participation requirement for studies that offer the participant potential educational benefits. An addendum to the APA guidelines suggests that other means of recruiting subjects should be used when the potential for educational gain is missing. For instance, in studies involving long-term participation, the potential for educational gain to subjects typically is exhausted at an early stage in the investigation. In such cases, it is common practice to offer other incentives, such as monetary ones, and to recruit participants from outside the university subject pool. This, in fact, was the approach used by Zimbardo and his colleagues to recruit subjects for what was originally

---

Box 7.3

Suggested guidelines for ethical use of the subject pool (American Psychological Association, 1982)

1 Students should be informed about the research requirement prior to their enrolling in a course.
2 Investigators who intend to use the university subject pool should obtain the prior approval of their research proposals (e.g. by a departmental committee or an Institutional Review Board).
3 Alternative options that are commensurate in time and effort to research participation should be provided so that students may choose the type of research experience, as well as the time and place of participation.
4 A description of the procedures to be employed and the right to withdraw at any time without penalty should be provided and consent sought before the student begins participation.
5 Steps should be taken so that students are treated with respect and courtesy.
6 Students should obtain some sort of reward in return for their participation, at least minimally consisting of as full an explanation of the research purposes as possible.
7 A mechanism should exist by which students may report any mistreatment or questionable conduct related to the research participation requirement.
8 The recruiting procedure should be under constant review, taking into consideration students' evaluations of the research participation requirement.

intended as a two-week prison simulation. Because the educational value of research participation is likely to drop considerably after participation in three or four short-term experiments, the number of total research hours for a research requirement should not be too high (Diener and Crandall, 1978).

Turning to the specific APA guidelines, it is suggested that students should be informed of a research participation requirement prior to their enrollment in the course. In essence, this point reflects the necessity for "truth in advertising" for students as they go about making their course selections (Hogan and Kimmel, 1992). Thus, the participation requirement might be stated in the course description in the university catalog. At a minimum, the requirement should appear on the course syllabus distributed at the beginning of the semester when students still have time to decide whether to continue with the class or find an alternative one.

The information that may be provided to students in a description of the subject pool requirement includes the following: the amount of participation required; the available alternatives to participation in research; the general nature of the kinds of studies from which students are expected to choose; the right of students to withdraw from a study without penalty at any time; the penalties for failing to satisfy the requirement or for not fulfilling one's obligations to the researcher; the researcher's obligation to treat subjects ethically and to provide an explanation of the research; procedures to follow if subjects are mistreated; and an explanation of the scientific purposes of research conducted in the departmental laboratories (APA, 1982).

With regard to alternative opportunities for research participation, it is recommended in the discussion accompanying the APA guidelines that students might observe and write about ongoing research or submit a short paper describing published research reports. Other useful suggestions for optional assignments which provide students with educationally beneficial experiences have been offered by researchers. Among the ideas considered at one American university, in addition to allowing students to sign up as observers rather than as participants in selected studies (subject to the investigator's approval), were: (1) scheduling a series of evening research presentations by graduate students and faculty for undergraduates to attend; (2) allowing the customary journal article summary, but with specific articles selected beforehand by faculty based on a determination that they report high-quality, recent studies; and (3) allowing students to assist in an ongoing research project for a specified period of time, by doing data entry and the like (McCord, 1991). These alternatives were selected on the basis of their being essentially equivalent to the research requirement in terms of time, effort, odiousness, and course credit available.

Another possible alternative to research participation would be to have students participate in community volunteer work, perhaps arranged through departmental student organizations, such as Psi Chi (the honors society for psychology students). Raupp and Cohen (1992), for example, have developed a model for incorporating community service volunteering as a course component. Their application involved a course option in which upper-level students were given the opportunity to assist in elementary school tutoring as an alternative to more traditional course requirements. This basic approach surely could be extended to a variety of community volunteer activities in order to provide students in introductory-level courses with an involving and educationally beneficial alternative to research participation. Such an approach would have to be carefully considered prior to implementation, however, given the potential problems inherent in sending inexperienced students into professional settings and the likelihood that some volunteer work would call for a substantially greater investment of time than participation in a research study. (See Table 7.1 for a summary listing of possible alternatives to the student research requirement.)

Sieber and Saks (1989) have argued that the educational value of subject pools can be maximized when an appropriate *quid pro quo* is established between researcher and student. That is, because the

Table 7.1 Alternatives to the student research requirement

1	The reading and summarizing of journal articles describing recent, high-quality research
2	Observations of selected, ongoing studies
3	Informal field observations of behavior
4	The viewing of videotapes of laboratory experiments or filmed presentations of how an investigation is carried out
5	Attendance at research presentations organized by graduate students and faculty
6	Assisting in an ongoing research project for a specified period of time
7	Engaging in volunteer community service activities
8	Readings (e.g. paperback books) about the research process

These recommendations were compiled from the following sources: APA, 1982; Diener and Crandall, 1978; McCord, 1991; Raupp and Cohen, 1992; Sieber and Saks, 1989.

student provides data that yield knowledge, the researcher should repay the student in kind, not with money or grades, but with interesting and worthwhile knowledge. Additionally, Sieber and Saks suggest that having an opportunity to observe a data collection session or to watch a videotaped presentation of how an investigation is carried out would represent far more inviting and educational alternatives to writing assignments and other course work for students who may choose not to participate in research.

Among the other ideas for enhancing the educational value of subject pools that emerged from Sieber and Saks' (1989) survey was the suggestion that subjects should receive, following their participation in a study, at least five minutes of instructional feedback devoted to a description of the research design, the hypotheses, and the relevancy of the data collected for testing the hypotheses. Other suggestions were to quiz research participants on what they learned through participation or to have them complete evaluation forms by rating the researcher on such qualities as respectful treatment of subjects, clarity and adequacy of informed consent and debriefing, and overall interest and value of the research experience.

As a means for protecting students from unfair coercion, some departments offer research participation not as a course requirement, but rather as one of several things that can be done to earn extra credit points. For example, Sieber and Saks (1989) described one psychology department where the completion of five hours of subject participation could raise a student's grade one third of a letter grade (e.g. from B+ to A–). Because students could obtain the extra credit through alternative means (such as by writing one-page reviews of articles on library reserve) and could still pass the course without feeling any pressure to participate in research, such a policy would avoid the problem of imposing a penalty on non-volunteers. As other protections against coercion, researchers should remain sensitive to the possibility that they are inducing research participation by implying that cooperation will be rewarded by more favorable perceptions of the participant by others. For example, subtle promises of potential benefits for subjects' major reference groups and suggestions that declining to participate reflects a sign of weakness or immaturity are tactics which, according to the APA (1982), represent coercive threats, especially during the process of recruiting subjects.

Taken together, these recommendations for the ethical use of subject pools suggest that (1) attention and effort should be devoted to ensuring that students gain educational value from research participation, and (2) research participation should be available to students as a means for enhancing their course grades rather than jeopardizing the grades in any way (Keith-Spiegel and Koocher, 1985). Towards these ends, students should be informed of the department's subject pool policy prior to enrolling in the course and then again as soon as the course begins. When research participation is made a requirement for successful completion of a course, the alternative means for meeting the requirement should be well-publicized to students, in case they object to participation for whatever reason (Hogan and Kimmel, 1992).

Students also need to be aware of their freedom to withdraw from participation in a study at any time, despite their agreement to serve as subjects. Prospective subjects may show up for a scheduled appointment only to learn that the research involves certain aversive procedures (such as the administration of low-intensity electric shocks) that were not apparent in the original description appearing on the subject sign-up sheet. Some subject pool participants may inaccurately assume that they have no choice but to go through with participation or else lose the promised incentives (such as course credit) as a penalty for withdrawing (cf. Orne, 1962). The freedom to withdraw from, or refuse to participate in, particular investigations should be explained when the research participation policy is first presented to students and they should be reminded of this right when they arrive for a study (Diener and Crandall, 1978). Investigators also are obligated to protect students from adverse consequences attributed to their having declined or withdrawn from research participation (APA, 1992).

Finally, the APA guidelines for the ethical use of subject pools emphasize the importance of having mechanisms in place for reviewing the research participation requirement and for enabling subjects to initiate complaint procedures when they feel that they have been mistreated. Most universities now require that investigators planning on using the subject pool first must have their research proposals approved by a campus subject pool committee or institutional review board. According to Sieber and Saks (1989), 71.6 percent of the respondents to their survey claimed to have a mandatory application policy for subject pool use, and 87 percent of the surveyed departments reviewed applications routinely. The subject pool application process can serve as a valuable reminder to researchers of their ethical responsibilities to research participants.

Research recruitment procedures can be reviewed and modified based on the regular monitoring of participants' attitudes toward the requirement and investigators' evaluations of the adequacy of the subject pool policy (APA, 1982). In some research settings, users of the subject pool are evaluated by subjects in terms of the extent to which ethical guidelines were followed and whether clear and informative feedback was provided. In order for participant evaluations to be fair and informative, it undoubtedly is necessary for students to receive some degree of preparation for subject pool participation, in terms of their own and the researcher's ethical obligations. Of course, it is important that this preparation does not merely serve to create false expectations among students about researcher misconduct, leading to self-fulfilling prophecies. However, formal channels should be in place so that subjects are able to address their complaints (should they feel that their rights have been violated by a researcher) and participation-related problems (such as an inability to attend a study for which they have volunteered). Some academic departments have subject pool committees for this purpose, or else designate a research assistant, subject pool administator, course instructor, or departmental chairperson to handle student concerns (Sieber and Saks, 1989). Repeated complaints about a particular investigator could result in a loss of subject pool usage or other penalties.

The ultimate goal of the subject pool policies described here is to maximize two desirable ends which need not be in conflict: the creation of psychological knowledge and the ethical treatment of student research participants. It is hoped that the ethical use of subject pools will serve to increase the likelihood that future generations of student research participants will view the research option as a valuable and attractive opportunity for learning, rather than as just another course requirement to be avoided.

Ethical responsibilities of research participants

To this point, our emphasis has been placed on the rights of research participants and their appropriate treatment by investigators. As we have seen, participants have the right to be informed of the general purpose of the study and what they will be expected to do; the right to withdraw from a study at any time; the right to receive promised benefits; the right to expect researchers to protect their anonymity and maintain the confidentiality of their responses when these are promised; and the right to know when they have been deceived in a study and why the deception was necessary. Additionally, it is clear that researchers also have certain rights and that research participation carries with it certain obligations. Korn (1988) has identified

Research Subjects

Table 7.2 Responsibilities of research participants

1 Participants have the responsibility to listen carefully to the experimenter and ask questions in order to understand the research
2 Participants should be on time for the research appointment
3 Participants should take the research seriously and cooperate with the experimenter
4 When the study has been completed, participants share the responsibility for understanding what happened
5 Participants have the responsibility for honoring the researcher's request that they not discuss the study with anyone else who might be a participant

Source: Korn (1988, p. 77).

five responsibilities that pertain specifically to the participant's relationship with the researcher (see Table 7.2). These responsibilities are consistent with the contractual nature of the researcher-subject relationship, recognizing that both parties have certain rights and obligations (Carlson, 1971).

Although the responsibilities listed in Table 7.2 are relatively straightforward, a few comments can be made regarding their rationale. For example, unless the participant carefully attends to the researcher and understands what the research involves, a truly informed consent cannot be given. By questioning the researcher where understanding is lacking both before and after the study is completed, the research experience can be a more rewarding experience for the subject. Researchers also stand to benefit from subjects' questions when it becomes clear that participants are misinterpreting certain instructions, explanations, and the like.

Subjects also should be aware of the inconveniences they can cause when they fail to show up for a research appointment at the agreed upon time. Such behavior is likely to disrupt the schedules of the researcher as well as other participants. Because the data collection stage of a study can be extremely time consuming for researchers, subjects at least should make an effort to inform the researcher in advance when an appointment must be canceled so that another participant can be recruited for the time slot. Once an individual has completed participation in an investigation, the importance of maintaining confidentiality about details of the study also should be respected. In chapter 4 we described the threats to validity that emerge when suspicious (or pre-informed) subjects participate in research. By reneging on the promise not to discuss a study with others, subjects may in the process invalidate their own and future subjects' performance, thereby resulting in a wasted endeavor for all involved.

Some writers have suggested that research participation can serve as a means for individuals to satisfy their broader social responsibilities. Kimble (1987), for example, suggested that research participation provides one with the opportunity to help solve practical problems and to contribute to knowledge. This view, of course, suggests the importance of providing meaningful opportunities for persons who may be interested in volunteering for research participation.

Issues Involved in the Use of Volunteers

On the surface, the use of all-volunteer samples would seem to eliminate the ethical problems that arise when subjects (such as students) are coerced into participating in research. After all, the assumption of voluntary participation lies at the heart of such fundamental ethical regulations as informed consent, freedom to withdraw, and the like. If only it were that easy. As is explained below, if all participation were strictly voluntary, researchers would have to cope with methodological problems concerning the representativeness of their samples. They also would be restrained from utilizing certain research approaches, such as nonreactive observation. Thus, the voluntary nature of participation represents yet another issue in which conflicts between methodological and ethical demands are likely to arise. Additionally, there is a gray line between applying undue pressure when recruiting volunteer participants and being an ethical and competent researcher (Blanck et al., 1992). In other words, there is the possibility that researchers may use coercive, sometimes subtle, inducements when recruiting volunteers for research participation, especially when subjects are selected from vulnerable populations. In short, the use of volunteers carries with it the potential for several ethical dilemmas.

Volunteers and the representativeness problem

One limitation to using only volunteers for research participation pertains to the fact that persons who volunteer for experiments are likely to differ in significant ways from persons who are forced to participate. As a result of an extensive comparison of studies involving both volunteer and non-volunteer samples, Rosenthal and Rosnow (1975) found it possible to distinguish volunteers from other research participants on the basis of several characteristics. As compared with captive participants or non–volunteers, volunteers tend to be better educated, higher in social class status, more intelligent, more sociable, and have a higher need for social approval. Although the evidence is less strong, it appears that volunteers may also have a greater need for new and exciting experiences, have less conventional attitudes, be less authoritarian, and be more non-conforming, altruistic, and self-disclosing than non-volunteers. Volunteers also are more likely than non-volunteers to be young, female, and from smaller towns. Table 7.3 summarizes these differences by classifying these (and other) distinguishing characteristics according to the degree of confidence we can have in each (Rosenthal and Rosnow, 1991). Thus, the first five characteristics warrant "maximum" confidence because they were based on more relevant studies and the highest percentage of supportive results; there is less supportive evidence for the remaining traits.

Table 7.3 Volunteer characteristics grouped by degree of confidence of conclusion

---

I. Maximum confidence
  1 Educated
  2 Higher social class
  3 Intelligent
  4 Approval-motivated
  5 Sociable

II. Considerable confidence
  6 Arousal-seeking
  7 Unconventional
  8 Female
  9 Non-authoritarian
  10 Jewish > Protestant > Catholic
  11 Non-conforming

III. Some confidence
  12 From smaller town
  13 Interested in religion
  14 Altruistic
  15 Self-disclosing
  16 Maladjusted
  17 Young

Source: R. Rosenthal and R.L. Rosnow (1991) Essentials of Behavioral Research: Methods and Data Analysis (2nd ed.). Copyright © 1991 McGraw-Hill, Inc. Reproduced by permission.

Given these differences between volunteers and non-volunteers, it is understandable that researchers may be reluctant to recruit subjects solely on the basis of voluntary participation. To do so would be to raise the possibility that the research conclusions are misleading as a result of volunteer bias. That is, the knowledge that research volunteers are likely to be brighter, less authoritarian, higher in approval need, and so on, than non-volunteers can lead one to expect conclusions that are either too liberal or too conservative as a result of the use of volunteer samples.

As an example, imagine a case in which a marketing researcher conducts a study to determine the persuasive impact of a new advertisement prior to recommending its use in a heavily funded television campaign (Rosnow and Rosenthal, 1993). Suppose the researcher finds it most convenient and practical to pilot test the advertisement by recruiting volunteer subjects at various shopping malls and randomly assigning them to an experimental group that is exposed to the new advertisement or a non-treatment control group. The problem with such a study is that its outcome may be contaminated by volunteer bias. Specifically, because volunteers are relatively high in need for approval, and people high in approval need are likely to be more easily influenced that those who are low in approval need (Buckhout, 1965), the combination of these tendencies may lead subjects in the experimental condition to overreact to the treatment. As a result, the possibility would exist that the anticipated effect of the new advertisement in the general population would be exaggerated.

As the previous example illustrates, the generalizability of data obtained from using voluntary subjects exclusively will be threatened to the extent that individual differences between volunteers and non-volunteers are related to the variables under investigation. Thus, because volunteers tend to be more intelligent than non-volunteers, a study involving the establishment of norms for a standardized intelligence test could result in artificially inflated values if only volunteer subjects were studied. Similarly, in prevention research, volunteer subjects might be better able to learn techniques for preventing interpersonal difficulties, depression, and the like than people with similar symptoms who choose not to volunteer (Kimmel, 1988a). The consequence would be an overestimation of the effectiveness of the prevention program that served as the focus of the study.

A recent illustration of the volunteer effect in an applied study was provided by Strohmetz et al., (1990), who examined differences in measures of problem severity among alcoholics who did or did not volunteer to participate in a treatment study. A positive relationship was revealed between patient volunteer status and severity of alcoholism problems reported during the pre-treatment period, such that those individuals who were more willing to participate in the treatment intervention tended to have more severe alcohol-related problems. According to Strohmetz et al., one plausible implication of this result is that patients who agree to participate in an intervention study may somehow be different from the population of interest. A similar concern regarding voluntary self-selection into intervention programs has been noted in research involving the adjustment of Vietnam War veterans (King and King, 1991). Awareness of this problem has led some researchers, where possible, to stratify the population of interest into respondents and non-respondents in order to assess the direction of the selection bias (cf. Blanck et al., 1992; Saks and Blanck, 1992).

In the context of research on prevention programs, the decision to limit participation solely to volunteers raises the possibility that individuals from certain societal groups might not have an opportunity to participate. For example, well-intentioned attempts on the part of researchers not to coerce people into prevention research programs could serve to withhold valuable resources from those most in need. This could occur as a result of recruitment techniques that fail to reach certain segments of the population, or are reaching them through different media (Muñoz, 1983).

Another way that differences between volunteers and non-volunteers can pose a threat to the validity of research conclusions has to do with the higher need for social approval in volunteers. Several studies have demonstrated that because of their higher approval need, volunteer subjects are particularly inclined to comply with an experiment's task-oriented cues (or demand characteristics) in order to play the role of good subject (Horowitz, 1969; Rosnow and Suls, 1970). This problem was apparent in a study carried out by Horowitz (1969) on fear-arousing communications in persuasion. Horowitz reasoned that conflicting findings obtained by attitude change researchers may have been due to whether or not their samples were comprised of volunteer subjects. In his view, volunteers might be more influenced in the expected direction by fear-arousing messages because of a greater sensitivity to demand cues and willingness to comply with them, whereas non-volunteers might respond in an opposite manner in response to forced participation. As predicted, Horowitz found a positive relationship between fear arousal and attitude change for volunteers but an inverse relationship for "captive subjects." Apparently, subjects' motivations to comply with the demands of the experiment were influenced by whether or not they had willingly agreed to participate.

The problem of inducements to participate

As we have seen, the use of volunteers can serve to upset the delicate balance between the methodological requirements and ethical imperatives of a research study by producing artifacts in the research data (Blanck et al., 1992; Kimmel, 1979; Suls and Rosnow, 1981). For this reason alone, recommendations to organize elite subject pools consisting of informed and well-paid volunteers would not be an advisable solution to the ethical problem of coercing people to participate in research (see Box 7.4). Instead, researchers have attempted to cope with threats to external validity due to limited participant samples (e.g. only volunteer subjects) by developing methods for enhancing the diversity of their subjects.

While respecting individuals' freedom to decline to participate, several empirically derived recruitment strategies for inducing more non-volunteers to enter the sampling pool have been proposed. These involve communicating the message to prospective subjects that participation in the research will be interesting, making the appeal for subjects as non-threatening as possible, emphasizing the theoretical and practical importance of the research, offering small courtesy gifts simply for considering participation, and avoiding experimental tasks that could be perceived as psychologically or biologically stressful (Rosenthal and Rosnow, 1975). Such suggestions are based in large part on research findings showing that volunteer rates tend to increase the greater the incentives offered for participation and the less aversive, more interesting, and more important the task. Such methods for stimulating research participation may have certain potential scientific benefits beyond drawing a more representative sample and thereby lessening the likelihood of subject selection bias. In addition, these recruitment strategies should make researchers more thoughtful and careful in their appeals for subjects and in the planning of their research, with more emphasis placed on doing important rather than trivial research (Blanck et al., 1992; Rosenthal and Rosnow, 1991).

Some researchers are not totally convinced that improvements in the techniques for recruiting subjects will eliminate the dilemmas inherent in using volunteer samples (e.g. Schuler, 1982). But perhaps the most serious ethical concern involves the point at which the researcher's offers of inducements to participate become coercive,

---

Box 7.4

"Volunteers for hire"

During the mid-1970s there were indications that graduate students in more than one American university had considered forming organizations along the lines of employment agencies like Manpower, Inc., which would offer volunteer subjects for hire (Rosenthal and Rosnow, 1975). While the author is not aware of any such groups actually having been formed, it is intriguing to consider the consequences of organized groups of professional volunteers who offer their services for research participation, albeit for a price. Although organizations of this kind might put to rest many ethical anxieties, the generalizability of data obtained from using voluntary subjects exclusively would be threatened to the extent that groups of volunteers differ from the population at large.

A recent article in the satirical magazine *Spy* helps put into perspective some additional problems that can arise when researchers find it necessary to utilize professional volunteers for their research (Zicklin, 1994). The article focuses on the experiences of professional volunteers who regularly consent to participate in experimental studies of new drugs or in other potentially risky tests of medical procedures in order to receive attractive monetary incentives. One self-proclaimed veteran of the medical-testing industry claimed to have participated in ten biomedical studies over a two-year period, sometimes involving painful procedures, but generally for good pay. For example, according to this participant, in one study he consented to have his forearms heated until they blistered so that an experimental skin treatment could be tested. While painful, he claims to have received US $500 for the eight-hour study.

An obvious problem inherent in offering large sums of money to willing participants in potentially dangerous medical studies is that there may be no limit to the kinds of risky procedures that some people will agree to undergo if the price is right. Another problem which has certain methodological implications also was revealed in the *Spy* article. Because partial completion of a drug study typically results in only partial payment, there apparently are some volunteers who choose not to divulge any adverse side-effects of a drug for fear of being dropped from the study. This tendency to hide the ill effects of an experimental drug could create a false confidence in its safety, ultimately resulting in the approval of a potentially harmful new medicine. Curie-Cohen et al., (1979) have noted a similar problem involving some sperm donors who attempt to hide illness and disease in order to maintain their status as paid volunteers.

Perhaps it is somewhat of a blessing in disguise that behavioral scientists generally do not have the resources to offer the kinds of monetary incentives that are used to attract research volunteers by major pharmaceutical companies or the medical research industry. Nevertheless, even when incentives are relatively small, the behavior of voluntary participants may be affected by the prospects of being dropped from a project or missing out on an opportunity to participate in a follow-up study at some later date.

thereby threatening the individual's right and freedom not to participate. To illustrate the kinds of ethical problems that might be encountered in recruiting research subjects, consider the following questions posed by Blanck et al. (1992, pp. 963–4):

1 ... Is it ethical to employ young, attractive, verbal, and intelligent assistants of ethnic backgrounds similar to the target population in order to recruit participants?
2 Is it ethical to have research assistants spend time in the recruitment site, building a positive reputation, so that the potential participants are familiar with the researchers?
3 May a researcher ethically reward children with a classroom party for returning parental consent forms for a research project, the reward contingent on return of the consent form whether or not the parent granted approval?

While each of these strategies is likely to increase participation rates, it is unclear to what extent they present situations involving undue pressure. As of yet, there are no clear ethical standards for obtaining guidance in such cases beyond the necessity for researchers to maintain a sensitivity to the possibility that they are exerting excessive pressure on targeted participants to enhance recruitment.

One subtle coercion that raises ethical issues occurs in the context of applied studies when researchers make promises in their recruitment efforts that cannot be carried out. For example, for some intervention programs, individuals from among the most powerless groups in society are invited to participate in voluntary programs that researchers consider to be beneficial to them. To ethically test the effectiveness of the intervention, researchers must be sensitive to the possibility that they are overselling their approach. When controlled studies actually demonstrating the effectiveness of the intervention techniques are not widely available, a researcher's confidence in the effectiveness of a particular technique in preventing mental health problems and the like might promote behavior that approaches a kind of "false advertising." False hopes may be created among volunteers, thereby unjustifiably raising their expectations about the potential benefits of participation.

As a means to circumvent the problem of raising false hopes, researchers might communicate to potential subjects the belief that an intervention program has a reasonable chance of positive effects, and little chance of damaging effects, while taking care not to oversell the potential benefits of the approach. For example, in his research on the effectiveness of interventions designed to prevent depression, Muñoz (1983) informs potential participants that the methods have been found helpful in controlling depression that interferes with daily functioning. However, he also warns that the techniques first must be learned by participants and will not necessarily work for all people in all situations. Potential volunteers thus are aware that research involvement may be beneficial to them, but that there is no guarantee that they can be successfully applied to prevent depression. This approach is consistent with the position that a researcher's attempts to influence recruitment are justifiable only when the results of the cost-benefit analysis are at least balanced for prospective subjects, or when subjects have all the prerequisites for performing the calculation themselves (Schuler, 1982).

Other important issues related to recruitment strategies pertain to the nature and size of the incentives offered to subjects. People may voluntarily choose to participate in research for several possible reasons, including intellectual curiosity, monetary payment, bonus points in a college class, and the advancement of science. As Diener and Crandall (1978) have pointed out, because all subjects are no doubt motivated by some kind of incentive, it would be foolish to argue that inducements for research participation should be disallowed. However, most ethical researchers would agree that incentives for research should not be excessive. The issue then becomes one of determining at what point an incentive may be considered "too strong" and thus coercive. One suggestion for assessing whether the strength of an incentive is unacceptable is to offer it to potential subjects in the context of investigations varying in risk. If nearly all agree to participate even when the research poses certain risks, one might conclude that the incentive is excessive (Diener and Crandall, 1978).

Another approach for determining whether incentives may be too strong is to empirically compare their effects on willingness to participate (Reiss, 1976). In a series of three such exeriments, Korn and Hogan (1992) investigated the influence of various incentives on college students' willingness to participate in experiments involving aversive procedures. Participants read descriptions of hypothetical experiments and then rated their willingness to participate in the studies. Different rewards were offered for participating; for example, one experiment compared a voluntary condition and three incentive conditions (US$10.00, 1 percent of the total possible points to be added to their class grade, or 5 percent of the total possible points). The aversiveness of the experimental procedure also was varied by comparing an innocuous control condition (a standard memory experiment involving nonsense syllables) with various aversive treatments consisting of the administration of a drug with possible side-effects.

The results of Korn and Hogan's experiments revealed that the incentive offered had an effect on willingness to volunteer for a research study, although it was not as powerful or consistent as the potential aversiveness of the experimental treatment. In the experiment described above, students were more willing to participate in a study (whether or not it involved an aversive treatment) for US$10 or 5 percent course credit than for 1 percent course credit or for strictly voluntary purposes. Although incentives increased willingness to participate in two of the three experiments, there were no significant differences related to kind of incentive (course credit or money). On the other hand, students were consistently less willing to participate in research as the aversiveness of the treatment increased.

Korn and Hogan limited their investigation to a range of incentives realistic enough to be considered plausible, rather than including extremely large incentives (such as 50 percent course credit or US$1,000). Their results may have been somewhat limited by the hypothetical nature of the experiments considered. However, the important issue may not be the incentives themselves, but rather the investigator's obligations to act responsibly (Diener and Crandall, 1978). For example, a college professor who indiscriminately offers 50 percent course credit for research participation would likely undermine his or her obligations as a responsible instructor. A fairer way to implement the incentive would be to first devise the class grade distribution without considering the extra credit option so that non-volunteers will not be at a disadvantage, and then to establish that bonus points will be given only when research participation will provide a learning experience for the students. Finally, a reasonable level of course credit should be established for research participation. As Korn and Hogan demonstrated, incentives need not be very large in order to influence willingness to participate.

It has been argued that there is nothing inherently unethical about offering people in need money in order to participate in research, and that participating for money is as legitimate a motivation for research participation as "to help science" (e.g. Diener and Crandall, 1978). While this author agrees that monetary incentives are acceptable for stimulating research participation so long as the amount offered is reasonable, such incentives may not be appropriate in certain circumstances, such as when the prospective subjects are students or professional volunteers. In the case of students, researchers might be tempted to view the monetary incentive as a justification for not having to provide educational feedback. And, as we have seen, attracting "volunteers for hire" through offers of relatively large monetary inducements can raise certain questions about the validity of the research results.

The ethical issues concerning the establishment of legitimate levels of incentives for research participation become somewhat more complex when subjects are recruited from special populations, as is the case with prisoner volunteers. In the remainder of this chapter, problems involved in the recruitment of subjects from special populations are considered.

Vulnerable Participants

While much of the focus of this chapter has been placed on the recruitment and selection of college students for research participation, it is important to recognize some of the issues underlying the use of other subject populations from which behavioral scientists often draw their samples. For example, many investigators have used prisoners as research subjects for some of the same reasons that students are so prevalent in research – the prison population represents a readily available, captive group of potential participants. Additionally, because of the nature of their investigative interests, behavioral scientists from such disciplines as clinical psychology, developmental psychology, and sociology are apt to select their subjects from certain well-defined, often vulnerable groups, such as children, the elderly, the mentally disabled and handicapped, and underprivileged persons. Individuals within these groups often have certain limited capacities to choose freely whether or not to participate in research, thereby raising some serious ethical issues.

Differing ethical positions on the involvement of research participants from special groups can be traced in large part to a consideration as to whether the research is therapeutic or nontherapeutic in nature (Ramsey, 1976). Therapeutic or clinical research focuses on the treatment of specific problems, and the subject typically can anticipate some direct or indirect benefits as a function of having participated. That is, if the methods studied prove to be effective, the participant can expect some improvement. By contrast, participants in non-therapeutic research do not expect to receive any direct benefits relating to their current state, beyond the incentives offered for entering the study (such as monetary payment).

Perhaps with the exception of experimentation with children, the problems inherent in research with vulnerable participants have received considerably greater attention in the context of biomedical research than in behavioral science disciplines (Schuler, 1982). As a result of extensive discussion and debate, special biomedical regulations were developed to protect individuals from special groups whose ability to give informed consent is assumed to be limited. Some of the recruitment and selection issues in using participants from special populations for behavioral science research are described below.

Prisoners

Most of the early research involving prisoner participants consisted of medical experiments on infection, the testing of medical drugs and cosmetics, and related studies, all of which offered minimal direct benefits to the subjects themselves (Jonsen et al., 1977; Reynolds, 1979). For example, one research approach that was commonly used in prison settings is the so-called "patch study" as a means of testing skin sensitivity (Hatfield, 1977). In a typical patch study, one-inch square bandages impregnated with a commercial product (such as perfume, deodorant, or soap) are taped to as many as ten sites on the back or forearm, along with one control site where no substance is used (in order to determine whether the subject is simply reacting to the bandage). A participant typically would wear the patches for up to three or four weeks to determine if the product caused any adverse skin reactions. In other large-scale studies, bath soaps and dandruff shampoos would be provided to volunteer prisoners and used for a period of about three months before the products would be released to retail outlets.

In recent years, such biomedical experiments in prison settings have been severely restricted (e.g. Herbert, 1978). This is largely a result of recommendations made in 1976 by the National Commission for the Protection of Human Subjects of Biomedical and Behavioral Research, a federal body which was particularly instrumental in bringing ethical issues to the attention of the public, researchers, and politicians (see chapter 2). There are now severe limits on all research with prisoners that does not contribute to their personal well-being or pertain to the circumstances and conditions of their previous lives (Reynolds, 1979: Schuler, 1982). This emphasis on therapeutic research is apt to raise certain role conflicts for behavioral scientists working within institutional settings where the objectives of institutionalization are unclear (Carroll et al., 1985). For example, the roles of prison have included punishment, rehabilitation, deterrence, and protection through detention (Gaylin and Blatte, 1975). In addition to carrying out research in prisons, behavioral scientists often function as therapists, evaluators, administrators, and consultants on matters of control within prisons, and often act as expert witnesses on such matters as parole, competence, sentencing, and dangerousness.

The American Psychological Association's Task Force on the Role of Psychology in the Criminal Justice System identified confidentiality as a central issue for psychologists working within the prison system (APA, 1978; Clingempeel et al., 1980). While the focus of the task force was placed on therapeutic practice, it is not difficult to recognize how dilemmas involving confidentiality may arise in research with prisoners. For example, during the course of a study with prisoners an investigator might learn that a subject was smuggling drugs into the prison (Steininger et al., 1984). As discussed in chapter 6, this kind of situation is common in applied studies, where a researcher is often faced with conflicting obligations. In this case, informing prison officials about the drug smuggling would constitute a breach of confidentiality to the prisoner. Although the prison officials are paying for the researcher's services and would desire the information, the prisoner may have volunteered for the research under conditions of anonymity and confidentiality. According to the APA Task Force, most psychologists surveyed tended to prefer not to break promises of confidentiality, citing concern for the potential harm to the prisoners (Clingempeel et al., 1980).

In other cases, such as those involving potential harm to others, the researcher's loyalties to his or her subjects are more seriously challenged. If, during the course of participation in a research study, a prisoner reveals plans to kill or seriously harm another person, some researchers probably would view the need to protect society at large as a more important obligation than any promises of confidentiality made to the prisoner. Such a decision to breach confidentiality when there is clear danger to others is consistent with recommendations in APA's (1982) *Ethical Principles*, its Task Force report (APA, 1978), and a California Supreme Court ruling (*Tarasoff* v. *Regents of the University of California*).

One additional issue to be considered in therapeutic research with prisoners pertains to the view that prisoners (as well as hospitalized patients) should be required to cooperate with research on programs oriented toward their own rehabilitation. The APA (1982, p. 45) advises against this practice on the grounds that the "position is too readily available as a rationalization for exploitation" and the possibility that the institutions involved may function more to incarcerate or punish than to rehabilitate inmates.

At the heart of typical objections to non-therapeutic research on prison participants is the concern that consent to participate may rarely be truly voluntary (see Box 7.5). Prisoners are extremely dependent on material rewards (Hatfield, 1977) and many have hopes that participation in research will lead to an easier or shorter imprisonment (Schuler, 1982). One early critic of the reduction of punishment for criminals who consent to participate in dangerous research was the philosopher Immanuel Kant (1797/1965). Kant

---

Box 7.5

Prisoners as experimental subjects: pros and cons

Pros: arguments for using prison volunteers in non-therapeutic Research

1 Being a subject in non-therapeutic research carries little risk.
2 Prisoners should not be denied the freedom to volunteer for experiments.

Cons: arguments against using prison volunteers in non-therapeutic research

1 Voluntary informed consent is impossible under prison conditions.
2 Nothing ethically justifies research not potentially beneficial for the prisoner.
3 Research results obtained using prisoners may not be valid for the free population.

Source: Shubin (1981).

believed that the practice would amount to no less than the destruction of legal justice, claiming that "if legal justice perishes, then it is no longer worthwhile for men to remain alive on this earth" (p. 100). While Kant's prediction may have been overly dire, there is wide agreement that early parole should not be given for involvement in a research study. Nonetheless, there is the risk that some prisoners will volunteer for dangerous experiments anyway, believing that in doing so their chances for release will somehow be improved. Undue pressure to participate also can stem from fears that refusing to volunteer for research will somehow lessen the possibility of parole. One recommendation for avoiding these threats to voluntary research involvement is to reward prisoner participants monetarily and to inform them that their participation will not be entered into parole records (Diener and Crandall, 1978).

Researchers must be especially sensitive to the possibility that incentives offered to prisoners are not so strong as to eliminate the freedom of choice in the decision to participate. However, it also should be noted that the opportunity to participate in research provides prisoners with an important means of contributing to society and for improving their self-esteem (Hatfield, 1977; Novak et al., 1977). The monetary incentive also allows them to obtain certain amenities through legitimate means and provides a source of contact with the outside world. These positive outcomes are evident in the following comments from an ex-prisoner who participated in several biomedical experiments during the early 1970s (Hatfield, 1977, p. 12):

> In the highly structured environment of prison, where I had no other opportunities to do anything that would give me a substantial monetary and emotional return, medical research gave me a sense of purpose ... Men in prison have few material resources and are denied many aspects of human dignity. Those few opportunities they do have should not be taken away, against their wishes, unless society is prepared to offer a meaningful alternative.

Children

Children represent one of the most widely used special groups for research participation in the behavioral sciences. A primary ethical (and legal) concern, as with other vulnerable groups with potentially limited understanding of information provided by the investigator, pertains to the adequacy of informed consent (Curran and Beecher, 1969) (see Box 7.6). That is, do children understand what they are

Box 7.6

Ethical principles for research with participants from special groups: some pertinent examples

American Psychological Association (APA, 1992)

6.11(e) For persons who are legally incapable of giving informed consent, psychologists nevertheless (1) provide an appropriate explanation, (2) obtain the participant's assent, and (3) obtain appropriate permission from a legally authorized person, if such substitute consent is permitted by law.

British Psychological Society (BPS, 1995a)

3.2 Research with children or with participants who have impairments that will limit understanding and/or communication such that they are unable to give their real consent requires special safe-guarding procedures.

3.3 Where possible, the real consent of children and of adults with impairments in understanding or communication should be obtained. In addition, where research involves any persons under sixteen years of age, consent should be obtained from parents or from those *in loco parentis*. If the nature of the research precludes consent being obtained from parents or permission being obtained from teachers, before proceeding with the research, the investigator must obtain approval from an Ethics Committee.

3.4 Where real consent cannot be obtained from adults with impairments in understanding or communication, wherever possible the investigator should consult a person well-placed to appreciate the participant's reaction, such as a member of the person's family, and must obtain the disinterested approval of the research from independent advisors.

3.5 When research is being conducted with detained persons, particular care should be taken over informed consent, paying attention to the special circumstances which may affect the person's ability to give free informed consent.

Canadian Psychological Association (CPA, 1991)

In adhering to the principle of respect for the dignity of persons, psychologists would:

I.27 Seek an independent and adequate ethical review of human rights issues and protections for any research involving vulnerable groups and/ or persons of diminished capacity to give informed consent, before making a decision to proceed.

I.28 Not use persons of diminished capacity to give informed consent in research studies, if the research involved might equally well be carried out with persons who have a fuller capacity to give informed consent.

I.29 Carry out informed consent processes with those persons who are legally responsible or appointed to give informed consent on behalf of individuals who are not competent to consent on their own behalf.

I.30 Seek willing and adequately informed participation from any person of diminished capacity to give informed consent, and proceed without this assent only if the service or research activity is considered to be of direct benefit to that person.

American Sociological Association (ASA, 1989)

B.5 Sociologists should take culturally appropriate steps to secure informed consent and to avoid invasions of privacy. Special actions may be necessary where the individuals studied are illiterate, have very low social status, or are unfamiliar with social research.

told about the nature and purpose of a study, and at what age can they be allowed to consent on their own, without the agreement of a legal guardian? A related question is whether children are able to weigh the risks that a study may entail for them. Additionally, children may be particularly prone to pressures to participate in research and have a lower tolerance for potentially risky or deceptive experimental manipulations.

Researchers clearly need to take special care to protect the rights and well-being of less powerful research participants, taking into account subjects' concerns and viewpoints when designing their studies. According to Thompson (1990), it is helpful for researchers to adopt a developmental perspective when planning research with children, by recognizing that while many risks to children decrease with increasing age, others are apt to increase, change in a curvilinear fashion, or remain relatively stable as the child matures. As with other vulnerable groups, the common approach to dealing with the issue of consent is to use parental or proxy consent as a substitute for, or in addition to, obtaining the child's consent. That is, if it is clear that the child cannot give personal consent but stands to benefit directly from the investigation, parents could be asked to provide consent for the child (Annas et al., 1977). As a straightforward example, it is not possible to inform a retarded adolescent as one could inform a normal adolescent. In such a case, it is up to the parents or other primary caregivers to decide whether the adolescent should enter an experimental treatment program, based on what is in his or her best interests (Steininger et al., 1984). By contrast, in an early legal decision a normal 19-year-old boy was considered capable and mature enough to choose a local anesthetic over general anesthesia for a tonsillectomy, despite the fact that his parents had chosen the general anesthetic (cf. Curran and Beecher, 1969).

The involvement of children in research generally entails fewer ethical problems when the research is therapeutic in nature. Most persons would agree that children should not be prevented from participating in studies that have a chance to improve their health or functioning, assuming that the investigations are well planned and focused. Examples of behavioral science investigations that offer potential therapeutic effects to children include preventive intervention studies oriented toward improving learning or interpersonal skills, research on techniques to rehabilitate juvenile offenders or drug abusers, and studies to prevent suicide or self-destructive behavior in at-risk youths.

One potential ethical concern in conducting research with children suffering from severe problems is the possibility that the promise of help or an irrational hope for a beneficial outcome will introduce coercive elements into the decision to have the child take part in the research (Carroll et al., 1985). Thus, in addition to clearly describing the anticipated benefits (and potential risks), investigators need to accurately present the likelihood of the experimental outcomes, based on current knowledge, to the adults responsible for the decision to have the child participate in the study. As we saw in the case of the Cambridge–Somerville study, even the most well-intentioned study can have unanticipated negative results (see chapter 6), a point that the researcher might also convey when obtaining consent. Additionally, when research participation is agreed upon in order to obtain promised therapeutic effects, it is essential that the study actually has the potential to help the individual participant rather than just advancing knowledge about the behavioral problem. Care also must be taken to provide effective procedures to children serving in untreated control groups as quickly as possible so that they do not experience unnecessary delay in receiving treatment.

More stringent safeguards need to be established for therapeutic research with children than for similar research with adults (Carroll et al., 1985). While exposure to certain risks may be tolerated in adult studies, it is generally believed that risks should be minimal or non-existent in studies with children. For example, an investigation involving a potentially painful procedure might be judged permissible among freely volunteering, informed adults (APA, 1982). The same procedure would be viewed as unacceptable for children because they are less likely to be familiar with such experiences and, as a result, more fearful of them. Alternative, risk-free procedures need to be explored prior to exposing children to untested research techniques that are intended for therapeutic purposes.

The views regarding children's participation in non-therapeutic research have been quite diverse, not only from an ethical standpoint, but from practical and legal perspectives as well. While some critics have argued against the involvement of children in any type of non-therapeutic research due to their questionable capacity to give consent (e.g. Ramsey, 1976), others have taken the opposite position by suggesting that children would consent if they had full capacities to do so and that research participation provides children with a rare opportunity to benefit others (e.g. McCormick, 1976). The debate is tempered, however, when the research in question clearly eliminates the possibility of harm as a possible consequence of participation. Most researchers would agree that a study of physiological reactions to aversive stimuli, such as physical pain versus negative criticism, would be unacceptable because of children's questionable ability to consent to such an investigation and the potential risks involved (Carroll et al., 1985). By contrast, a perception study involving young children's preferences for colorful and novel stimuli would no doubt be a stimulating and enjoyable experience for them, assuming that the research is conducted in a comfortable, non-fearful environment and parental consent has been freely obtained.

With regard to the issue of consent, parents (or legal guardians) are generally allowed to provide consent for a child in therapeutic studies, but the child's consent also must be obtained when the research is non-therapeutic (BPS, 1995a; Reynolds, 1979). For non-risk, basic research like the perception study just described, researchers often use a consent procedure involving "proxy and parallel consent" (Carroll et al., 1985). This involves first obtaining the fully informed consent of the parents and then the voluntary consent of the child, regardless of his or her age. In other words, the parents or persons legally responsible for the prospective participant are provided with full information about the nature and goals of the investigation. Following their agreement, the child is informed that participation is voluntary, participation can be stopped at any time, and that involvement in any phase of the research is not required.

In order for procedures like the proxy and parallel consent approach to be ethically sound, the child must be able to understand the researcher's description of the study and should not feel undue pressure to consent simply because parental consent has been obtained. Researchers have wrestled for some time with the question of when a child becomes capable of providing informed consent. Several recommended solutions have focused on the criterion of age, with ages ranging from 6 to 14 years offered as minimal limits for obtaining a child's consent (Curran and Beecher, 1969; Reynolds, 1979). In a recent review of empirical findings on children's capacity to consent, Abramovitch et al. (1991) reported that most children who were asked to participate in psychological research understood what they would be required to do. However, few children under the age of 12 years believed that their performance would be kept confidential and many felt there would be negative consequences if they ended their participation prematurely.

Prior to carrying out non-therapeutic (as well as therapeutic) investigations with children, researchers should be aware of legal requirements regarding the issue of consent (Annas et al., 1977). As Reynolds (1979) has pointed out, it may be that in some research contexts the investigator will be held responsible for determining whether the child has the intelligence and maturity to understand the research protocol and potential consequences. Some investigators have recommended novel procedures for providing information about a study to children. For example, Levy and Brackbill (1979) have suggested showing to children during the consent procedure a narrated video which depicts an investigator and child model going through the various steps in the study. Such a video could serve to clarify the procedure and at the same time reduce any fears the child might be experiencing about the unfamiliar experience. As long as the research in question does not appear overly dull or threatening, most children are apt to be curious about the experience and willing to participate, especially when it is described by a friendly investigator (Diener and Crandall, 1978).

Institutionalized individuals

The issues related to research with institutionalized persons, such as mental patients and the mentally handicapped, are similar to those described above in relation to children (Carroll et al., 1985). Of foremost concern are potential problems relating to coercion and restrictions on the voluntary decision to participate, protections against research risks, and questions related to competence to give fully informed consent. As in the case of children, it is easier to justify the carrying out of therapeutic research with institutionalized persons than research offering them no direct benefits. According to one critic, non-therapeutic research conducted with mentally infirm participants often is meaningless, with many such studies simply intended to demonstrate that there are differences between people who are institutionalized and people who are not (Haywood, 1976). Certainly, much meaningful basic research can be conducted with institutionalized subjects; however, researchers must carefully evaluate the potential merit of their investigations in order to ensure that they are not taking advantage of "captive" participants for the sake of trivial research.

Summary

The ethical researcher is one who recognizes that the power differential *vis-à-vis* potential (and actual) research participants can result in unfair coercion and the unacceptable exploitation of persons from vulnerable groups, and who takes steps to avoid these outcomes. Every person who participates in research is deserving of humane treatment and clearly specified benefits, whether or not the goals of the investigation are therapeutic in nature. Although all potential subjects need to be pre-tested and screened for a study which poses certain risk of harm to them, this is particularly true for persons with obvious vulnerabilities. For example, despite the ethical criticisms that were raised about the Stanford prison experiment (see Box 4.2), it must be said that the researchers took special steps to select only those volunteers who fell within the normal range on all measures in an extensive battery of psychological tests. By doing so, the researchers hoped to reduce the possibility of destructive effects (Zimbardo, 1973a, 1973b).

In view of the numerous ethical dilemmas that can arise in the recruitment and selection of research subjects, perhaps the best approach for researchers to take is one that regards the researcher-participant relationship as a collaborative one. Several researchers have argued for such a model in the past, though it has been rarely implemented in practice (Adair, 1973; Carlson, 1971; Jourard, 1967). A collaborative relationship would be one in which both the researcher and subjects work together as equals on mutually interesting behavioral research questions. The following passage from Carlson (1971, p. 216) effectively describes the essence of a truly *collaborative* research approach:

[A collaborative model] demands more candor and more thought on the part of the investigator in posing research problems, in engaging appropriate subjects, and in interpreting the nature of the experience; it demands more involvement from the subject, and offers the important reward of having his experience taken seriously.

Despite the fact that the collaborative model may represent an unattainable ideal in light of certain practical constraints (cf. Korn, 1988), the approach argues for a more balanced relationship between investigator and research participants than is currently found in the behavioral sciences. In the view of Rosenthal and Rosnow (1991), by treating subjects as though they were another granting agency, granting researchers their time instead of money, the research relationship can become more reciprocal, research procedures more humane, and the investigations more meaningful for all parties involved.

8

Ethical Issues in Research with Animals

My first exposure with animal rights was someone who came to the lab and said that the chickens I was working with were suffering anguish. I just dismissed them as silly. I said, "That's emotional nonsense. This is science. This is important," and I just dismissed it.
Animal rights activist, as quoted in H. A. Herzog, Jr.,
"The movement is my life"

In this chapter we turn our attention away from human subject research in order to focus on ethical issues involving research with animals. The use of animals for research purposes has long been considered a legitimate scientific activity. Millions of animals are studied in research laboratories each year in order to test the effects of experimental treatments that ethical issues prevent from first being tested on humans, and because the behavior of animals is interesting in and of itself. In addition to the widespread use of rodents, such as rats and mice, researchers have carried out investigations on monkeys, dogs, cats, birds, shrews, and fish, because these animals can serve as effective models of many human responses (Rissman, 1995; Shaughnessy and Zechmeister, 1990). Chimpanzees, for example, currently represent the only species of non-human primates in which progress of the HIV virus that causes AIDS can be appropriately studied. Rodents and birds now comprise roughly 90 percent of all non-human research subjects (Bowd and Shapiro, 1993).

A wide range of important questions have been investigated through animal research, including the effects of new drugs intended for human use and of substances meant for release into the environment. Tests of new surgical techniques often are first carried out on animals, and some animal subjects are exposed to diseases so that researchers can observe resulting symptoms and investigate potential cures (Shaughnessy and Zechmeister, 1990). In addition to research on the treatment of patients recovering from brain and spinal cord injuries, drug dependencies, and nervous disorders, behavioral scientists have carried out a number of animal studies that have advanced our basic understanding of the relationship between the brain and behavior. There is general agreement that these investigations have yielded valuable contributions not only to human welfare but to the welfare of animals as well (e.g. Azar, 1994b; Baldwin, 1993; Miller, 1985).

As Shaughnessy and Zechmeister (1990) have pointed out, the use of animals as subjects in controlled laboratory studies has often been taken for granted. While it is possible to invoke the biblical reference to human "dominion" over lower species of animals (e.g. Rollin, 1985) as a justification for animal research, perhaps the strongest argument from a scientific perspective pertains to the need to gain knowledge about phenomena that affect the human condition while at the same time protecting subjects from serious risk of harm. It is difficult to imagine much progress in the fight against such diseases as cancer, AIDS, or muscular dystrophy without first being able to investigate the course of the disease and potential cures in animal subjects.

Despite the numerous scientific advances that have emerged from research on animals, controversy over animal rights and research ethics has given rise to a number of important questions about the appropriateness of using animals for research purposes. Critics have charged researchers with mistreating animal subjects and causing unnecessary suffering, and also have questioned the value of research involving laboratory animals altogether. Behavioral scientists have been especially targeted by these critics, in large part because the benefits of behavioral research involving animals may not be as obvious as in other fields. The public debate over animal research is an emotional one, and logic and reason have often been overshadowed by rhetoric (see Box 8.1). The quotation that appears at the start of this chapter (attributed to a researcher turned animal activist) suggests that even staunch animal researchers ultimately may become conflicted about these issues. As a result of the passions that have been aroused on both sides of the issue, one thing is certain: no longer can research on animals be taken for granted.

In this chapter, the ethical questions pertaining to whether animals

---

| Box 8.1 |

Reflections from the animal research debate

The animal research debate has given rise to passionate reactions from both supporters and detractors of the use of animals in science. What follows are some representative comments from both sides of the debate.

I worked in a research lab for several summers during college, and I remember that it was a natural tendency to lose all empathy with one's animal subjects. My supervisor seemed actually to delight in swinging rats around by their tails and flinging them against a concrete wall as a way of stunning the animals before killing them. Rats and rabbits, to those who injected, weighed, and dissected them, were little different from cultures in a petri dish: they were just things to manipulate and observe.

                                                            Steven Zak, 1989

I can still remember the picture of that little monkey. They had severed his nerves, and he couldn't use his arm. They taped the other arm and made him use the handicapped arm. The monkeys were terrified of the experimenters.

                          Animal rights advocate, as quoted in Herzog, 1993)

Animal liberationists do not separate out the human animal. A rat is a pig is a dog is a boy.

              Ingrid Newkirk, Director, People for the Ethical Treatment
                                                              of Animals

How can one accept the eating of animals for the betterment of mankind but censure research on them for the betterment of mankind? It doesn't make sense to me.

              Animal researcher, as quoted in Keith-Spiegel and Koocher, 1985

These people are either stupid or too rigid to accept proven facts. Research studies with animals were the keys to polio, diphtheria, hepatitis, immunization, and to the development of antibiotics, insulin, arthritis medication, chemotherapy agents, hypertension control medication, joint replacement techniques, kidney dialysis, heart surgery, and organ transplants, to name a few. And yet they either deny that these facts are true or insist that such research should not have been done in the first place or should have been done on prisoners or on the mentally retarded. I think they hate people.

              Animal researcher, as quoted in Keith-Spiegel and Koocher, 1985 should be used for scientific investigations are assessed through a consideration of the philosophical arguments used to support the various positions on the issue, the potential benefits that have been accrued from animal research, and the ethical guidelines and regulations that have been established for protecting the welfare of animal subjects. In addition, the alternatives to animal research that have been proposed are described and evaluated. As we have strived to do throughout this text, an attempt is made to find a reasonable balance for assuring that sound research can be carried out in the most ethical way possible.

Background: Psychology and the Animal Research Controversy

Among the behavioral scientists who conduct research with animal subjects, it is psychologists who are most likely to study animals outside of the species' natural habitats. In fact, certain lower species (such as mice and pigeons) are especially bred by some researchers for the sole purpose of having them serve as laboratory subjects in psychological studies. While it is difficult to obtain accurate assessments of the percentage of researchers who conduct animal research, it appears that at least within psychology, a minority of researchers are actively engaged in such work. According to one estimate, animals are used in about 8 percent of psychological research (cf. Shapiro, 1991). Neal Miller (1985), a strong defender of behavioral research on animals, revealed that only 7 percent of the pages in American Psychological Association research journals reported research primarily on animals, in comparison with 93 percent reporting primarily on human subject research. However, one recent estimate based on surveys of psychological departments, analyses of *Psychological Abstracts*, and extrapolations from other countries is that roughly one to two million animals may be used in psychological research each year (Bowd and Shapiro, 1993). Despite this estimate, many more millions of animals are studied in research involving safety tests of a variety of products, including industrial chemicals and cosmetics, and other non-psychological research (Eckholm, 1985). For example, as many as 50,000 laboratory animals are killed or maimed in the European Community nations alone each year in safety tests carried out by cosmetic firms on shampoos, sun creams, and other beauty products. Although the European Community has agreed to ban such cosmetics testing on animals beginning in

1998, the deadline may be postponed if it appears that the beauty industry will suffer economically ("Animal testing," 1992).

Psychologists as vulnerable targets

Animal research in psychology has been criticized for its invasive procedures and apparently trivial or repetitive results (Bowd and Shapiro, 1993; Pratt, 1976; Ryder, 1975; Singer, 1975). Psychologists who conduct laboratory research on animals are frequently confronted with mistrust and hostility from the public, the media, students, and even fellow researchers (Devenport and Devenport, 1990). Rollin (1981), for example, criticized experimental psychology as "the field most consistently guilty of mindless activity that results in great suffering" (p. 124). At times, the hostility toward psychologists has taken a violent turn as research laboratories have been vandalized and researchers' lives threatened by extreme animal rights activists.

In contrast to the allegations of some activists, psychologists do not routinely subject their research animals to extreme pain, deprivation, or other inhumane treatment. Yet psychological research has been an early and easy target of animal rights activists. This may be due in large part to the fact that for most people, behavioral research has few easily explained links to specific human health benefits (Baldwin, 1993). Compared with some of the dramatic advances in biomedical animal research, the scientific findings and applications of behavioral research may not be perceived by the lay public as valuable. For example, psychological research that is focused on the development of new therapies for stress disorders likely will not appear to be as effective or as important as medical studies devoted to the discovery of new treatments for a life-threatening disease, such as cancer (Larson, 1982). Behavioral research also has lacked the powerful support of the medical establishment (Archer, 1986).

Other reasons have been suggested for the distinct vulnerability of psychologists to critics of animal research. Because psychological research involves the whole living organism, rather than single cells or specific anatomical structures, researchers often are required to maintain animal subjects for long periods of time and to observe them under a variety of conditions. Some of these conditions can in fact be stressful to the animals involved, and so the duration and circumstances of the studies may contribute to the appearance that they are abusive in nature (Larson, 1982).

Various constraints to the research process itself add to the vulnerability of psychologists who conduct animal studies. There often are a number of hypotheses that must first be tested in behavioral research before scientific advances become evident. Studies are conducted to eliminate the less fruitful hypotheses as well as to discover the more promising ones. As a result, progress in psychology and other behavioral science disciplines tends to be slow and perhaps not readily apparent. Behavioral researchers typically are unable to predict with certainty what the outcomes of their studies will be or what benefits can be expected. For these reasons, researchers choose to maintain a low profile, a tendency that can contribute to the public's misunderstandings and suspicions about the nature of the research that is being carried out (Larson, 1982). Even so, there is evidence that a majority of people generally recognize the need to use animals in experimental research (see Box 8.2).

Box 8.2

Should animal research be allowed?

According to recent polls conducted in the United States, most Americans accept the need for using animals in experimental and medical research. In the mid-1980s, a Media General-Associated Press survey found that eight out of ten people questioned believed it was necessary to use animals in medical research on serious diseases and seven of ten people agreed that animal research was needed for basic scientific research ("Poll finds," 1985). Nearly half of the 1412 respondents believed that laboratory animals were treated humanely, although another 30 percent did not. Nearly nine in ten respondents had no objection to using rats in research. Similar findings were obtained in a national poll conducted by the National Science Foundation (National Science Board, 1991). In that survey, half of the respondents answered affirmatively to the following question: "Should scientists be allowed to do research that causes pain and injury to animals like dogs and chimpanzees if it produces new information about human health problems?" This finding is particularly interesting in light of the explicit mention of "pain and injury" to two species of popular animals (cf. Kallgren and Kenrick, 1990). Taken together, these survey results suggest that most people do not share the goals of animal rights activists who believe that all research involving animals should be eliminated (Baldwin, 1993).

Emergence of the animal rights movement

Although the modern animal rights movement began to have a clear impact on psychology shortly after publication of Australian philosopher Peter Singer's influential book *Animal Liberation* in 1975, concern about the use of animals in scientific experimentation can be traced back to the Victorian anti-vivisection movement in Britain (Sperling, 1988). This nineteenth-century movement appears to have emerged as a reaction to the rapid scientific and technological changes of the period, which had begun to challenge the prevailing notions of humans' relationship to nature (Baldwin, 1993). The most significant event in the British movement came in 1876 with the passage of the Cruelty to Animals Act, which required licenses for conducting animal research and introduced a "pain rule" to protect domestic animals from cruel and improper treatment. Still in effect today, the law states that there is a limit to the amount of pain that an animal should be allowed to endure for the sake of science (Larson, 1982). There were several challenges to animal research in the ensuing years, but scientists were able to successfully protect their interests and continue using animal subjects.

In the United States, early colonists had introduced a law in 1641 to protect animals from cruelty; however, it was not until 1966 that the first federal legislation intended to safeguard the well-being of research animals was passed in the form of the Animal Welfare Act (see "Regulation and review" below). This legislation, along with subsequent revisions, focused primarily on requirements for housing and breeding research animals rather than on experimental procedures which could cause pain and suffering in the animals.

Public concern over the ethical treatment of animals has increased dramatically since the mid-1970s, apparently sparked by the publication of Singer's (1975) *Animal Liberation* (Baldwin, 1993; Bowd and Shapiro, 1993; Herzog, 1993). A second influential book, Ryder's *Victims of Science*, also appeared in 1975. Both books similarly targeted behavioral researchers for conducting painful and unnecessary experiments on animals. In his book, Singer presented a utilitarian position against animal research and laid the philosophical groundwork for later arguments that animals have rights, including the right not to be used as research subjects. One indication of the impact of these publications on the animal rights movement can be seen in the rapid growth of the leading animal rights organization in the United States, People for the Ethical Treatment of Animals

(PETA). The membership of PETA increased from 18 members in 1981 to more than 250,000 members in 1990 (McCabe, 1990).

PETA was founded by Alexander Pacheco, an animal rights advocate who in 1981 surreptitiously infiltrated Edward Taub's laboratory in order to obtain interviews and photographs relevant to Taub's studies involving deafferented monkeys (see chapter 1). The ensuing controversy over Taub's research also did much to bring animal research issues to the forefront of public awareness and debate (Boffey, 1981; Keith-Spiegel and Koocher, 1985; Larson, 1982). Earlier, considerable media attention had focused on a long-term investigation carried out by an experimental psychologist at the American Museum of Natural History involving the sexual behavior of domestic cats (cf. Shaughnessy and Zechmeister, 1990; Wade, 1976). In order to determine behavioral changes in the cats following the loss of various sensations or brain function, the researchers removed parts of the brains or the testicles of some of their animal subjects. Public awareness of this procedure in 1976 brought a fierce reaction from animal rights groups who believed the experimental manipulations to be cruel and unethical. As a result, activists demonstrated against the museum, pressure was placed on museum contributors to withhold their support, and death threats were received by the researchers.

In addition to PETA, other organized animal protection groups have emerged in recent years. Among these is a more radical group, the Animal Liberation Front (ALF), whose members have harassed and disrupted the work of animal researchers whom they believe to be engaged in cruel and unethical experiments. During a period lasting from early 1984 to May 1985, ALF members carried out raids on more than 20 research centers, freeing experimental animals and destroying laboratory equipment (Eckholm, 1985). It should be noted, however, that only a small minority of animal rights proponents belong to groups like ALF, which advocate and sometimes carry out illegal activities on behalf of research animals (Herzog, 1993). In recent years, the tactics of animal rights activists have become more sophisticated, with activists resorting to the courts and legislative lobbying to stop animal research, and challenging the scientific merit of university animal studies (Azar, 1994c). The existence of animal rights groups is not restricted to the United States. In France, for example, the French National Society for the Defense of Animals (Société Nationale pour la Défense des Animaux) has actively fought against the suffering inflicted on animals since it was founded in 1972. A recent study of participants in animal rights groups suggests that there are strong parallels between involvement in the animal rights movement and involvement in religious movements (see Box 8.3).

The response from psychologists

It is somewhat ironic that at the same time that the animal rights campaign against behavioral scientists began to accelerate, psychologists themselves were beginning to express concern over the welfare of their experimental animals and the use of animals in research and toxicity tests had begun to decline (Boffey, 1981; Larson, 1982). During the 1970s, several discussions concerning the use of animals in research were sponsored by the APA and local psychological associations, current policy statements were reviewed, and an ethics principle devoted exclusively to animal experimentation was developed and included in the 1981 *Ethical Principles*. This principle has since undergone minor revisions and now appears as standard 6.20 in the 1992 APA code (see Box 8.4). Psychologists in other countries, such as Canada, Great Britain, and Australia have engaged in similar activities. A survey conducted by the Institute of Laboratory Animal Resources of the National Academy of Sciences revealed a 40 percent decrease in the number of animals used in research during the period, from 33 million birds and mammals in 1968 to 20 million in 1978. While this decline can be attributed in part to the role of animal welfare groups in sensitizing scientists to the importance of the humane treatment of research animals, economic considerations and the emergence of alternative research approaches not requiring the use of animals also help to explain the trend (Gallup and Suarez, 1980; Holden, 1982).

Despite the fact that psychologists have responded to concerns about their animal experiments, some critics have argued that discussions within the professional literature have been one-sided and that psychologists in general have reacted in markedly defensive ways (Bowd and Shapiro, 1993; Phillips and Sechzer, 1989). According to Baldwin (1993), however, when activists initially targeted individual psychologists or areas of research that were particularly vulnerable to attack, little or no institutional support was forthcoming; thus, researchers had little choice but to assume a defensive posture or else ignore the activists in hopes that they would make their point and then go away. In short, the research community failed initially to communicate its position effectively and, as a result, public attitudes were shaped by the dramatic images, catchy slogans, and one-sided

Box 8.3

Animal rights activists: who are they?

Activists within the animal rights movement seem to come from a diverse range of backgrounds, including scientists who had themselves conducted research on animals at one time and many other individuals who never had expected to become involved in the cause. Only recently have attempts been made to characterize the members of animal welfare groups (Galvin and Herzog, 1992; Jamison and Lunch, 1992; Jasper and Nelkin, 1992; Plous, 1991), and there is scant information available about the social and psychological consequences of involvement in the movement.

The results of a recent participant-observation study conducted within the animal rights movement provide an informative look at the individuals who become active in the movement and the consequences of their activism. Psychologist Harold Herzog (1993) completed extensive interviews with 23 activists in the southern United States, none of whom claimed to belong to radical organizations such as ALF, and accumulated more than 400 pages of transcribed interview data. Sixteen of those interviewed were females, a disparity that reflects the greater representation of women in the movement (Jamison and Lunch, 1992; Jasper and Nelkin, 1992; Plous, 1991). Researchers have found that females tend to be less willing than males to see non-human species used in painful research (Kallgren and Kenrick, 1990). For the three-year period during which Herzog conducted the interviews, he also attended meetings and protest demonstrations in an attempt to learn more about the activists and how their lives were affected by their beliefs.

All of the participants in Herzog's study were committed individuals who had made major lifestyle changes as a result of their beliefs that there is a fundamental equality between humans and other species and that people do not have the right to use animals for their own purposes. Despite this shared world view, however, Herzog found that there was a surprising diversity of attitude and behavior among the activists. Some had come to the movement as a result of the careful evaluation of philosophical arguments, while others were motivated by more emotional reasons; some approved of civil disobedience, while others did not; and a similar diversity characterized the extent to which the activists felt morally superior to non-activists.

One of the most striking findings that resulted from Herzog's investigation was the extent to which the animal rights movement had become the central focus of the lives of nearly all those he interviewed. Most of the activists strived to maintain consistency between their ideals and their actions by refraining from eating meat, boycotting consumer products that had been tested on animals or made from leather, and attempting to spread their message to others. Some participants claimed that the changes in lifestyle brought on by their activism presented serious problems in their family relationships. Interestingly, Herzog noted that some of these findings, such as the change in fundamental beliefs evident in most activists, the dramatic alterations in their lifestyle, and the strong desire to spread their message suggest parallels between involvement in the animal rights movement and religious conversion. The portrait that emerges from this research nonetheless helps to break down the stereotype of animal rights activists as overemotional and irrational individuals. According to Herzog (p. 118):

> I found the people I interviewed to be intelligent, articulate and sincere. Though some were obsessed with their cause, none were "crazy" or irrational. The vast majority of the activists I have met were eager to discuss their views about the treatment of animals, and they were pleased and surprised that a member of the scientific community was interested in hearing what they thought.

arguments provided by the animal rights activists. Some animal rights supporters nonetheless have argued that the defensive posture taken by psychologists was in part intended as a strategic attempt to trivialize the issue of animal protection.

Psychologists also were faced with the difficult task of having to explain to an uninformed public the nature of the scientific method and the fact that one could not expect every animal experiment to produce beneficial results (Baldwin, 1993; Orem, 1990). The inability to respond effectively to growing criticism also failed to allay public fears that psychologists were incapable of self-regulating the care and use of research animals. In recent years, psychologists and other scientists have recognized the need to respond more forcefully and publicly to charges leveled by animal rights activists and have been able to persuasively refute misleading allegations (such as the false claim that animal subjects are routinely subjected to painful electric shocks) (Coile and Miller, 1984; Johnson, 1990).

Compared to the extensive attention devoted to issues involving human subjects in recent decades, psychologists have shown relatively little concern about the ethical issues pertaining to the treatment of animals (Carroll et al., 1985). However, if the rapidly growing psychological literature on ethical issues in animal research is any indication, this claim is less true today than it once was. It is interesting to note, however, that whereas researchers have expressed far more concern about the protection of human subjects in psycho-logical research, it is the public that has been most vocal about animal experimentation (Keith-Spiegel and Koocher, 1985).

Arguments For and Against Animal Research

In order to more fully examine the arguments for and against conducting research with animals it first is important to clarify the

---

Box 8.4

APA's Principles for the Care and Use of Animals

a. Psychologists who conduct research involving animals treat them humanely.

b. Psychologists acquire, care for, use, and dispose of animals in compliance with current federal, state, and local laws and regula-tions, and with professional standards.

c. Psychologists trained in research methods and experienced in the care of laboratory animals supervise all procedures involving animals and are responsible for ensuring appropriate consideration of their comfort, health, and humane treatment.

d. Psychologists ensure that all individuals using animals under their supervision have received instruction in research methods and in the care, maintenance, and handling of the species being used, to the extent appropriate to their role.

e. Responsibilities and activities of individuals assisting in a research project are consistent with their respective competencies.

f. Psychologists make reasonable efforts to minimize the discomfort, infection, illness, and pain of animal subjects.

g. A procedure subjecting animals to pain, stress, or privation is used only when an alternative procedure is unavailable and the goal is justified by its prospective scientific, educational, or applied value.

h. Surgical procedures are performed under appropriate anesthesia; techniques to avoid infection and minimize pain are followed during and after surgery.

i. When it is appropriate that the animal's life be terminated, it is done rapidly, with an effort to minimize pain, and in accordance with accepted procedures.

Source: *American Psychologist* (1992). 47, p. 1609. Copyright © 1992 by the American Psychological Association. Reprinted with permission.

difference between the goals of animal welfare groups and animal rights groups (Baldwin, 1993; Jasper and Nelkin, 1992; Johnson, 1990). Those individuals whose concerns pertain to animal welfare are primarily interested in improving laboratory conditions and reducing the number of animals needed to conduct research studies. These mainstream concerns for the humane treatment of animals are shared by most researchers (Baldwin, 1993). By contrast, the animal rights movement is not mainstream, holding that animals have rights equal to humans and thus should not be used at all as laboratory subjects. The primary goal of animal rights groups is to abolish all laboratory research on animals (Plous, 1991), a goal that researchers and, if responses to national polls are any indication, the general public do not agree with (see Box 8.2).

The philosophical roots of the animal welfare movement can be traced to the Judeo-Christian belief that humans are the stewards of the planet and, as such, are privileged to use its plants and animals as they deem fit (Johnson, 1990). However, this privilege bears a corresponding responsibility to serve as caretakers for that which they have been entrusted. An important element of Judeo-Christian thought is the belief that humans are qualitatively different from other animal species. In this view, a hierarchy of value exists among the planet's life forms, with human life and welfare having the highest value. By virtue of their unique capacity for moral choice, it is argued that humans can and should be held responsible for their actions towards animals (Robb, 1988). Stemming from these notions are the efforts of animal welfarists to prevent the cruel treatment of animals and to improve their health and well-being. This perspective has a long history in Judeo-Christian thought, in contrast to the relatively recent positions held by animal rights activists.

While "welfarists" often belong to local humane societies and tend to express sympathy for the plight of animals, they are unlikely to undergo the sorts of lifestyle changes that can be seen in the more committed rights advocates (Herzog, 1993). The latter have been labeled "fundamentalists" by Jasper and Nelkin (1992) to refer to individuals who believe that humans do not have the right to use animals for their own purposes, regardless of the benefits. A similar categorization has been offered by Herzog (1990), who suggests that animal protectionists groups can be distinguished as "reformers," who admit the necessity to utilize animals for research purposes and the more radical "animal liberators," who believe that animal research is immoral and should be abolished.

The case against animal research

Underlying the various arguments promulgated by animal rights advocates is the recognition that the ethical issues and problems involved in conducting research with animals are different from those involved in human subject research. One essential difference has to do with the fact that animals cannot be informed of the risks posed by a research study or give their consent to participate, and thus they are at the mercy of investigators (Carroll et al., 1985). Because of the near absolute power that researchers hold over their animal subjects, consent requirements and privacy issues essentially are rendered irrelevant in the context of animal research (Keith-Spiegel and Koocher, 1985). In addition to the design and procedure of the investigation, researchers determine and control the conditions under which animals live, including what and how much they are provided to eat. Also, unlike research with humans, the risks of research participation may be far greater for animals and, in some cases, may necessitate the termination of their lives.

Numerous arguments have been put forth by those individuals who believe that animal research should be abolished, although many activists often are unable to articulate the philosophical perspectives that form the basis of their positions (Sperling, 1988). Two of the most influential philosophical positions used to present the case against using animals in research are the utilitarian and the rights arguments. The major points related to these two positions are summarized below.

The utilitarian argument

Described earlier as the basic perspective underlying the development of research principles for human subject research, the utilitarian position has also been applied to animal research. The basic focus of the utilitarian position is on the consequences of one's actions. According to this perspective, ethical conduct is that which results in the greatest pleasure and happiness, relative to pain and suffering, for the most people. In utilitarian considerations, no one person's happiness is to be held as more important than another's.

In his book *Animal Liberation*, Singer (1975) extended this argument by maintaining that considerations of the greatest good should include animals. At the heart of this view is the "principle of equality" (or "equal consideration of interests") which holds that all living creatures with the capacity to sense have an equal interest in avoiding pain and suffering. Consistent with this principle is the conclusion that there is no basis for elevating the interests of one species, such as humans, above any other sentient species. In Singer's view, the only relevant moral criterion for distinguishing between species is the capacity to suffer. Because all sentient animals have that capacity, they are subject to equal moral consideration. Thus, humans and all other sentient animals have an equal interest not to be used in experiments that could cause them harm and suffering. These arguments are apparent in the following passage from Bowd and Shapiro (1993, pp. 136–7):

> The most morally relevant factor in a decision to cause suffering to others is their ability to experience it ... We feel methods involving inescapable pain, deprivation, or fear are unacceptable because each sentient being, regardless of its other capabilities, has an interest in being spared suffering. Modern-day society rejects the notion of performing painful experiments on humans who are incapable of granting consent, regardless of the benefits which might accrue to others. In the absence of morally relevant distinctions between ourselves and other animals, painful research on sentient non-humans should be rejected for the same reasons.

The critical aspect of the utilitarian argument against the use of animals in research is the claim that it is inappropriate to elevate humans above all other species on the basis of any criterion other than suffering. To do so would be to commit a form of *speciesism*, which is the animal rights supporters' equivalent of racism. In Singer's (1985) terms, speciesism is a prejudicial attitude biased toward the interests of members of one's own species and against the members of other species. Research on animals would be permitted in some circumstances, according to the utilitarian view, but only when the potential value of the study is high and it could also be carried out using human subjects (cf. Herzog, 1990). One problem with this utilitarian position has to do with the difficulty of having to determine where one draws the line between species that suffer and those that do not. On this matter, for example, Singer draws the line at the phylogenetic level of oysters. A second problem with the position is that it does not consider qualitative differences in pain and pleasure across species (cf. Baldwin, 1993).

As is discussed below, another variation of the utilitarian perspective often is used as an argument to support the use of animals in research. Researchers have pointed to the numerous examples of animal research applications that have improved human welfare and they have maintained that the benefits to humanity accrued from such research clearly outweigh the costs imposed on the research animals (Miller, 1985). In this view, animals are not given the same degree of consideration as people because of significant differences that exist between humans and other species (e.g. higher intelligence and the capacity for language). Some critics believe that such a cost-benefit approach for justifying animal research is unsatisfactory because any analysis of costs to animals presumes that they are willing participants in the research enterprise (Bowd and Shapiro, 1993). Further, it is argued that whereas the benefits from laboratory research with animals often are uncertain, the costs to animals are obvious and real.

The rights argument

The fundamental assumption of the rights argument against animal research, as articulated by Regan (1983), is that animals possess certain rights based on their "inherent value." That is, it is believed that humans and other sentient animal species have inherent value in equal measure which entitles them to certain fundamental rights. Included among these rights is the right to moral consideration, the right to be treated with respect, and the right not to be harmed. Consistent with this argument, it is held that animals have a right not to be used for research by humans, even if the research promises certain direct human benefits. Proponents of this position suggest that animal research should be abolished because scientists tend to treat research animals as renewable resources rather than as creatures with inherent value and rights. Failure to recognize the rights of other species, according to Regan, constitutes a violation of a principle of respect. In contrast to the utilitarian argument for using animals in research, Regan has further argued that the sacrifice of even a small number of innocent creatures should not be permitted even if such an action should lead to benefits for many individuals. In his view, such cost-benefit considerations do not provide an adequate moral justification for animal research.

Some proponents of the rights philosophy rely on certain criteria – such as language, higher intelligence, and self-consciousness – for determining who is entitled to fundamental rights. Because non-human species generally do not meet such criteria, it is believed that the rights should be restricted to humans. This philosophical position, however, runs into difficulties when confronted with considerations of the moral status of humans who do not meet the criteria (such as infants and the mentally infirm) and animals that do appear to meet some of the criteria (such as members of primate and cetacean species that have demonstrated some language skills, etc.) (Herzog, 1990). By broadening the criteria to include members of sentient animal species as being morally entitled to fundamental rights, animal rights advocates avoid these concerns. Nonetheless, questions relevant to how the rights in question are defined and who is entitled to them remain at issue for critics of this approach (Baldwin, 1993).

Despite the differences between the utilitarian and rights-based positions against animal research, it is possible to identify some commonalities. At the outset, it is apparent that aside from their somewhat divergent philosophical underpinnings, both positions lead to the same conclusion: that research with animals as it is now conducted should be eliminated. Both views acknowledge the ethical significance of certain similarities between humans and other species (e.g. the capacity for suffering), but downplay the importance of apparent differences (e.g. intelligence level). In so doing, the two positions represent anti-speciesism perspectives on animal research. For many people, however, the utilitarian and rights arguments that human life *per se* is not more important than that of other species are difficult to accept (Herzog, 1990).

Other issues
In addition to the two general philosophical arguments described above, other points have been made to bolster the position that animal research is immoral and should not be allowed. Consistent with the philosophical argument that sentient species have the right to be protected from pain and suffering is the accusation that many research animals do in fact experience much pain and suffering during the course of laboratory studies. The controversial test cases involved in charges of animal mistreatment, such as the research on monkeys conducted by Taub, are often cited as evidence supporting accusations of animal abuse inflicted during the course of research.

Despite arguments to the contrary (see "The case for animal research" below), critics have maintained that many animal studies are highly invasive in nature, involving pain, stress, punishment, social and environmental deprivation, as well as experimentally induced emotional and intellectual deficits (Bowd and Shapiro, 1993). Typically cited among the specific research examples are studies of visual deprivation in kittens, infant maternal deprivation and drug addiction in macaque monkeys, brain studies of eating behavior in rats, and the like. While it is impossible for humans to know with any degree of certainty what members of non-human species experience or feel, animal research critics have argued that considerations of the mistreatment of laboratory animals should include their pre- and post-experimental care (i.e. "husbandry"), in addition to solely focusing on the invasive experimental procedures (Bowd and Shapiro). For instance, although much of the public outcry over Taub's research focused on the experimental procedure itself (i.e. the deafferentation of limbs), the actual charges against Taub centered on abusive husbandry practices, such as the inadequate care, feeding, and housing of his monkeys (Shapiro, 1989). Among the customary husbandry practices currently under scrutiny is the issue of housing animals in individual cages which, according to one survey, is the method used by 84 per cent of investigators to house their primate subjects (Bayne, 1991). Animal critics believe that housing animals of the same species apart deprives them of essential social interaction (Bowd and Shapiro, 1993).

Issues related to the invasive treatment of research animals are exacerbated by additional concerns, such as a lack of apparent value in purely theoretical studies, non-generalizability of animal findings to humans and other species, and the diversion of funds from treatment programs. For example, Bowd (1980) has claimed that much of the findings derived from animal research performed by psychologists is not published, implying that the studies have been deemed insignificant by the investigators' peers. Moreover, one rather consistent theme running through the critiques of animal research has to do with the apparent redundancy in the research that *is* published. According to Bowd (1980, p. 207), "a significant proportion of published research is repetitive or deals with problems to which the answers are self-evident." As an example, Drewett and Kani (1981) noted that aversive stimulation research on animals has been excessively replicated and that the human applications obtained from the early animal studies no doubt could have been developed from ethically sound human subject research. The latter point implies that the use of animals in research intended to benefit humans may not be the only, or most effective, means of gathering the necessary information (Shapiro, 1983). In the assessment of many research critics, the potential benefits of animal research are indeterminate, due to such factors as the questionable applicability of the findings to human welfare and the strong possibility that the research will not be published (because of the high rejection rate of mainstream psychology journals) (Bowd and Shapiro, 1993). Such notions are intended to strike at the heart of the argument used most often to justify animal research: namely, that such research leads to significant, practical benefits for promoting the well-being of humans.

Finally, the case against animal research also consists of some arguments that pertain to issues underlying the politics of the debate. While many supporters of animal research would agree that animal rights advocates have successfully mounted a "calculatedly deceptive and seductive campaign to win sympathy and financial support from those who have affection for animals" (Johnson, 1990, p. 213) through the use of shocking posters and the like, the research profession has been similarly criticized for presenting a sanitized and misleading picture of the animal research enterprise (Bowd and Shapiro, 1993). For example, it is argued that the basic media through which students obtain their exposure to animal research (i.e. popular psychology textbooks and audiovisual materials) tend to minimize descriptions of the suffering involved in invasive animal studies, showing instead pictures of cute animals with accompanying text that focuses on the human benefits derived from the research (Field, 1989, 1990). Surprisingly, however, a recent analysis of leading introductory psychology textbooks revealed that authors rarely make explicit the contributions stemming from research with animals and frequently present major findings from animal research as if they had been obtained from humans (Domjan and Purdy, 1995). Other critics have complained about the scientific jargon used to refer to invasive procedures in animal research. Bowd (1980), for example, has criticized research psychologists for using language that belies the true nature of the procedures, such as referring to starvation as "deprivation," to the animal participant as an "organism," and describing research animals as being "sacrificed" rather than killed. In Bowd's view, this use of jargon enables many researchers to maintain an insensitivity to their treatment of animals.

The case for animal research

As previously described, in recent years researchers and other supporters of animal research have become more forceful in their rebuttals of the arguments summarized in the preceding section. While each of the points raised in favor of eliminating animal research have been countered by researchers (see Table 8.1), the arguments most typically emphasized in defense of that activity are the following: (1) that many benefits to humans and animals have accrued from animal research; (2) that researchers are sensitive to the pain and suffering experienced by animals and use procedures ensuring that animals are humanely treated; and (3) that researchers have developed alternative procedures which have led to a reduction in the use of animals for research purposes. Additionally, researchers have pointed to the practical advantages of using animals rather than human participants in various research contexts.

The philosophical approach underlying most attempts to justify animal research stems from historically early views of humans as being distinct from other species. These views are rooted in Judeo-Christian belief, as well as Cartesian notions found in the philosophy of the seventeenth-century French thinker René Descartes, which differentiated humans from non-humans on the basis of human possession of a "soul" and the capacity for moral choice. Because of their capacity for moral choice, it is maintained that humans can enter into contractual situations with binding rights and responsibilities for both parties (Baldwin, 1993; Robb, 1988). In light of

Table 8.1 Some arguments for and against experiments using animals

Point	*Counterpoint*
* Animals have rights, and these should be protected by the [US] Constitution	* To claim that animals have the same rights as people is an elitist stance
* The interests of human beings should not overwhelm the interests of animals	* Animal research has achieved major benefits for both humans and animals
* Animal research is an ethical travesty that justifies extraordinary and even illegal measures to prevent it	* Animal research is declining, but it is still necessary and justified for some uses
* If researchers were not allowed to use animals in research, more alternative research methods would be explored	* Less invasive animal research procedures have been developed and are being used in animal research
* Animal protection laws should be replaced by a rights act that would prohibit the use of any animals to their detriment	* There are laws and research guidelines that protect animals used in research

Adapted from *Taking Sides: Clashing Views on Controversial Psychological Issues, Seventh Edition* edited by Brent Slife and Joseph Rubinstein. Copyright © 1992 by The Dushkin Publishing Group, a unit of Brown & Benchmark Publishers, a division of Times Mirror Higher Education Group, Inc., Guildford, CT 06437. Reprinted by permission. Taking Sides ® is a registered trademark of The Dushkin Publishing Group.

neurological studies demonstrating that there are in fact continuities between human and animal brain structures and functions, it is now acknowledged even within the scientific community that animals experience a broad range of emotions and feelings, and possess psychological features previously ascribed solely to humans (cf. Blumberg and Wasserman, 1995). Nevertheless, one vestige of the early philosophical views of animals which conceptualized them as merely mechanical systems is the utilitarian argument that the use of animals in research is justified by the potential benefits to human welfare. Indicative of this utilitarian perspective is the comment from Herzog (1990), that "If it were possible to transplant an organ from a healthy sheep into a dying infant, most of us would readily approve of the operation" (p. 91). In short, this view is consistent with the belief that scientists have a right to attempt to improve human lives through research.

Turning next to the specific arguments that have been raised in support of research on animals, some researchers have emphasized the practical advantages of using animal subjects. In contrast to human participants, it is easier in the sense of methodological control to maintain similar manipulated experiences or genetic make up in animal comparison groups. It also is possible to study long-term changes with greater facility in animals due to easier access, more rapid developmental changes, and shorter life spans in various species. It is generally agreed, however, that these practical considerations do not alone serve to justify the use of animal research participants (Shapiro, 1984).

Perhaps the most persuasive argument cited within the research community in defense of animal research is that the research has led to numerous benefits, not only to humans but to other animals as well. Most supporters of animal experimentation contend that these benefits have gone largely unrecognized by animal rights activists. Among the contributions to animals that have accrued from behavioral research are improvements in the environments of captive animals, including modifications in the conditions under which research animals are studied, thereby leading to a reduction of suffering and discomfort during the research experience (Novak and Petto, 1991). Other benefits include biomedical developments in veterinary procedures, vaccines for deadly diseases (e.g. rabies, Lyme disease, feline leukemia), and captive breeding programs that have resulted in saving some animals from the brink of extinction (Baldwin, 1993; King and Yarbrough, 1985). Miller (1983, 1985) has provided several examples of how traditionally lethal methods of controlling animals that damage crops or flocks (e.g. by shooting or poisoning the destructive animals) have begun to be replaced by the application of more humane techniques derived through experimental studies of animal behavior (see Box 8.5).

Research benefits for humans

Although animal rights activists often argue that animal experiments are not really applicable to humans and thus are without any value, the numerous benefits to humans that have come about as a result of research on animals have been clearly documented by others (Baldwin, 1993; Cole, 1992; Comroe, 1983; King and Yarbrough, 1985; Miller, 1983, 1985; Randall, 1983). The fact that animals have

---

Box 8.5

Animal research benefits for animals

Psychologist Neal Miller has for more than a decade emphatically presented the case for animal research by documenting the many benefits to animals and humans that have been derived from the research. Among the examples of research benefiting animals described by Miller (1983, 1985) is the application of the so-called "taste-aversion effect," an example of aversive conditioning that causes animals to avoid a nausea-inducing flavor, even if it was tasted several hours before the onset of the nausea. One application of this learned effect involves the administration of a moderately bad-tasting chemical to birds that is reinforced by non-harmful nausea if the birds continue to eat. The administration of the chemical, methiocarb, has proven to be a humane approach to protecting fruit crops and lawns from Canadian geese, and researchers have begun to focus on other possible applications (e.g. to train coyotes to avoid the taste of sheep meat; to better understand the food aversions in children receiving chemotherapy for cancer).

Other behavioral research studies on such topics as imprinting, animal sexual behavior, genetic and experiential factors that control the timing and course of migration, and animal requirements for habitat and for territory have proven essential in efforts to save endangered species from extinction. In one example derived from imprinting research (see page 269), artificially incubated condor chicks are exposed to and fed by a puppet surrogate of an adult female condor rather than by an undisguised human caretaker. This procedure has been successful in facilitating the young chicks' safe association with wild condors when they are released into their natural habitat.

played a pivotal role in improving the human condition represents a primary reason that most people claim to support the continued use of animals for research purposes (e.g. National Science Board, 1991). It is not our intent here to enumerate all of the benefits that have been cited beyond simply to acknowledge some examples of important contributions that have emerged through animal experimentation. Within biomedical areas alone, the list of contributions attributed to animal research is impressive, including the development of vaccines for a number of diseases (such as typhoid, diphtheria, and polio), antibiotics, antidepressants, local anesthetics, insulin, cancer therapies, and most surgical procedures (Cole, 1992).

The human benefits of behavioral research with animals also are many, although, as described earlier in this chapter, they may not be as apparent to the lay public as the contributions attributed to research in medical fields. Many of the contributions of animal research conducted by behavioral scientists are indeed significant advances, including the rehabilitation of persons suffering from spinal cord injury, stroke, head injuries, and other neuromuscular disorders; the discovery and greater understanding of Alzheimer's disease and eating disorders; improvements in communication with the severely retarded; the discovery of drugs for treating anxiety, psychoses, and Parkinson's disease; the control of chronic anxiety through non-drug therapies; advances in the treatment of premature infants, eye disorders in children, and the problem of enuresis; and improved knowledge about and treatment of alcoholism and other types of substance abuse, obesity, hypertension, memory deficits in the aged, and insomnia (Baldwin, 1993; Miller, 1985). Non-human primate research also has enabled researchers to investigate a variety of complex behaviors, such as aggression, social organization and communication, learning and memory, and growth and development (King and Yarbrough, 1985).

Studies conducted by Roger Sperry during the 1960s provide an illustrative example of the scientific advances that can be derived from behavioral research with animal subjects. Sperry won a Nobel Prize for his work on the "split brain," a term which pertains to the functioning of the brain after the corpus callosum (a massive bundle of fibers connecting the left and right hemispheres of the brain) is severed. This research was of particular interest because of its potential applications in the control of epileptic seizures. For patients suffering from severe epilepsy, seizures typically begin in one hemisphere and then spread to affect the whole brain. Thus, it was believed that by surgically isolating the two hemispheres, it would be possible to greatly diminish the debilitating effects of the seizures. Because the consequences of such a radical surgical procedure in humans were unknown at the time, Sperry's early investigations were carried out on cat and monkey subjects. The animal studies revealed that the two hemispheres of the brain can operate independently for a wide range of mental activities, including learning and memory functions. This finding provided researchers with a basis for conducting subsequent studies on human participants (e.g. Gazzaniga and LeDoux, 1978; Sperry, 1968). Today, many epileptic patients who otherwise would have to be confined to hospitals are able to lead relatively normal lives because of the split-brain procedure. Had it not been for the initial animal studies, this consequence may not have been possible.

Given the nature of such benefits, animal research supporters understandably react with great consternation when barriers to animal research arise or when research laboratories are trashed by radical animal rights activists. Thus, the destruction of one psychologist's sudden infant death syndrome (SIDS) laboratory was seen not as a victory for animals, but rather as "a postponement of the day when we will know how to stop babies from dying of SIDS." The loss of an addiction research grant as a result of pressure from animal rights groups was viewed not as a victory for the cats that were to be used in the research, but as a delay in our learning "how to save human beings from lives of drug dependence" (Johnson, 1990, p. 214). Many researchers no doubt would agree with Johnson's overall assessment that "The victories of the animal rights movement can be properly counted only in terms of the human pain and death they have guaranteed."

Reducing the pain and suffering of research animals

In countering the charge by animal research critics that scientists fail to recognize the pain experienced by their laboratory animals, Gallistel (1981) has argued that if that were the case, researchers would not take the care that they do to limit it during experimental procedures. Among the recent steps taken to reduce the possibility that research animals will undergo unnecessary pain or suffering is the establishment of policies at some large research facilities prohibiting the use of muscle-paralyzing agents during experimental surgery.

So-called "paralytics" (such as curare) have been used to immobilize experimental animals without rendering them unconscious during surgical procedures. Scientists also have become progressively aware that, unless the focus of the research is to study the effects of trauma *per se*, causing animal suffering does not represent good science because the research ultimately will yield invalid, non-generalizable results (Zola et al., 1984).

Further evidence that researchers have grown increasingly sensitive to the concerns of animal rights activists is apparent in statistics indicating that the number of animals used in research, at least in the United States, has sharply declined in recent years. While this reduction also can be attributed somewhat to the increased costs of conducting animal research, it is clear that researchers have been successful in developing methodological alternatives to animal research, such as computer simulations, mechanical models, and the use of tissue and cell cultures (see "Alternatives to animal research" below). Research animals now have greater protections as a result of the emergence in recent years of ethical standards and guidelines, such as those developed by professional associations. The establishment of ethical review committees at a growing number of research facilities serves as an additional mechanism for overseeing the procedures used for housing animals and for controlling pain, training personnel, and determining the adequacy of the experimental design in terms of costs, benefits, and alternative approaches which do not require the use of animals.

Despite these advances, there are some who contend that animal rights activists are not concerned so much with the humane treatment of animals in research as they are with eliminating their use in research entirely (e.g. Johnson, 1990). This is apparent, say some critics, in the misleading and manipulative images in campaigns against animal research, which depict animals apparently undergoing extremely painful experimental manipulations. For example, Baldwin (1993) has argued that graphic descriptions of animals receiving severe electric shocks lead the public to believe that the administration of intense shocks capable of inducing convulsions is routine in animal research when in fact that is not at all the case. Rather, when electric shock is used in behavioral research, its level tends to be relatively mild and the procedure typically must first be approved by a review committee. Baldwin also points out that when stronger shock is used, the purpose is to study medical problems such as epilepsy and other convulsive disorders and their treatment.

Regulation and Review of Animal Research

In a relatively short period of time, much has been done to regulate the care and use of animals in behavioral and biomedical research. In the United States, for example, an elaborate system exists for safe-guarding the well-being of research animals, consisting of federal regulations, state laws, and professional guidelines and review mechanisms. These safeguards are briefly summarized below.

Governmental regulations

In recent years, the American Congress has played an increasingly involved role in the animal research debate, as is evidenced by the considerable legislation that has been passed. Federal regulations under the Animal Welfare Act and Public Health Service (PHS) policy are now in place, as are state laws governing the availability of pound animals for research. To a great extent, the campaign against animal research launched by animal rights groups during the 1980s was responsible for the emergence of these federal and state requirements.

At the center of federal policy regulating the use of animals in research is the Department of Agriculture's (USDA, 1989, 1990, 1991) Animal Welfare Act. Enforced by the USDA's Animal and Plant Health Inspection Service (APHIS), the law consists of guide-lines for the care and use of research animals and specifies that unannounced inspections of animal research facilities be carried out periodically by APHIS inspectors. The inspections, which are intended to ensure compliance with the regulations, include assess-ments of feeding schedules, adequacy of housing, exercise require-ments for certain animal species, promotion of the psychological well-being of non-human primates, and so on.

In addition to the APHIS inspections, two Public Health Service agencies (the National Institutes of Health and the Centers for Disease Control) require institutions receiving federal funds to estab-lish a standing Animal Care and Use Committee (ACUC) to review all proposed and ongoing animal research conducted at the institu-tion. These review committees are to be comprised of at least five members, including a veterinarian, an experienced animal researcher, a non-scientist (e.g. a lawyer or ethicist), and a person who is not affiliated in any way with the institution (Office for Protection from Research Risks, 1986). Each ACUC review of a research grant proposal is to include a consideration of the justification for the species to be used and number of animals required to carry out the research; the adequacy of the procedures for controlling pain, housing animals, carrying out euthanasia, and training personnel; any available alternatives to the research; and evidence that the study will not represent an unnecessary replication of previous research. ACUCs also are expected to carry out local inspections of research facilities in order to assess ongoing compliance with state research requirements.

There are some obvious similarities between ACUCs and the National Institutes of Health regulations requiring the establishment of institutional review boards (IRBs) to review federally funded research with human subjects. Thus, it was not surprising that when legislation was proposed to create ACUC review at PHS grantee institutions, animal researchers reacted with some of the same concerns as those expressed by human subject researchers when IRBs were first created (during the 1970s). In fact, while researchers have feared that federal regulations would inhibit research, this appears not to be the case in either the human or animal research context. Rather, institutionalized review seems to have had the effect of improving research design by reducing risk and discomfort for subjects (Mishkin, 1985).

Recently, several animal rights organizations sued the USDA, arguing that current regulations were not strict enough and that more detailed regulations should be issued ("Ruling on animals," 1994). The focus of the lawsuit was on so-called "performance-based" regulations that had been adopted by the USDA in 1991, which placed responsibility for determining the appropriate environment and exercise program for laboratory animals onto each research facility. A judge ruled in favor of the activists in 1993 and ordered the USDA to rewrite the regulations. However, this ruling, which would have required more rigid standards for animal research, was overturned one year later by an unanimous appeals panel decision. It was determined that the animal rights groups did not have legal standing to challenge USDA rules because it could not be shown that they had been directly harmed as a result of USDA actions. The determination also applied to a challenge of USDA's decision not to include rodents and birds in research under the Animal Welfare Act. This recent court decision has been viewed as an important victory for the research community, which had generally supported the reinstated 1991 guidelines ("Ruling on animals," 1994).

Professional developments

Various steps also have been taken within the professions to provide guidelines for the use and care of research animals. Earlier, we alluded to the development of an ethical principle specific to animal research which was first added to the APA code of ethics in 1981. Similar principles have been adopted by the British and Australian psychological associations. Some examples of the principles included in the British Psychological Society's (1995b) *Guidelines for the Use of Animals in Research* appear in Box 8.6.

Despite the fact that a separate principle devoted to animal research was not added to the *Ethical Principles* until fairly recently, the APA has long been involved in promoting the well-being of animals in psychological research. For example, Principle 10 was heavily adapted from an earlier APA policy statement (APA, 1968), which was to be posted in all animal psychology laboratories where animal research was conducted. Guidelines for the use of animals in schools also were detailed in another early APA policy (Committee for the Use of Animals in School Science Behavior Projects, 1972).

The APA Committee on Animal Research and Ethics (CARE) publishes another useful document, *Guidelines for Ethical Conduct in the Care and Use of Animals* (APA, 1993a), which is used widely by researchers in psychology and other scientific fields. Established in 1925, CARE is responsible for reviewing the ethical issues involved in animal research and revising guidelines for the care and use of laboratory animals. The APA also promotes the well-being of animals in research through membership in an organization known as the American Association for the Accreditation of Laboratory Animal Care (AAALAC), the only such accrediting body recognized by the PHS (Baldwin, 1993). For an institution to receive AAALAC accreditation, its animal research facilities must exceed the requirements established by federal policy. Finally, Psychologists for the Ethical Treatment of Animals (PsyETA), an independent organization of American psychologists who share the concerns of animal welfare groups, was created during the 1980s. Among the goals of this organization are the improvement of the care standards and conditions for animals used in psychological research, the development of educational approaches that contribute to the well-being of research animals, and the maintaining of liaisons between scientific and animal welfare groups (Keith-Spiegel and Koocher, 1985).

While these professional developments have done much to raise ethical sensitivities within the disciplines, they have not eliminated

---

Box 8.6

The British Psychological Society's Guidelines for Animal Research

The BPS (1995b) *Guidelines for the Use of Animals in Research* are based on standards originally published in 1981 by the Association for the Study of Animal Behaviour and the Animal Behaviour Society. The guidelines supplement an earlier BPS statement published as a Report of the Working Party on Animal Experimentation (*Bulletin of The British Psychological Society*, 1979, 32, 44–52). The intent of the 1995 document is to provide a checklist of points for investigators to carefully consider prior to conducting their experiments with animals. Regardless of the nature of the research, an overriding principle is that animal researchers have a "general obligation to avoid, or at least to minimize, discomfort to living animals" (p. 12). Additionally, investigators are encouraged to seek independent advice from experts as to whether the anticipated scientific contribution of the proposed research justifies the use of living animals.

The BPS guidelines are presented in 15 separate sections. Although the document is too lengthy to reproduce here in its entirety, a sample of statements from several of the sections are presented below.

3 Species

Choosing an appropriate subject usually requires knowledge of the species' natural history as well as its special needs.

4 Number of animals

Laboratory studies should use the smallest number of animals necessary. The number cannot be precisely specified, but ... can often be greatly reduced through good experimental design and the use of statistical tests such as the analysis of variance and multiple regression that enable several factors to be examined at one time.

7 Caging and social environment

Caging conditions should take into account the social behaviour of the species. An acceptable density of animals of one species may constitute overcrowding for a different species. In social animals caging in isolation may have undesirable effects.

9 Aggression and predation including infanticide

The fact that pain and injury may come to animals in the wild is not a defence for allowing it to occur in the laboratory ... wherever possible, field studies of natural encounters should be used in preference to staged encounters.

10 *Motivation*

Whenever arranging schedules of deprivation the experimenter should consider the animal's normal eating and drinking habits and its metabolic requirements . . . differences between species must be borne in mind: a short period of deprivation for one species may be unacceptably long for another.

11 *Aversive stimulation and stressful procedures*

Procedures that cause pain or distress to animals are illegal in the UK unless the experimenter holds a Home Office licence and the relevant certificates. The investigator should be satisfied that there are no alternative ways of conducting the experiment without the use of aversive stimulation. If alternatives are not available the investigator has the responsibility of ensuring that any suffering is kept to a minimum and that it is justified by the expected scientific contribution of the experiment.

12 *Surgical and pharmacological procedures*

It is illegal to perform any surgical or pharmacological procedure on vertebrates in the UK without a Home Office licence. In pharmacological procedures, experimenters must be familiar with the literature on the behavioural effects and toxicity of the drugs being used. It is essential that low dosage pilot studies are done to determine behavioural effects and toxicities when new compounds are used or when a drug is being used for the first time in the particular laboratory.

13 *Anaesthesia, analgesia and euthanasia*

The experimenter must ensure that animals receive adequate post-operative care and that, if there is any possibility of post-operative suffering, this is minimized by suitable nursing and the use of local anaesthetics where appropriate.

14 *Independent advice*

If an experimenter is ever in any doubt about the condition of an animal, a second opinion should be obtained, preferably from a qualified veterinarian, but certainly from someone not directly involved in the experiments concerned.

concerns held by many critics that the scientific community is unable to self-regulate animal research. Reactions to APA's animal research principles are a case in point. According to the principles listed in Box 8.4, the investigator who uses animals in research has an ethical obligation to protect their welfare and to treat them humanely (principle 10a); those persons involved in the research project must be qualified to manage and care for the particular species that are used (principles 10c and 10d); and procedures that expose the animals to pain or discomfort must be justified by the potential scientific, educational, or applied merits of the study (principle 10g). However, these principles have met with some of the same criticisms that were voiced in reaction to the APA's principles for human subject research. For example, Carroll et al. (1985) cautioned that because researchers are always likely to see some value in their own research, they may be prone to rationalize painful procedures as necessary and make little effort to find alternative, painless procedures.

In a more general sense, others have pointed out that the APA's animal research principles deal almost exclusively with the care and treatment of research animals rather than providing restrictions on their use (Bowd and Shapiro, 1993; Shaughnessy and Zechmeister, 1990). Further, Bowd and Shapiro (1993) noted that only one animal welfare case was considered by the APA Ethics Committee from 1982 to 1990, despite charges by the animal rights movement that several research laboratories had committed specific animal welfare violations during that period. (The case that *was* heard by the Ethics Committee involved Taub's research on deafferentation in macaque monkeys.)

Critics of animal research also have claimed that the primary interest of psychology bodies (such as CARE) responsible for overseeing animal welfare is not so much in defending and protecting animals as it is in promoting animal research, suggesting that the professional response to ethical challenges has been to form "advocacy groups rather than impartial and balanced review panels" (Bowd and Shapiro, 1993, p. 138). Whether or not this is a fair assessment, it is clear that behavioral scientists now take a much more active role in safeguarding their research interests. Groups representing behavioral and biomedical research have begun to form alliances with civil liberties groups and federal agencies to advocate for legislation and other protections for animal researchers. These efforts recently paid off in the form of passage into law of The Animal Enterprise Protection Act of 1992, more commonly referred to as the "break-in" law ("Animal research," 1992). Among other things, this law makes it a federal offense to commit acts of vandalism and thefts at animal research facilities on an interstate or international basis that result in more than US$10,000 in damages. The law also provides for strict penalties for situations in which anyone is injured or killed during an attack on an animal enterprise. This legislation is intended to protect animal researchers from costly and destructive attacks on their laboratories and to reduce fears among researchers whose lives have been threatened by animal rights extremists in recent years.

Alternatives to Animal Research

In response to the ethical concerns raised by the use of animals in laboratory research, researchers have been encouraged to develop alternative approaches for carrying out their investigations. While many behavioral scientists continue to believe that a strong defense of current practices is the best policy for responding to their critics, others have questioned whether the use of animals in laboratory research is even the most appropriate approach for studying many behavioral research questions (Devenport and Devenport, 1990). In recent years, a diverse set of alternatives to laboratory animals have emerged and, as a result, the number of animals used for behavioral research purposes in general has continued to decline. Within the behavioral sciences, researchers have employed mathematical simulations, mechanical models, anthropomorphic "dummies," and simulated tissue and body fluids in research situations where at one time animals would have been used as a matter of course.

Several of the more widely utilized alternatives to animal research have particular applicability to biomedical research, such as tissue and cell cultures to test and produce some vaccines, nuclear magnetic resonance to visualize internal organs without necessitating surgery on an animal, and computer models to mimic biological processes. Although these methods have been employed by behavioral scientists as well, some researchers believe that for most investigations on the behavior of animals it is necessary to study fully functioning and whole animals (Drewett and Kani, 1981).

A growing number of researchers have begun to consider whether research on commonly used animal species could be conducted instead with the use of human subjects or, at the other extreme, lower organisms. Because a major proportion of animal studies are carried out in order to serve as more acceptable alternatives to research situations in which issues of practicality and ethics preclude the use of humans, lower organisms have been increasingly studied in recent years (Holden, 1982). For instance, questions relating to behavioral genetics have been investigated through the use of *Drosophila* fruit flies (e.g. Pruzan, 1976; Pruzan et al., 1977), which provide researchers with certain methodological advantages (e.g. accelerated lifespan and generational breeding; ease with which the genetic material can be observed) in addition to alleviating ethical concerns about pain and suffering. Researchers also have begun to identify other alternative species whose nature more closely mirrors human functioning. As an example, Rissman (1995) recently made the case for the study of the musk shrew (*Suncus murinus*) as an ideal model for investigating the neural and hormonal bases of female sexual behavior.

Other ideas for research alternatives when animals represent the only appropriate subject population have been suggested. Lea (1979) recommended that whenever possible researchers could replace animal stimulation (e.g. of nervous system processes) with mere recording; similarly, stimulation could provide a substitute for lesioning techniques. Such alternative approaches would serve to reduce the amount of discomfort experienced by the animal and limit possible damage to the animal's nervous system. Lea also provided several suggestions for reducing unnecessary degrees of deprivation and other punishments routinely used in learning studies: a food deprivation period of only six hours seems to be as effective for studying learning in rats as a 23-hour deprivation period; rewards can be used instead of punishments in avoidance studies; aversive (but less painful and traumatizing) stimuli, such as loud noises or bright lights, can reduce the need to administer electric shocks. While such recommendations might not be applicable in physiological studies, Lea observed that the more painful procedures common in physiological research may be justified by the greater likelihood of medical applications derived from the findings.

The reduction in numbers of animal subjects used and the study of animals in their natural habitats represent two of the most basic steps available to researchers for reducing some of the ethical concerns that surround their work. For example, the use of animal subjects in classroom demonstrations can be replaced with films and videotapes, or by having students carry out naturalistic, non-invasive behavioral animal studies (Azar, 1994a). As an alternative to requiring students to conduct repetitive experiments that do not stand to advance science, videotapes could be made for teaching purposes at a relatively low cost and reused for several years (Bowd, 1980). Such recommendations are especially appropriate for situations in which the research would put the animal at some discomfort or require that it be killed following the demonstration or exercise. Alternatives to animal research requirements should be provided as a matter of course in teaching contexts, in light of indications that a growing number of students are sensitive to animal rights issues and object to working with animals (Azar, 1994a; Hogan and Kimmel, 1992).

The investigation of animal behavior in its natural habitat has been recommended by those who have questioned the validity of scientific findings derived from laboratory studies (Bowd and Shapiro, 1993; Devenport and Devenport, 1990; Drewett and Kani, 1981). Such an approach is consistent with the long-standing ethological tradition of observing animal species in their natural environments in order to determine how environmental cues affect behavior. Ethological studies, for example, have provided numerous examples of how physical characteristics and behavior patterns that are adaptive to survival have evolved in species through the process of natural selection (Bernstein et al., 1994). A classic example of natural habitat observation is evident in the work of Dian Fossey (1981, 1983), a researcher whose life and tragic murder were depicted in the film *Gorillas in the Mist*. Fossey journeyed to Africa in 1967 in order to document a vanishing breed of mountain gorillas for *National Geographic* magazine. Through a procedure known as desensitization, she was able to adapt her animal subjects to the presence of an observer and put them at ease. Over a period of time she gradually moved closer to the gorillas, imitating their movements (such as munching on foliage and scratching herself), until it was possible to sit among them and freely conduct her observations. This enabled Fossey to obtain some fascinating insights into the behavior of her animal subjects.

Ethologists also have used structured observations to investigate the effects of systematic manipulations of various aspects of an animal's natural habitat (Shaughnessy and Zechmeister, 1990). The Nobel-Prize-winning ethologist Konrad Lorenz employed this method in his well-known research on imprinting, the attachment shown by some animal species to objects they encounter during critical or sensitive periods early in their lives (see Eibl-Eibesfeldt, 1975). Lorenz studied the imprinting response in young graylag goslings, who have a tendency to imprint on their mothers (or almost any moving object in their environment) immediately after hatching.

Box 8.7

Studying animals in our own backyards

Researchers continue to make progress in their attempts to develop new approaches to the behavioral study of animals in order to avoid some of the ethical conflicts that have emerged in the context of invasive laboratory research. One intriguing alternative to laboratory experimentation recently suggested by Devenport and Devenport (1990) involves the use of domestic pets as subjects, with experiments carried out at owners' homes. This approach requires the cooperation of local veterinarians, from whom researchers can obtain large, homogeneous samples of pets with the required combination of characteristics relevant to the intended research purpose.

In one application of this approach, Devenport (1989) carried out several experiments to study the effects of contextual changes of experience in a food-seeking task, using sixty adult sporting breed dogs tested in their owners' backyards. Various dimensions of the dogs' behavior were assessed across training, extinction, and contextual change conditions in studies similar to ones previously run in the laboratory with rats. This research apparently brought little disruption to the families (who received a minimal payment for each session and a bonus for completing all sessions) or to their pets. Although daily feedings had to be postponed until after testing, a desirable reward (Vienna sausages) motivated the animals without requiring the use of deprivation.

According to Devenport and Devenport, home testing of animals has several advantages. For example, the animals do not have to be physically confined to unfamiliar and threatening settings as is the case in the laboratory; as a result, they are calmer, more tractable, and are maintained in a stimulating environment in relation with humans. An advantage that the home environment has over the study of animals in the wild is that the former provides an experimental setting in which variables of interest often can be controlled as efficiently as in the laboratory. Outside interference can be kept to a minimum, thereby allowing the measurement of subtle differences in temperament and attitude variables under stable conditions. The approach is potentially adaptable to the study of a variety of behavioral questions, on topics such as development and aging, learning, and social behavior.

One important additional benefit of pet research is the apparent effect it has on improving attitudes about animal experimentation. Based on responses to post-experimental questionnaires, Devenport and Devenport (1990) found that having a pet participate in research appears to increase interest in science among the owners and provides a better understanding of the research process. Despite obvious drawbacks to this approach (such as limitations in the range of equipment that can be used and the type of research questions that can be studied) the home-testing of pets appears to have potential as a harmless method for studying animals and for improving the image of animal experimentation.

In order to investigate the role of parental signals thought to be naturally attractive to the baby geese, Lorenz created models of male and female parents and tested the young animals' response to the models' movement and honking sounds. In one humorous example of this work, Lorenz squatted and imitated goose noises in front of newborn goslings whose mother was not present; as a result, he himself became an irresistible figure to a gaggle of geese who diligently followed him wherever he went.

Despite some practical limitations inherent in the study of animals in natural settings, as the work of Fossey and Lorenz demonstrates, this approach can provide a more accurate picture of the true capacities and normal, instinctual lives of many animal species than that obtained from laboratory-based invasive research. Another interesting application of this methodology involving the study of domesticated animals outside of the laboratory setting has been suggested by Devenport and Devenport (1990; see Box 8.7).

Summary

Within the behavioral and biomedical sciences, there is general agreement that until additional alternatives to animal research can be found, research on live animal subjects will continue. Indeed, there are many researchers who believe that the complete elimination of animal research will never be possible (Feeney, 1987; Gallup and Suarez, 1985; Miller, 1985). As a result, researchers must continue to balance the need to carry out potentially valuable investigations against the need for adequate protections in the way animals are treated (Shaughnessy and Zechmeister, 1990). As we have documented in this chapter, much has been done to add safeguards to the treatment and care of laboratory animals, and there is evidence of a steady reduction of animals used for research purposes. Nonetheless, for extreme animal rights groups, these developments have done little to allay their call for a complete halt to animal research. This being the case, it may be that there is no common ground between animal researchers and the radical proponents of animal rights, regardless of how humanely animals are treated by researchers (Johnson, 1993).

As Herzog (1990, p. 93) has noted, "the issue of animal rights is philosophically and psychologically complex . . . mired in a milieu of rationality, emotion, convention, ethical intuition, and self-interest." Thus, it is in the best interest of those on both sides of the issue to refrain from perpetuating stereotypes, disseminating propaganda, and engaging in other destructive activities which are likely to do little to resolve the debate. Further research into the variables influencing moral judgments about animals could provide a better understanding of the different reactions to animal research that have been observed. For example, it has been found that judgments about animals are influenced by such factors as their phylogenetic distance from humans, their physiognomic similarity to humans, their "cuteness" and perceived intelligence, the affective valence (i.e. perceived goodness or badness) of a species, and the labels we assign to animals (Burghardt and Herzog, 1980; Herzog, 1988; Kallgren and Kenrick, 1990).

Most researchers reject the notion that animals have rights that preclude their use in research, but recognize that scientists have certain responsibilities to the research animals that are studied (Baldwin, 1993). Among these responsibilities are the need to ask whether each investigation necessitates the use of animals in the first place and, if so, to use as few animals as possible in the research. Animal researchers must be sensitive to the ethical issues surrounding their work as well as to the professional and governmental requirements that currently are in place. They must proceed in ways that lead to the humane treatment and care of research animals, and to the avoidance of invasive and painful procedures whenever possible. These are common goals of both scientists and animal welfare advocates alike.

9

Ethical Review and the Communication of Results

Scientists must write, therefore, so that their discoveries may be known to others.

R. Barrass, *Scientists Must Write*

The only ethical principle which has made science possible is that the truth shall be told all the time. If we do not penalise false statements made in error, we open up the way, don't you see, for false statements by intention. And of course a false statement of fact, made deliberately, is the most serious crime a scientist can commit.

C. P. Snow, *The Search*

When aspiring researchers contemplate what is involved in carrying out an investigation in the behavioral sciences what no doubt first comes to mind are such activities as recruiting subjects and collecting data through implementation of whatever methodological procedures seem most appropriate to the research questions at hand. As a result, the most salient ethical issues may appear to be those directly pertaining to the treatment of the participants during the research process. Is it ethical to deceive subjects? How will subjects be debriefed? Will the privacy of the research participants be protected? Will the experimental procedure inflict unnecessary discomfort on human (or animal) subjects? In preceding chapters, we have attempted to elucidate the scope of these questions and to offer possible solutions to the ethical dilemmas they present.

When we consider the researcher's responsibilities both before and after carrying out an investigation, other ethical issues become apparent which need to be addressed with the same level of serious-ness as those that arise in the treatment of research participants. In this final chapter we turn our attention to those additional issues which pertain to research activities that take place both before and after the data are collected. In particular, problems involved in the ethical review process and in the communication of research findings are considered. While these research-related activities may at first appear to have little in common, certain similarities can be noted. Most importantly, both activities contribute to the system of "checks and balances" that is built into the scientific research enterprise. That is, the processes of ethical peer review and the public communication of one's research findings provide, in addition to the sharing of knowledge, means by which research can be evaluated on both methodological and ethical dimensions and a way in which the honesty and integrity of the investigator can be checked.

In the first part of this chapter we shall consider the review process by focusing on the importance of ethical review, the nature of ethical boards, and threats to objective peer review in light of recent research on ethical decision making. We then will conclude the book by turning our attention to some of the issues relating to integrity in the analysis of data and in the reporting of research results. Although these latter issues do not directly deal with the protection of human rights, they are compelling nonetheless because of their potential impact on science as a body of knowledge (Reese and Fremouw, 1984). Among the specific ethical responsibilities to be discussed are those that arise once an investigation has been completed, such as the accurate reporting of results, assignment of appropriate credit for those who contributed to the research, and ethical conduct in attempts to publish and to protect against the misuse of research.

The Ethical Review Process

Prior to the recruitment and testing of subjects, a thorough search of the relevant literature must be conducted, the investigation must be planned and designed, and the research proposal that emerges from these activities typically must be subjected to review. As we have discussed in previous chapters, the ethical review of research pro-posals now is required at many research institutions before the researcher is given the go ahead to carry out a study; review also may occur at various intervals once the data collection process has begun. Some researchers, however, have voiced their concerns about the adequacy of external review in light of subjective biases and vested interests in the decision-making process. Additionally, institutional review boards (IRBs) have been criticized as being overly stringent and conservative in carrying out their gate-keeping function, by unnecessarily restricting scientists in their attempts to investigate sound research questions (see Ceci et al., 1985; Christakis, 1988; Rosnow et al., 1993).

In satisfying their responsibilities to ensure the protection and welfare of subjects, ethical review boards typically attempt to ascertain that the anticipated benefits of an investigation are greater than any risks posed to the research participants (and to society at large) (Stanley and Sieber, 1992). At one time, such systematic evaluations of research studies were not conducted and researchers did not have to concern themselves with ethical standards or IRB review. The IRB concept was first realized during the mid-1960s following recommendations by the National Commission of the United States Department of Health and Human Services (see chapter 2). Prior to that time, there was very little evidence of any type of committee review of research projects in the biomedical or behavioral science fields. The widespread existence of committees to review research with human participants was clearly apparent a few years later, even before enactment of the National Research Act of 1974. The results of a study conducted for the National Commission by the Survey Research Center at the University of Michigan revealed that by 1975 IRBs had been established in 113 medical schools and 424 other research settings (see Gray et al., 1978).

Whereas formal review and external monitoring originally were limited mostly to large, funded research projects, nowadays most studies are subjected to some kind of external review process. (Table 9.1 presents the sorts of questions commonly asked of investigators during a research review.) Universities and other institutions where research with human or animal subjects is conducted now tend to require some form of ethical review for the approval and monitoring of research. For example, a psychology student who intends to conduct an experiment using human subjects might first have to satisfy the requirements of a departmental review committee (Shaughnessy and Zechmeister, 1990). Depending on the nature of the study and the degree of risk it entails, the study then might be referred to a university-wide IRB. However, should the study qualify as involving only minimal risk to subjects, it could be approved by the student's course instructor who has the responsibility to expedite the review process for research falling into certain predefined categories. Because serious consequences can arise when university, federal, and local review procedures conflict (see Tedeschi and

Table 9.1 Sample questions for ethical review

Investigator
 1 Who is the primary investigator, and who is supervising the study?

Research participants
 2 What are the general characteristics of the research participants to be used (e.g., age range, sex, projected number of subjects, affiliation of subjects)?
 3 What, if any, are the special characteristics of the research participants (e.g., children, pregnant women, racial or ethnic minorities, mentally retarded people, prisoners, alcoholics)?
 4 Are other institutions or individuals cooperating in or cosponsoring the study?
 5 What is the general state of health (mental and physical) of the research participants?
 6 How will subjects be selected for, or excluded from, participation in this study?

Procedure
 7 What will the subjects be asked to do, or what behaviors will be observed by the researcher?
 8 Will deception be used? If the answer is yes, why is it necessary?
 9 What is the nature of the deception, and when will the debriefing take place?

Material
 10 If electrical or mechanical equipment will be used, how has it been checked for safety?
 11 What standardized tests, if any, will be used? What information will be provided to the subjects about their scores on these tests?

Confidentiality
 12 What procedure will you use to ensure the confidentiality of the data?

Risks
 13 Are there any immediate risks to the subjects, including possibly causing them embarrassment, inconvenience, or discomfort?
 14 Are there any long-range risks to the subjects?
 15 If there are risks, what is the necessity for them, and how will subjects be compensated for facing such risks?

Source: Rosnow and Rosenthal, *Beginning Behavioral Research: A Conceptual Primer*, ©1993, p. 58. Reprinted by permission of Prentice Hall, Upper Saddle River, New Jersey.

Rosenfeld, 1981), many officials believe it is a wise policy to have such formal review procedures in place at their institutions. Thus, despite the fact that professional standards tend to place the responsibility for ethical conduct on the individual researcher, the trend in recent years has been for research institutions to adopt the approach typically taken by governments, which is to rely on external enforcement for the approval and monitoring of studies (Schuler, 1982).

Over the years, membership requirements and review criteria for IRBs have undergone several changes in order to increase their effectiveness and comprehensiveness (Gray and Cooke, 1980; Rosnow et al., 1993). The initial federal regulations requiring IRB review for human subject research in the United States specified that the composition of a review committee was to consist of at least five members with varying backgrounds and fields of expertise (e.g. an IRB could not consist solely of behavioral scientists) (*Federal Register*, March 13, 1975). The committee was also to include at least one member who was not affiliated with the institution, and it could not be comprised of either all men or all women (*Federal Register*, January 26, 1981). Such review committee membership requirements are typical in many university settings, whether external review is required to conform to federal policy or is established on a voluntary basis. In fact, the composition of ethical review boards has broadened over the years to the point at which about one-third of the members are not researchers (Gray and Cooke, 1980). It is now common to find responsible individuals from the community, including members of the clergy, lawyers, and nurses serving on such boards.

As a personal example, the author served as an original member of the human studies committee at the college where he is employed as a psychology professor and, as it turned out, was the only representative from the social or behavioral sciences serving on the committee. The committee was formed after a faculty member at the college received federal funding to carry out a research project involving human subjects. Other members of the original committee consisted of two physical education professors (one of whom regularly conducts research), two nursing faculty, a college administrator, and a person from the community. Prior to the establishment of the college-wide review committee, an ethical review committee had been formed by the college's behavioral sciences department to review research proposals within the department and to recommend and support campus ethics programs (see Hogan and Kimmel, 1992). The composition of this voluntary departmental committee was consistent with governmental requirements for IRBs: the minimum representation on the committee included three full-time behavioral science faculty members, one student representative, one faculty representative from another college department, and one professional person from off campus.

The operation of the human studies committee was such that the investigator first would be expected to complete an Application for Approval of Research Involving Human Subjects, which then was discussed by committee members (see appendix 3). If clarifications regarding certain details of the application or the proposed study were needed, the investigator would be invited to meet with the committee. This would not be necessary if the committee was in agreement that the study represented an example of "minimal risk research" according to governmental criteria. As is usually the case with such review boards (cf. Gray and Cooke, 1980), studies rarely were disapproved outright; rather, further details regarding the procedure (such as how the subjects were to be debriefed) were often requested or else additional safeguards were suggested (such as the inclusion of a more complete informed consent process). Revised proposals were then reconsidered by the committee.

It probably is not surprising that some behavioral scientists have criticized the usual composition of IRBs, given that committee members are likely to consist of persons who are not researchers themselves and who may not be familiar with the research process (see Box 9.1). That is, it is difficult for some researchers to accept having their proposals reviewed by members of the clergy, lawyers, professors of English, history, physical education, and so on, who may appear to have little appreciation of the potential merits of scientific research and who are likely to be unaware of the methodological difficulties in conducting well-controlled human research while maintaining a strict adherence to ethical regulations. However, a review board entirely staffed by people who are within the same field may be less inclined to ask certain basic questions and may be prejudiced in favor of approving the research of their peers. In this light, it is best that IRBs be comprised of a mix of people.

The changing nature of IRB review

Since their inception, the primary functions of IRBs have consisted of assessing risks and benefits, protecting the rights and welfare of research participants, and assuring that informed consent is obtained. Over the years, the responsibilities of review boards have

Box 9.1

Reactions to ethical committee review

The following comments were written by an American psychologist in response to a questionnaire intended to assess systematic biases in ethical decision making (Kimmel, 1991; see pp. 283–4). These excerpts reflect a high degree of personal frustration with the external review process; growing evidence suggests that this researcher's experiences may not be uncommon among behavioral scientists.

I am not very happy with university ethics committees. I had a three-year running feud with the committee at [name of university]. I considered the research I proposed to be risk-free, which they refused to accept. The objections ranged from the ignorant to the absurd. The research involved simple psychophysical tasks. I noted in my proposal that subjects would take frequent rest intervals to reduce the effects of boredom and fatigue on performance. One committee member felt that boredom was a risk. (We're talking about listening to tone bursts at threshold for periods of up to three minutes). Since the response box the subjects used was manufactured in the laboratory, the same committee member ... insisted there was shock hazard and demanded that I indicate this to my subjects. The committee in general wanted me to inform my subjects that listening to tones no louder than ordinary speech entailed a slight risk to their welfare. I was not and am not now willing to so stipulate. I have since ... switched to animal research where the composition of animal welfare committees is a bit more reasonable. I fully recognize and support the need for review of all research involving living subjects. However, when there is no appeal of committee decisions, the review process quickly loses its protective value and the committee instead begins to attempt to *direct* research from their meetings. My chief antagonist on [the university's] committee was a professor in the family resources department who clearly had no idea whatsoever of what laboratory research entailed.

Another candid view of the external review process is taken from an informal talk delivered by Herbert Kelman at the 1987 meeting of the Eastern Psychological Association (cf. Kimmel, 1988b, p. 157). Unlike the preceding comments, however, Kelman's remarks represent the perspective of an IRB member and reflect the kinds of unintended biases that might enter into evaluations of research in one's own discipline.

> I remember my own experience on an IRB once when we had a proposal for a study, one of those field experiments that involved somebody falling down in the middle of the street and clutching his heart and all of that. I was much more conflicted about saying no to that study than some other people for the simple reason that I saw the study within a tradition. I mean, that's what we've been doing in social psychology and I can understand why this person would want to do this . . . this is part of the traditional way in which work is being done in the field . . . and I had some appreciation for the value of this work. But there were some other people whose immediate reaction was, "This is ridiculous research. Why should one even consider the risks or the ethical violations or the legal problems that might arise out of this for the sake of this worthless kind of enterprise?" Well, I think it's useful to have that corrective there but, on the other hand, I wouldn't want it to be entirely those people. I think you need a kind of dialectic.

expanded and the criteria for ethical review have continued to evolve. These developments have been traced in part to a rise in pressures applied by certain advocacy groups, such as groups lobbying for the quicker release of drugs for dying patients, drug companies attempting to gain support for changes enabling research participants to have access to experimental treatments, and researchers pressing for the education of people whose understanding of informed consent is deficient (Mitchell and Steingrub, 1988). According to Rosnow et al. (1993), such pressures can lead to changes in the criteria ultimately used by IRBs to evaluate research.

Within the research community there are growing sentiments that IRBs have begun to consider more than is necessary to adequately assess the physical and mental risks to prospective research participants (Ceci et al., 1985). Indeed, there is mounting evidence that the role of IRBs has changed in recent years to include issues not specifically related to the rights and welfare of research participants in evaluations of research proposals. For instance, the discussion of research design and other methodological considerations is no longer uncommon among IRB members (Rosnow et al., 1993). This development is not necessarily a bad one. Because a poorly designed study can have serious negative ramifications, one might argue that the technical aspects of a proposed investigation in fact should be included as an aspect of external review. To illustrate, Meier (1992) suggested that the design of a large-scale study to examine the efficacy of an experimental drug intended to halt the progression of a fatal illness must provide for clear conclusions to be drawn. This would require that the investigation meet certain standards of control and sample size, considerations that might be incorporated into an IRB decision. In essence, a failure to meet these technical standards likely would result in a significant waste of institutional, public, and other resources.

There remains widespread disagreement over whether technical aspects of investigations ought to be taken into account by IRBs and whether the composition of IRBs should be expanded to include persons with scientific expertise. Some people believe that scientific issues ought to be resolved before a research proposal is submitted to an ethical review committee. Additionally, there are potential negative consequences in allowing a consideration of scientific issues into the review process. For example, it has been noted that some IRBs may be prone to apply overly rigorous design standards, such as those that are required for drug trials, to studies in behavioral and social research areas (Rosnow et al., 1993). Further, adding another dimension to the evaluation of research proposals would no doubt be seen by some researchers as an additional barrier in their pursuit of complex research questions or intention to use less conventional research approaches. Such fears eventually could serve to restrict the kinds of studies that are carried out in certain areas (Suls and Rosnow, 1981; West and Gunn, 1978).

Adequacy of the IRB process

Evidence regarding the effectiveness of ethical review committees in protecting the autonomy of research participants and the impact of such committees on research is somewhat mixed. Some informative early results regarding the performance and impact of IRBs were revealed by the aforementioned survey conducted by the Survey Research Center (Gray et al., 1978). More than 2,000 researchers and 800 review board members at 61 research institutions were interviewed about various positive and negative aspects of the functioning of IRBs during a one-year period ending in 1975. One major finding of the survey was that although the IRBs commonly required modifications in the proposed research (usually regarding some aspect of the procedure to obtain consent), the outright rejection of research was rare. Most of the researchers interviewed considered the IRBs' judgments regarding their research proposals as sound; however, approximately half of the researchers reported that they had experienced some delays in undertaking their study or that their research had been impeded in a way that was not balanced by the benefits of the review process. Only 8 percent of the researchers believed that the overall difficulties imposed by IRB review outweighed the benefits of the review process in protecting research participants.

Taken together, the results of the Survey Research Center study do not lend support to the fears often expressed by researchers that the external review process will lead to overly stringent judgments and the frequent rejection of research proposals. Nonetheless, since that study, a growing body of evidence has begun to emerge suggesting that there may be important deficiencies in the performance of some IRBs (Ceci et al., 1985; Goldman and Katz, 1982; Prentice and Antonson, 1987). Specifically, substantial inconsistency has been found in the application of decision standards and in the subsequent recommendations put forth by ethical review committees.

Inconsistencies in ethical decision making
Variability in the evaluation of research proposals among IRBs can stem from several sources, including the composition of the review boards, subjective biases in the assessment of risks and benefits, the nature of interaction among committee members, the types of studies evaluated, committee members' review experience, and so on. Given these considerations, it is not surprising that inconsistencies in IRB decision making have been reported by researchers who have systematically studied the review process. In one study, Goldman and Katz (1982) investigated the adequacy of peer review by asking 22 IRBs at major American medical schools to review three medical research protocols and consent forms according to their standard procedures. The three proposals each posed serious ethical issues, flaws in the research design, and a consent form that was incomplete and in violation of federal guidelines. While outright approval of the reviewed proposals was rare, there was substantial inconsistency in the reasons offered for non-approval (including "modify" and "disapprove" decisions) and in the application of ethical standards. Some of the boards requested greater specification of the procedures for administering the experimental treatment and for obtaining measurements of the effects, while others sought clarification of confidentiality procedures, more complete explanation of potential benefits and risks, or a more explicit comparison of alternative treatments. There also were wide-ranging differences in specific ethical and methodological objections to each of the three proposals, variations in reactions to the inadequate consent forms, and a substantial number

of review boards approved the intentionally flawed research designs. Goldman and Katz concluded that these variations can be explained in part by systematic differences in approval guidelines and community standards, and that their findings reflect a failure of institutions to obtain uniformity throughout the external review system.

The results of an investigation utilizing archival records suggest that while there may be variability among IRBs, the decision making *within* IRBs may be fairly consistent. Grodin et al. (1986) conducted a 12-year audit of one IRB (at the Boston City Hospital in Massachusetts), analyzing more than 700 research protocols that were considered between the years 1973 and 1984 for factors related to consistency of decision making in requiring changes to the proposed research designs. The findings showed that the review standards remained constant despite changes in the composition of the committee, the committee's experience over time, types of studies, funding sources, and political climate. Of course, the applicability of these results to other institutional reviews is quite limited, given the fact that only one IRB, which was functioning in a hospital setting, was investigated. Indeed, in Goldman and Katz's (1982) study, individual review boards demonstrated little internal consistency in their interpretation of guidelines or application of standards for approval, as evidenced by their varying reactions to three similarly unacceptable consent forms.

Although the preceding examples pertain to the institutional review of medical protocols, similar inconsistencies in ethical decision making have been observed among behavioral scientists as well. For example, Eaton (1983) found that experienced reviewers were in agreement about the ethicality of 111 psychological research proposals (in terms of the way subjects were to be treated) only 8 percent of the time. The reviewers who were studied were members of the university psychology deparment in which the proposals were prepared.

In an investigation of biases in ethical decision making, the author asked a sample of 259 American psychologists to evaluate several hypothetical studies varying in costs and benefits (Kimmel, 1991). Consistent with previous research on ethical decision making (e.g. Forsyth, 1980; Hamsher and Reznikoff, 1967; Schlenker and Forsyth, 1977), the results were such that it was possible to distinguish between conservative decision makers (i.e. those who emphasized research costs) and liberal decision makers (i.e. those who emphasized research benefits) on the basis of certain individual background characteristics. Respondents who tended to reflect greater conservatism in their ethical evaluations (emphasizing research costs) were primarily female; had recently received the doctoral degree in an applied psychology area, such as counseling, school, or community psychology; and were employed in service-oriented contexts. By contrast, those respondents who tended to be more liberal in their ethical assessments (emphasizing research benefits) were more likely to be male; had held their doctoral degrees in a basic psychology area (such as social, experimental, or developmental psychology) for a longer period of time; and were employed in research-oriented settings. To the extent that these results are generalizable to members of IRBs, there may be predictable biases in the way certain review committees appraise costs and benefits and ultimately decide on the fate of particular research proposals.

In another investigation of review board decision making, Ceci et al. (1985) demonstrated how personal values can play a critical role in decisions involving socially sensitive research. The researchers had 157 university IRBs review hypothetical proposals differing in terms of their degree of social sensitivity and their level of ethical concern. The proposals described an investigation of employment discrimination and were characterized by a procedure involving varying levels of deception. The results of the study revealed that socially sensitive proposals (intended either to examine discrimination against minorities or reverse discrimination against white males in corporate hiring practices) were twice as likely to be rejected by the review committees as were non-sensitive proposals. These outcomes did not appear to be influenced by the presence or absence of ethical problems, although the use of deception was most frequently cited as the reason for non-approval of the sensitive proposals. According to Ceci et al., this latter finding seems to suggest that, at least for some studies, IRBs may use whatever reasons are most convenient to justify their review decisions. These results clearly have implications for emerging research in other socially sensitive areas, such as AIDS research (Thomas and Quinn, 1991).

Given the findings described here, it is apparent that similar research proposals involving identical levels of risks and benefits may be approved at one institution but not at another. To some extent, local laws, community standards, and institutional restrictions that limit the types of information and degree of acceptable risk to research participants provide indirect sources of inconsistency among IRBs (Rosnow et al., 1993). As a result, luck might sometimes be involved in the approval of a research proposal, especially one that is socially sensitive. A favorable decision might simply be a matter of where one happened to be working when the proposal was submitted for review (Ceci et al., 1985). Rigorous review constraints placed on tests of potential drugs in AIDS vaccine trials already have tempted some AIDS researchers to carry out their studies in developing regions where standards are less strict than in industrialized countries (Rosnow et al., 1993; Thomas and Quinn, 1991).

Recommendations for review
Although most researchers appear to accept the concept of external review in principle, a growing number have expressed their dissatisfaction with the review process in actual practice. As an essential mechanism for protecting research participants from major risks of violation of their rights and welfare, whatever steps can be taken to ensure that the review process is adequate and evenhanded would be worthwhile. In the words of one ethicist, "researchers should not be burdened by being victims to an overly zealous IRB any more than subjects should be victims of overly lenient ones" (Veatch, 1982, p. 180). In short, a process is needed for maximizing consistency among IRBs and keeping them informed of emergent issues and shifting standards governing research.

One recommendation that has been offered for achieving these goals is to provide IRB members with a casebook of actual research protocols that have received extensive review and analysis by both investigators and subjects. According to Rosnow et al. (1993), such a casebook could serve as an aid to help sensitize IRB members to relevant costs and benefits of doing and not doing research, while providing them with a means for scrutinizing their own ethical biases. For unique research cases that require novel analysis, Rosnow et al. also suggested that an advisory board could be created within a discipline's professional association. The purpose of the advisory board would be to analyze and review an IRB decision when it is requested due to disagreements among committee members, investigators, and institutions. The advisory board's analysis could either serve an educative function or be binding (assuming the parties involved have agreed to accept the determination as such).

Another approach to improving the review process is to take steps to minimize problems of communication between IRB members and investigators. For example, members of review committees could clarify unclear guidelines and keep investigators fully informed of unfamiliar regulations. This, of course, would require that IRB members be sufficiently prepared for the review process through training sessions and regular communication with other IRBs (Tanke and Tanke, 1982). It is unlikely that ethical training can produce unanimity in particular judgments, but at least it can provide IRB members with some understanding of the sources of the systematic differences in their ethical decision making. The education of IRB members also can be provided through attendance at regional and national conferences and by exposure to sample review cases in journals such as *IRB: A Review of Human Subjects Research* (Veatch, 1982). Similarly, investigators can aid the review process by familiarizing themselves with current ethical standards and principles of review. They also need to be straightforward in providing complete and adequate information about the risks and benefits of their proposed studies, especially when potentially sensitive responses are sought from subjects.

Ethical Issues in the Communication of Research

The research process does not end once the data have been subjected to statistical analysis and the results have been interpreted by the investigator. In fact, ethical responsibilities extend beyond the actual carrying out of a study (i.e. the collection and analysis of data) to include issues involved in the reporting of research and the prevention of the misuse of research results. At the final stage of the research process, peer review comes not from ethical review boards, but rather through the careful scrutiny of the completed research by other members of the scientific community. The nature of this review, however, is fundamentally different from that of ethics committees. Whereas the concerns of review boards focus on the ethical treatment of research participants and the protection of their rights, the nature of peer review following completion of a study emphasizes the scientific quality of the study and professional issues related to honesty and accuracy in the reporting of the research findings. These latter concerns are fundamental to the protection of science as a body of knowledge.

In order for an investigation to have any sort of impact on scientific progress within one's discipline or utility in terms of application within specified settings, the research findings must be communicated to others. The details of a study may be communicated to other scientists in various forms, including articles published in scientific journals, chapters in edited books, technical reports, presentations at scientific conferences, and so on. Research findings often are transmitted to the general public through radio or television interviews, newspaper articles, magazine features, public lectures, and the like. Whichever medium is used to report a study to others, certain ethical responsibilities are necessarily imposed on the researcher, who may be held accountable for the consequences of what is reported. As a result, in recent years professional associations have begun to formulate ethical standards for clarifying professional conduct in the reporting and publishing of scientific information. Standards such as those developed by the American Psychological Association (see Box 9.2) can serve as a useful guide for our consideration of some of the ethical issues related to the communication of research results, such as accuracy in the reporting of results and professional issues in research publication.

Honesty and accuracy in reporting research

If there is one ethical guideline for which there is complete consensus it is the one that maintains that scientists must not falsify their research findings. In fact, the basic rule requiring honesty and accuracy in the conduct and reporting of research is such a "given" in science that it almost may seem absurd to have to mention it at all. After all, one of the basic purposes of conducting research is to build on current scientific knowledge through incremental discovery. Within the research context, the pursuit of truth – in the sense of veracity in the conduct of inquiry – represents the dominant value (Weinstein, 1981). In other words, the norms of scientific inquiry demand honesty throughout the truth-seeking process. Despite the fact that researchers are assumed to be honest and unbiased in their pursuit of scientific truth, instances of fraud and other related forms of scientific misconduct have been revealed throughout the history of science. These issues thus present legitimate ethical concerns for investigators within the behavioral sciences.

Categorizing scientific dishonesty

One of the earliest attempts to delineate various forms of scientific dishonesty came in 1830, when Charles Babbage, a British professor of mathematics at Cambridge University, published a book entitled *Reflections on the Decline of Science in England*. Although Babbage is sometimes referred to as a prophet of the electronic computer, he may be best known for distinguishing in his book between three forms of scientific dishonesty: trimming, cooking, and forging. In contemporary terms, "trimming" refers to the smoothing of irregularities in one's data to make them look more accurate and precise. In essence, trimming is the manipulation of data to make them look better (Babbage, 1830/1969):

Box 9.2

Ethical standards for the reporting and publishing of scientific information

The following standards, which deal with the communication of scientific information, were extracted from the American Psychological Association's *Ethical Principles of Psychologists and Code of Conduct* (1992), *American Psychologist*, 47, 1597–1611.

6.21 *Reporting of Results*

(a) Psychologists do not fabricate data or falsify results in their publications.
(b) If psychologists discover significant errors in their published data, they take reasonable steps to correct such errors in a correction, retraction, erratum, or other appropriate publication means.

6.22 *Plagiarism*

Psychologists do not present substantial portions or elements of another's work or data as their own, even if the other work or data source is cited occasionally.

6.23 *Publication Credit*

(a) Psychologists take responsibility and credit, including authorship credit, only for work they have actually performed or to which they have contributed.
(b) Principal authorship and other publication credits accurately reflect the relative scientific or professional contributions of the individuals involved, regardless of their relative status. Mere possession of an institutional position, such as Department Chair, does not justify authorship credit. Minor contributions to the research or to the writing for publications are appropriately acknowledged, such as in footnotes or in an introductory statement.

(c) A student is usually listed as principal author on any multiple-authored article that is substantially based on the student's dissertation or thesis.

6.24 *Duplicate Publication of Data*

Psychologists do not publish, as original data, data that have been previously published. This does not preclude republishing data when they are accompanied by proper acknowledgment.

6.25 *Sharing Data*

After research results are published, psychologists do not withhold the data on which their conclusions are based from other competent professionals who seek to verify the substantive claims through reanalysis and who intend to use such data only for that purpose, provided that the confidentiality of the participants can be protected and unless legal rights concerning proprietary data preclude their release.

6.26 *Professional Reviewers*

Psychologists who review material submitted for publication, grant, or other research proposal review respect the confidentiality of and the proprietary rights in such information of those who submitted it.

... in clipping off little bits here and there from those observations which differ most in excess from the mean and in sticking them on to those which are too small.

A second type of dishonesty involves the "cooking" of data, which consists of retaining only those results that best fit one's hypothesis or theory and discarding those that do not. "Forging" pertains to the act of inventing some or all of the research data that are reported, and may even include the reporting of experiments that never were conducted. Because it involves instances of complete fabrication of results, forging typically is considered one of the most blatant forms of scientific misconduct.

Of these three research abuses, forging is the most vulnerable to exposure; because it is based on completely fictitious data, it is the most likely to look fake to the knowing eye or to be revealed as a result of unsuccessful attempts to replicate the reported results. Trimming and cooking, on the other hand, may long go undetected, especially if the researcher reaches correct conclusions by illegitimate means. In fact, cooking may be the most difficult to notice given that what it yields is genuine, though only a part of the truth.

One of the most notorious cases of data cooking is the experimental work of the physicist Robert A. Millikan (Sigma Xi, 1984). Early in this century, Millikan was involved in an intense controversy with the Viennese physicist Felix Ehrenhaft on the nature of the electronic charge. Following several years of inconclusive and conflicting experiments, Millikan published a major paper in 1913 in which he reported the results of a series of experiments on liquid droplets. Millikan explicitly stated in the paper that the results were based on all of the droplets observed over a period of 60 consecutive days. However, an examination of his laboratory notebooks nearly half a century later revealed that the observations presented in his 1913 article were in fact selected from a much larger total: only 58 out of 140 observations were reported (Holton, 1978). Interestingly, reviewers of Millikan's research have pointed out that his selectivity was unnecessary in that the totality of the research ultimately would have had great scientific importance even if the anomalous values had been reported (Franklin, 1981; Holton, 1978). It is not so much that Millikan's selection of data *per se* was at issue; the unacceptable conduct in this case was his assertion that the published data were based on all the observations.

Although Babbage's categories of scientific dishonesty represent an early typology, they still have relevance today. For example, the debate surrounding the work of Sir Cyril Burt, which was described in chapter 1, involves charges that Burt may have been guilty of engaging in each of the forms of unacceptable conduct identified by Babbage (cf. Green, 1992). Burt's critics have contended that some of his statistical results were "too good to be true," and thus had to have been either trimmed or forged. His studies consistently and overwhelmingly supported his strongly held belief in the heritability of intelligence, suggesting the possibility of data cooking. Further, because he reported some identical correlation values (to three decimal places) in different studies varying in sample size (from 21 to 66), it appears that Burt may have fabricated at least some of the values (Broad and Wade, 1982). It may be recalled from our earlier discussion of this case that Burt even may have "invented" at least two of his research assistants.

The Millikan and Burt cases serve to illustrate how many instances of scientific dishonesty often are neither quickly discovered nor damaging to the perpetrator's ongoing career (Sigma Xi, 1984). Millikan went on to become one of the most influential and renowned scientists in the United States, having received the Nobel Prize in 1923, partly as a result of his early work on the electronic charge. Similarly, Burt acquired a reputation as a pre-eminent researcher and leader in his field, was the first psychologist to be knighted in England, and was the first foreigner to receive the prestigious Thorndike Award from the American Psychological Association. Questions about dishonesty in the work of Millikan and Burt did not emerge until some years after their deaths and, as a result, these men did not have a chance to respond to the charges leveled against them. Similar circumstances characterize some of the well-known early cases of cheating that have been perpetrated during the history of science (see "Cases of scientific fraud" below).

In addition to Babbage's categories of scientific dishonesty, other subtle forms of misconduct in the sciences have been identified. These include the suppression of negative evidence or unsuccessful results; the selective citing of the literature or selective reporting of others' data to bolster one's own position; misrepresentation of others' ideas; the incorrect analysis of research data; misleading descriptions of methodologies; and plagiarism in proposing or reporting research (Mahoney, 1976; Swazey et al., 1993). These forms of abuse often are less blatant than some of the more notorious cases of outright cheating and, as a result, may be more prevalent and difficult to detect. A much grayer area consists of a variety of unprofessional behaviors, such as not sharing resources, maintaining sloppy notebooks or research records, nastiness, non-collegiality, not helping younger colleagues, and professional secrecy (see Box 9.3). While such misbehavior does not represent scientific misconduct *per se*, its effects may be such that they serve to undermine the scientific endeavor in significant ways (Herman, 1993).

Cases of scientific fraud

The occurrence of fraud in science is hardly a recent phenomenon. Throughout the annals of science, even some of the most influential scientists of their day have been suspected of falsifying data or engaging in other forms of scientific misconduct. In fact, at one time or another, all of the scientific disciplines have been affected by scandals resulting from the unethical behavior of researchers within their ranks. Many of the cases have appeared in the physical sciences where, because of greater replication and precision, the faking of data may be more readily detectable (Mahoney, 1976).

As interest in scientific honesty has increased in recent decades, historians of science have begun to uncover previously undetected examples of misconduct that may have occurred many years earlier.

Box 9.3

Professional secrecy and the "passionate scientist"

The stereotyped image of the scientist is one that includes such character-istics as objectivity, flexibility, personal disinterest in fame or recognition, and a communal interest in the open sharing of knowledge and coopera-tion with fellow researchers. In his 1976 book, *Scientist as Subject: The Psychological Imperative*, psychologist Michael Mahoney contends that this "storybook" image is far from an accurate representation. In Maho-ney's view, the portrayal of the scientist as a virtuous and "passionless purveyor of truth" should be replaced by one that acknowledges the subjective and fallible aspects of human nature. Thus, in contrast to the list of scientific virtues often used to describe scientists, Mahoney argues that the scientist is perhaps the most passionate of professionals: he or she is often dogmatically tenacious in maintaining opinions; behaves in selfish and ambitious ways; is frequently motivated by a desire for personal recognition; and occasionally is secretive and suppresses data for personal reasons. A compelling illustration of these tendencies is provided by the research leading up to the discovery of the molecular structure of DNA by James Watson and Francis Crick, which was wrought with secrecy, personal rivalries, animosity, and professional jealousies (Watson, 1969).

In partial support for his views, Mahoney describes how some scien-tists have gone to great lengths to conceal their discoveries for fear of being "scooped" by professional rivals or of having their ideas stolen by others. For example, Leonardo Da Vinci apparently had a penchant for coding his personal notes by writing them backwards, in mirror image. Similarly, Galileo is said to have used anagrams (scrambled letters containing a hidden message) as a protection against having his priority claims undermined by others. (In Galileo's case, one might say that personal dishonesty begets paranoia [see text].) Interviews with contem-porary scientists suggest that secrecy is prevalent in the sciences, although more discreet means of protecting one's work have replaced the rather crude methods used by early scientists (Gaston, 1971; Mahoney, 1976; Mitroff, 1974). One reason for such secrecy, in addition to the fear that one may miss out on an important publication, is the fact that many scientists are concerned about the possibility of having their work stolen. Gaston (1971) reported that more than one-sixth of the high energy physicists he interviewed believed that some of their work had been stolen, and an average of 14 percent of the physical and behavioral scientists interviewed by Mahoney and Kimper felt that their colleagues had plagiarized other's ideas (see Mahoney, 1976).

For instance, more than 1,800 years after the appearance of his influential publication on celestial motions, the renowned Greek astronomer Claudius Ptolemy was accused of having carried out one of the most successful frauds in the history of science. According to the physicist Robert Newton (1977), very few of Ptolemy's observations were his own and those were either fictitious or incorrect. Newton has argued that Ptolemy used a technique that is common among intellectual cheats, which was to work backward to prove the results he wanted to obtain. Ptolemy's key astronomical work (now known as *Almagest*, which is Arabic for "The Greatest") consisted of 13 volumes which placed the earth at the center of the solar system and detailed his theory about the motions of the planets and stars. Newton maintains that evidence now points to the likelihood that Ptolemy came up with his theory based on incorrect measurements made by Hipparchus almost 200 years earlier. Despite Ptolemy's claim that he worked for several years to confirm the measurements made by Hipparchus, it appears that he merely took the measurements on faith and never made any of his own. Newton has uncovered other discrepancies in Ptolemy's work to support his charges against the astronomer.

Galileo, the famous Italian scientist who is largely credited with overturning Ptolemy's theories of the solar system, has himself been the target of charges that he engaged in scientific misconduct, including the likelihood that he faked some data in order to bolster his own theory of gravity (Cohen, 1957). Galileo's great discovery, based on his use of the newly developed telescope, was that the planets revolve around the sun rather than the earth. Recently, the noted historian of science Richard Westfall has charged that Galileo stole the proof for this discovery from a student and used it without acknowledgment to obtain the favors of a patron, the Grand Duke of Tuscany (Broad, 1983). In December 1610 Benedetto Castelli wrote a letter to his former teacher Galileo suggesting that observations of the planet Venus might settle the long-running debate about the structure of the solar system. Shortly after receiving the letter, Galileo wrote to the Medicis asserting that he had made detailed observations of Venus over the course of three months and in so doing had proven the controversial theory proposed by Copernicus that the sun was at the center of the universe. According to Westfall, however, there is no evidence that Galileo ever made the observations or that he had ever thought out a serious program of observation prior to receiving Castelli's letter.

With each challenge to Galileo's reputation as a scientific genius, historians have come to his defense and have attempted to duplicate the experiments he claimed to have conducted. Nonetheless, the accumulation of evidence against Galileo's scientific integrity and questions regarding his personal motives seem to suggest that the greatness of the man perhaps can more readily be found in his ability to conduct "thought experiments," imagining his experimental outcomes instead of actually doing the work to obtain them (Broad, 1983).

During the twentieth century, one can find numerous examples of scandals involving the faking of data that have received widespread attention both within the scientific professions and among the general public. One of the most dramatic cases was the Piltdown hoax, which involved the discovery in 1912 of a primitive skull and jawbone near Piltdown Common, England by an amateur geologist. The fossil remains, which were referred to as "Piltdown man," were so different from anything previously known to anthropologists that it quickly threw their discipline into disarray, forcing a rethinking of existing theories of human evolution. Piltdown man effectively dismayed two generations of anthropologists until the conclusion was reached in the 1950s that a hoax had been perpetrated (Weiner, 1955). Modern tests revealed that a human skull and an ape jawbone from different periods had been altered to simulate the bones of primitive man; for instance, the teeth had been filed in order to give the appearance of having been worn down by chewing. To this day, it is unclear who was responsible for this sophisticated hoax, although one recent theory points to none other than the creator of Sherlock Holmes, Sir Arthur Conan Doyle, as a prime suspect ("Arthur Conan Doyle," 1983; Winslow and Meyer, 1983).

Another famous case involving the altering of a scientific specimen was that of the Austrian biologist Paul Kammerer, who was engaged in research to study the Lamarkian hypothesis of inheritance of acquired characteristics (Weinstein, 1979). Kammerer spent many years attempting to demonstrate that the physical skills and developments acquired by one generation of animals are passed on to their offspring through hereditary processes. Support for this view came when Kammerer reported that salamanders he had raised on a yellow or black background acquired more of the color in their own pigmentation and then passed this acquired characteristic on to their offspring. After World War I, however, when a suspicious investigator examined the preserved specimen of a toad which had purportedly acquired a black patch in one of Kammerer's experiments, it was revealed that the specimen had been tampered with. Apparently, the specimen's visible coloration had been produced by India ink, not by pigmentation. The forgery was widely publicized by the media and, as a result, Kammerer's reputation was disgraced and he committed suicide. It never has been established with any certainty whether Kammerer or some other person, such as a research assistant or jealous colleague, was responsible for the fabrication (see Koestler, 1972). In a similar, more recent example, a scientist at the renowned Sloan-Kettering Institute for Cancer Research fabricated his research results after several unsuccessful attempts during the 1970s at grafting skin from black mice onto white mice. In what has come to be known as the "painted mice affair," William Summerlin painted black patches on two white mice with a felt-tipped pen so that they would appear to be successful transplants from other animals (Culliton, 1974).

The behavioral sciences have not been immune from cases of scientific fraud. One well-known example surrounds some of the long-term research on extrasensory perception and other psychic phenomena carried out at J. B. Rhine's Institute for Parapsychology in North Carolina. Rhine was a respected researcher on psychic phenomena for many years. As he approached retirement, he appointed a bright young physician named Walter Levy to direct the Institute. Levy proceeded to automate the research laboratories with computers and sophisticated devices in order to eliminate human error in the data collection procedures. Ironically, it eventually was discovered that Levy himself had altered the data in some studies of psychokinesis in laboratory rats. Electrodes had been implanted in the "pleasure centers" of the rats' brains in order to study whether the animals could psychically influence a random generator to deliver pleasurable stimulations. The fraud was revealed when Levy's assistants found that he was surreptitiously unplugging the automated data counter at certain times so that it would appear that the rats had psychokinetically influenced the generator (Asher, 1974). As a result of this revelation, Levy's promising research career abruptly came to an end and parapsychological research, which had begun to gain acceptance within a skeptical scientific community, was once again cast under a cloud of suspicion.

In a lesser known, but perhaps more representative case of scientific fraud in behavioral research, Stephen Breuning, a psychologist who was participating in research on the effects of psychotropic medication in mentally retarded patients, admitted to having falsified some of his research reports. Breuning had been working on a research grant awarded to Robert Sprague, a clinical psychologist at the University of Illinois. Sprague's suspicions about the integrity of his promising young colleague were first aroused by a comment from one of Breuning's research assistants (Sprague, 1993). The assistant stated that Breuning was obtaining perfect agreement between nurses in their independent ratings of the abnormal involuntary movements of patients suffering from a neurological disorder (tardive dyskinesia) due to the long-term use of the medication under investigation. However, such perfect agreement among independent raters of complex movements in patients is highly unlikely.

Subsequently, during his own investigation of Breuning's research, Sprague found clearer evidence of dishonesty. This came in the form of an abstract Breuning had submitted for a symposium on tardive dyskinesia in which he claimed to have examined 45 patients over a two-year period at the research facility where he previously had been employed. Yet Breuning no longer had assistants at that facility and had not traveled there himself during that time to conduct the research. Unable to provide documentation of the examinations, Breuning confessed to having falsified the abstract. After a prolonged investigation conducted by the federal agency that had funded the research, Breuning was indicted for scientific misconduct and sentenced by a federal judge in 1988. According to Sprague (1993), had the misconduct not been revealed, Breuning's fraudulent research reports could have had major policy and medical implications for numerous mentally retarded persons receiving the medication about which he wrote.

In recent years, cases of scientific misconduct have come to light in a variety of contexts. Disturbing questions about a 1986 research paper on immunology written in part by the Nobel laureate David Baltimore resulted in the conclusion that crucial data provided by a leading molecular biologist who co-authored the paper had been faked (Hilts, 1991). Federal investigators found Thereza Imanishi-Kari guilty of 19 cases of falsifying data and recommended that she be barred from participating in any federally funded research for ten years. Although Baltimore was not himself accused of fraud, his reputation as a brilliant biologist was sadly tarnished by criticisms that he was more intent on squashing the investigation of his colleague's conduct than getting to the bottom of the charges ("A scientific Watergate?," 1991).

In another highly publicized case during the 1980s, Robert Gallo, a leading AIDS researcher, was found to have falsely stated in a 1984 publication that the French version of the virus that causes AIDS had not been successfully grown in a cell culture, but that his laboratory had achieved success in growing another version of the virus. Actually, both viruses had been grown in cultures in his laboratory, and Gallo was held responsible for having "impeded potential AIDS-research progress" by failing to reveal the utility of the French virus (Herman, 1993). The federal panel that investigated Gallo's laboratory reported a pattern of conduct characterized by a "propensity to misrepresent and mislead in favor of his own research findings or hypotheses" (Herman, 1993, p. 7).

What is particularly disconcerting about the cases of scientific misconduct described here is that a leading (or up-and-coming) scientist was at the center of each scandal. It is unclear whether this can be attributed to the greater pressures placed on the scientist to maintain his or her lofty status within the field, the fact that research by successful investigators is more likely to be subjected to careful scrutiny within the research community, or the possibility that greater media attention is drawn to celebrated cases involving well-known figures (cf. Bechtel and Pearson, 1985).

Extent of scientific dishonesty

The faking of one's data is a phenomenon that is considered rare by most scientists, yet as we have seen in the preceding section, scandals continue to be revealed. The question remains as to to what extent the cases that have come to light represent a preponderance of all of the incidents that have occurred or are merely the tip of an iceberg (Weinstein, 1981). While there is general agreement that honesty and accuracy are critical elements of the research enterprise, there are disparate opinions about the extent and significance of scientific misconduct in the research disciplines. It is difficult to estimate just how pervasive fraud and other forms of misconduct are, although concerted efforts have been made in recent years to determine the extent of such practices.

In one attempt to ascertain the extent of cheating in science, anonymous questionnaires were made available to scientists subscribing to the British journal *New Scientist* (St James-Roberts, 1976). A significant percentage of the respondents (92 percent) reported that they had knowledge of "intentional bias" in their scientific disciplines. Another survey on scientific conduct within four disciplines (physics, biology, psychology, and sociology) carried out by Mahoney and Kimper found that 42 percent of the participants knew of at least one instance of data fabrication in their field and nearly half reported knowledge of at least one instance of data suppression (see Mahoney, 1976). Because the survey also revealed much lower estimates of the prevalence of such behavior, it may be that these data were influenced by some well-known scandals. It also must be noted that the generalizability of the results obtained from both of the surveys described here is limited by methodological problems inherent in the data collection procedures.

More recently, researchers from the Acadia Institute Project on Professional Values and Ethical Issues in the Graduate Education of Scientists and Engineers surveyed 2,000 doctoral candidates and 2,000 of their faculty in one of the first major studies of misconduct in science (Swazey et al., 1993). Anonymous questionnaires were returned by about 2,600 of the scientists from 99 graduate departments in chemistry, civil engineering, microbiology, and sociology who were asked about their experiences with 15 different types of ethically questionable behaviors. As is typical in surveys of this kind, the results did not measure the actual frequency of misconduct; rather, rates of exposure to perceived misconduct were obtained. The findings revealed that scientific misconduct may be more pervasive than many scientists believe: 43 percent of the students and 50 percent of the faculty members reported direct knowledge of more than one kind of misconduct in their laboratories. The forms of reported misconduct ranged from faking results to withholding research findings from competitors.

Other findings from the Acadia study revealed that from 6 to 9 percent of the respondents said they had direct knowledge of faculty members who had falsified or plagiarized data; one-third of the faculty members reported that they had direct evidence of such misconduct among students. Reports of questionable research practices (such as inappropriate assignment of authorship in research papers and misusing research money and equipment for personal use) were far more common than reports of outright misconduct. In addition, striking differences were found between scientists' espoused values and their actual practices. For example, in principle, nearly all of the faculty members endorsed some degree of collective responsibility for the behavior of their graduate students and also claimed some responsibility for their faculty colleagues. Yet, in practice, a relatively small percentage of faculty members believed that they actually exercise their responsibilities to their students (27 percent) or their colleagues (13 percent) to any great extent. Taken together, these results suggest that although misconduct is not rampant in the four scientific disciplines studied, it also cannot be said that it is rare.

One additional finding from the Acadia project is noteworthy. A significant percentage of students (53 percent) and faculty (26 percent) revealed that they were unlikely to report scientific misconduct because they were fearful of reprisals. This finding is consistent with the results of a recent survey of psychology interns and university faculty in which fear of retaliation emerged as the most frequently given reason for failure to report ethical violations (Holaday and Yost, 1993).

Some early research conducted by Azrin and his colleagues suggests that the falsification of data by young scientists and research assistants may be fairly widespread (Azrin et al., 1961). In one study, the researchers assigned a field experiment on verbal conditioning to 16 students who were to either reinforce the opinions stated by others (by agreeing with them) or to extinguish the opinions (by remaining silent). The procedure of the study, however, was intentionally designed by the researchers to be impossible to carry out in the prescribed way. Despite this fact, all but one of the students reported to their professor that they had successfully completed the experiment. After the one honest student described how it was impossible to run the experiment as intended, eight of the other students admitted to having significantly deviated from the prescribed data collection procedure. In a replication of the verbal conditioning study, 12 out of 19 students approached by a student confederate admitted to him that they had faked at least part of their data and 5 other students revealed that they had changed the required procedure to make it work (Azrin et al., 1961). Both of these studies were based on an earlier published study (Verplanck, 1955) in which students had collected data. Given the results obtained by Azrin et al., it may be that the original study was comprised of falsified data (Diener and Crandall, 1978).

Student researchers are especially susceptible to dishonest activities – such as falsifying data, making up data for subjects without having run them (a procedure commonly referred to as "dry labbing"), and altering the prescribed research procedure – for some of the same reasons that give rise to cheating in the classroom. Determinants of classroom cheating include pressures for good grades, student stress, ineffective deterrents, a belief that cheating is a normal part of life, and an overall diminishing sense of academic integrity (Davis et al., 1992). Further, there is evidence that cheating rates among students are as high as 75 percent to 87 percent (Baird, 1980; Jendreck, 1989), whereas detection rates are as low as 1.30 percent (Haines et al., 1986), suggesting that academic dishonesty is more likely to be reinforced than punished. Dishonest behavior that is reinforced in one academic context (the classroom) may readily generalize to other academic activities, such as the carrying out and reporting of research projects in laboratory courses. Thus, students who are pressured for time to complete their coursework may find it relatively easy to invent data that are consistent with the hypothesis or are similar to data collected from some actual subjects. Similarly, pre-scribed research procedures can easily be simplified to save time; for example, rather than taking the time to carefully administer written questionnaires to randomly selected participants, student researchers might instead obtain responses over the telephone from a few friends and relatives.

Because student research is generally unlikely to be published, the consequences of faking data are potentially less destructive than when such misconduct is perpretrated by a research professional. Nonetheless, success at falsifying data or other forms of misconduct when one is a student may increase the temptation to engage in such behaviors at a later time should one go on to work in a research field. It thus is important for students and research assistants to recognize the potential impact their work can have on scientific progress and the necessity for maintaining honesty in the reporting of their research investigations (see Box 9.4). Of course, it also is essential that young researchers be made aware that they could lose the opportunity to pursue a professional career if they are caught cheating. Just as students may be expelled from the university for cheating, professional researchers run the risk of personal disgrace and expulsion from the scientific community when they are caught falsifying their results.

Given the dire consequences that being found out may have on one's professional career, it stands to reason that the rewards can be great for the successful cheat in science. As suggested in Box 9.4, the motivations for these abuses are rather obvious, such as professional advancement, the awarding of research grants, and publications in respected scientific journals (Carroll et al., 1985). The rather low probability of being caught and the ease and opportunity with which data manipulation can be carried out also increase the likelihood of scientific misconduct. It should be added that the tendency to fake or invent data may be attributed to more subtle influences, such as self-deceit and rationalization. For example, researchers might unwittingly convince themselves that the data they have collected cannot possibly be accurate if they vary greatly from what was expected (McNemar, 1960); as a result, they may simply change the data to

---

| Box 9.4 |

Enhancing the accuracy of student research

Several suggestions have been offered for discouraging cheating in academic settings, including simply informing the students not to cheat, explaining why they should not cheat, distributing separate forms of exams or assignments, physically separating students (e.g. by rearranging seating assignments), and maintaining a close vigilance over students (Davis et al., 1992). Through a serendipitous finding, the author may have stumbled upon a method that enhances to some extent the accuracy of undergraduate student research reports. The method consists of carefully explaining to students prior to their carrying out a required research project that their grades will be independent of their results; that is, a research report could receive a high grade even if the hypotheses tested were not supported by the obtained data. It may be that many students incorrectly assume that in order to obtain a good grade for their research projects it is necessary to confirm their hypotheses.

The potential efficacy of this approach became apparent one semester following the assignment of a research project to two separate sections of an undergraduate course in social psychology that I was teaching. The study represented a simple between-subjects design intended to assess the effects of physical appearance on judgments of guilt and severity of recommended punishment for a fictitious student who had been accused of cheating on an exam. Students in the two class sections were provided with the stimulus material for carrying out the study, including two rudimentary drawings of an "attractive" and "unattractive" female student, a single description of the case of suspected cheating, and rating scales for assessing judgments of guilt and recommended punishment. The two drawings, which were intended to manipulate the physical appearance of the fictitious student, were somewhat crude and thus probably did not represent a very effective operationalization of the attractiveness variable. (In retrospect, actual photographs probably would have comprised a stronger manipulation.) Consistent with previous research (e.g. Efran, 1974), the student researchers were asked to test three hypotheses: (1) the attractive defendant will be judged as less likely to have cheated; (2) subjects will rate the attractive defendant as less likely to commit the transgression again; and (3) a less severe punishment will be recommended for the attractive defendant.

At the end of the semester, as I graded the completed research projects, I noticed that an interesting pattern was emerging. It appeared that the student researchers from one class were much more likely to have obtained support for the project's hypotheses than were students from the other section. Once I had finished grading all of the reports I examined this apparent difference more closely. In fact, my initial suspicions were confirmed: 75 percent of the students in one section obtained differences in the predicted direction (i.e. consistent with the hypotheses) as opposed to the remaining students who reported no apparent difference or differences in the opposite direction. In stark contrast, these percentages were reversed for students in the other section – only 25 percent reported differences in the predicted direction. Given the fact that both class sections had carried out the same research project, it was interesting to ponder what could have caused the varied results.

It eventually occurred to me that the two sections in fact had not received identical instructions for carrying out the study. For one class I recall clearly stating how grades for the research report would not at all be affected by the obtained results. I explained that the evaluation would be based on the clarity of the write-up, the adequacy of the literature review and discussion, and the procedure used to carry out the study. I also added a brief discussion of the importance of honor in science and how cheating (by falsifying research results) represents a serious breach of research ethics. Because of time constraints, this explanation was not provided to the other section and I failed to mention it during subsequent class meetings. Not surprisingly, the students who were not privy to the discussion on honesty and the caveat about the grading criteria comprised the section that was more likely to have obtained support for the hypotheses. Given the weak attractiveness manipulation used, it may thus be inferred that at least some of the students may have falsified or invented their data in the direction of the hypotheses in anticipation of receiving a better grade for the project. Of course, such an inference is merely tentative because of the *post hoc* nature of the comparison. It may be that the sections differed in important ways prior to assignment of the project, such that students in the class that tended to find support for the hypotheses were less honest to begin with than students in the other section. Another possibility is that the hypothesis confirming section may have taken greater care in carrying out the study and thus obtained the predicted effects despite the weak manipulation. Nonetheless, the serendipitous finding suggests a possible direction for future research on the effects of teaching students the importance of accuracy in research.

In practice, graduate students and professional researchers are well aware that rewards in the scientific community, such as publications and career advancement, are more likely to be forthcoming when research outcomes are statistically interesting. In essence, professional rewards are clearly linked to meaningful research outcomes (Altman, 1986). As a result, the strategy for enhancing scientific honesty described above would not be very useful for research professionals. However, for younger researchers at the undergraduate level, it could serve as a useful socializing mechanism for encouraging ethical scientific conduct.

what was expected. In short, scientists may become so committed to their pet hypotheses that they accept the hypotheses rather than the contradictory data when, in fact, they should do the opposite. This possibility is reflected in the following comment from a biochemist at the Max Planck Institute in Germany who admitted to having invented data which were published in various journal articles: "I published my hypotheses rather than experimentally determined results. The reason was that I was so convinced of my ideas that I simply put them down on paper" ("Researcher admits," 1977). A similar explanation has been offered to explain Burt's motivations for likely altering the results of his research on the heritability of intelligence (Green, 1992; Hearnshaw, 1979).

Whatever the causes of scientific fraud, it is clear that the consequences of such forms of misconduct can be great. In addition to the risks imposed on the unethical researcher (such as professional ostracism if caught or the possibility that the conduct will become more frequent if it is not exposed), violations of scientific honesty can harm science as well. Because the data reported in an investigation are assumed to have intellectual integrity and trustworthiness, falsified data can undermine science by adding purported facts to the existing body of knowledge within a discipline (Reese and Fremouw, 1984). Thus, false results are not merely worthless; they can contaminate scientific knowledge and mislead others working in the same field. In this light, it is no wonder that fraud perpetrated by scientists is considered a very serious offense within the scientific community. In the words of one eminent researcher, "any kind of falsification or fiddling with professedly factual results is rightly regarded as an unforgiveable professional crime" (Medawar, 1976).

Safeguards against scientific misconduct

Scientific research ideally is self-correcting; that is, there are many checks and balances that can lead to the detection of mistakes and falsified results. The potential harm caused by violations of scientific honesty can be undone by ethical researchers who attempt to replicate the reported results and fail. In effect, the replication of research findings (i.e. the repeating of an earlier study to see if its findings can be duplicated) and peer review (i.e. a critical evaluation of an investigation by qualified persons within a field) represent the standard means for guarding against dishonesty and error in science. In recent years, however, there has been growing skepticism about the ability of this system to safeguard against scientific misconduct (e.g. Altman, 1986; Sigma Xi, 1984; Weinstein, 1981). In short, it has been argued that current protections against scientific misconduct often fail because findings are infrequently checked in other laboratories and the peer review system frequently depends on misplaced trust ("Fraud in medical research," 1989).

Replication

It has long been assumed that replication provides a kind of "self-cleansing" function in the scientific process, quickly removing error and revealing faked results soon after they appear. Presumably, when important findings are published, they are promptly checked in other laboratories. Such was the case following the initial reports of high temperature superconductivity and low temperature nuclear fusion. Researchers in dozens of laboratories immediately checked and confirmed the initial superconductivity evidence, but failed to confirm the fusion results. Although there was no suspicion of fakery in the latter case, the failure to replicate the nuclear fusion findings suggested that there were mistakes in the work. Given this system of checking major studies, it is believed that researchers will refrain from fakery because they recognize that if their published findings are important, they will promptly be caught and exposed when attempts to replicate their work fail.

In theory, replication ought to serve as a deterrent against scientific dishonesty; in actual practice, there appear to be serious flaws in the effectiveness of this safeguarding mechanism. Not only are replication attempts relatively rare in science, but when they are carried out they may not have much of an impact on exposing inaccurate results. On the one hand, faked research findings indeed might be successfully replicated in cases in which the unethical researcher reached the correct answers through inappropriate means. On the other hand, replications of other scientists' work have become increasing less likely because fewer researchers have the competence to replicate (due to greater specialization in the sciences), funding agencies are unwilling to support "merely" replicative research, and one can expect little recognition as a result of devoting a substantial amount of time to replication research (Mahoney, 1976; Weinstein, 1981). Additionally, Fisher (1982) has suggested that replications may be especially rare in cases in which a single study has a strong impact but is so large in scale that a replication would be financially prohibitive. Another factor working against replication research is the reluctance of journal editors and reviewers to favorably consider studies for publication that do not demonstrate new effects. The results of a recent survey of 80 social science journal reviewers revealed just such a bias against replication studies, with several respondents explaining that they viewed replications as a waste of time, resources, and journal space (Neuliep and Crandall, 1993).

Replications are seldom easy to perform and their results may not be straightforward. Few researchers would be willing to accuse their colleagues of incompetence or dishonesty when the evidence of misconduct is not obvious (Sigma Xi, 1984), such as when the failure to replicate can be explained by other factors. After all, skepticism must be tempered by trust in the honesty of fellow researchers in order for scientific progress to proceed in a relatively smooth fashion (Bronowksi, 1956).

Peer review

As in the case of replication, the effectiveness of the peer review system as a mechanism for safeguarding science from fraud and other forms of research misconduct has been subject to increasing criticism. The conclusions of an investigative panel formed by the Institute of Medicine to study misconduct in biomedical research are typical of recent assessments of the quality of peer review at major research institutions. In its 1989 report, the panel identified an "excessively permissive" attitude by institutions which tended to allow the carrying out of careless and sometimes even fraudulent medical research ("Fraud in medical research," 1989). Few institutions were found to have explicit guidelines for evaluating research; as a result, individual investigators often failed to observe generally accepted practices and academic institutions were found to be limited in their ability to carry out objective evaluations of misconduct.

Peer review plays a central role in ensuring that the articles published by scientific journals are accurate, methodologically sound, and high enough in quality and innovativeness to merit communication to a larger scientific audience. In most cases, when a research manuscript is submitted to a scientific journal for publication consideration, it is evaluated by one's peers. Assuming the journal editor deems the research worthy of review, he or she will send the manuscript to a few experts (or "referees") for their criticisms and opinions on whether the work represents a contribution to the field and should be published. Such reviews typically are carried out under anonymous conditions; that is, the author's name and institution are left off the reviewed manuscript and the identity of the reviewers are kept secret from the author. Based on the experts' recommendations, the editor makes a final decision as to whether the manuscript should be accepted outright, accepted with revisions, or rejected (see Fogg and Fiske, 1993). It is assumed that this system of peer review prevents the publication of research errors and false claims, improves the quality of accepted papers, prevents duplication among scientific journals, and provides a fair means for the editor to select the most worthy submissions (Altman, 1986).

Unfortunately, adequate measures are unavailable for assessing the effectiveness of the peer review system in achieving the aforementioned goals in research publication (Altman, 1989). Thus, it is not possible to obtain accurate indications as to how many fraudulent research articles are exposed through peer review, how many actually survive the review process, and the extent to which good research is prevented from being published. Despite the fact that about three-fourths of the major scientific journals utilize peer review for assessing at least some of the articles they publish, there have been numerous calls within the sciences for improvements to the system. Researchers have complained about the unevenness of the review process (e.g. an article rejected by one journal may well be favorably evaluated by another) and the tendency for the typical journal review to be excessively time consuming (see Box 9.5). Another concern has to do with the possibility that personal, institutional, and gender biases may enter into the decisions of editors and reviewers. For example, in a recent survey of psychology and management research reviewers, Campion (1993) found evidence suggesting that reviewers employed in predominantly applied settings (such as consulting and management firms) tend to weight the practical importance of research articles somewhat higher and the theoretical importance somewhat lower than their counterparts in academic settings. These results are consistent with a commonly held assumption that academic-oriented scientists have a greater interest in the theoretical implications of research, while practice-oriented scientists tend to focus on the applied implications of research.

Other drawbacks attributed to the journal review process include the lack of systematic assessment of the adequacy of reviewers' comments; a tendency to publish only studies with findings that support the researcher's hypotheses; and a proclivity to publish research by prestigious authors (Altman, 1986; Cummings and Frost, 1985; Dickersin et al., 1992; Mahoney, 1976). A revealing anecdote relevant to the influence of a prestigious researcher's name on the likelihood of publication was conveyed by the son of the eminent nineteenth-century scientist, Lord Rayleigh (Strutt, 1924). Apparently, Rayleigh's name had either been omitted or accidentally detached from an article on electricity he had submitted to the British Association for the Advancement of Science. The paper was rejected by the committee as having been written by an eccentric theorist. When the authorship of the article was discovered, however, its merits suddenly became apparent and the publication decision was reversed.

Perhaps most troubling from an ethical perspective is the concern that the review system is open to abuse by unscrupulous reviewers who may appropriate ideas from rejected articles or make use of a paper's contents prior to its publication. Also, journal referees who are competing with the authors of reviewed papers for research

Box 9.5

To publish and not to publish

It is well known that a research manuscript may be negatively evaluated by one journal, only to be accepted for publication as an important contribution to the field by another journal (Bowen et al., 1972; Mahoney, 1976; Zuckerman and Merton, 1971). Perhaps even more surprising is evidence suggesting that research articles also may be differentially evaluated by the *same* journal on different occasions. This possibility was uncovered in an intriguing study carried out by Peters and Ceci (1982). In their investigation of the reviewing process within psychology, Peters and Ceci resubmitted previously published articles to each of twelve prestigious psychology journals, respectively. The articles included only cosmetic alterations, including author and institutional affiliation changes. Of the twelve articles resubmitted, only three were detected as resubmissions. Further, eight of the remaining nine articles actually were *rejected* by the journals that previously had published them a short time earlier. The researchers found no evidence that the editorial policies or rejection rates had changed for these journals since the original publication of the articles.

Peters and Ceci's findings, along with research reporting low inter-referee agreement for manuscript review, add credence to the often-voiced charge that the peer review system is capricious and inequitable. In his indictment of the publication process, Mahoney (1976) aptly asserted that the comments made by journal reviewers are often so divergent that one wonders whether they actually had read the same research manuscript. In light of recent research, we can now ask whether journal reviewers read the same manuscript the same way twice.

grants may unfairly evaluate their rivals' work or utilize the knowledge gained in the review process to their own advantage (Altman, 1986). Again, the extent to which reviewers are guilty of such conduct is impossible to gauge; nonetheless, such activities represent clear abuses of the trust bestowed on those who are invited to take part in the peer review system.

Several recommendations have been offered for improving the peer review process. One suggestion is that journals should develop more explicit guidelines for those who take part in the reviewing of manuscripts, including the establishment of criteria for the selection of referees and for the review of manuscripts. A clear procedure for authors to contest criticism of their work also has been called for. For some journals, explicit guidelines for review already have been established, along with steps to speed up the review process and better protect the anonymity of authors and reviewers. Another recommendation is to have more reviewers judge each manuscript (rather than the more typical two or three) in order to overcome possible biases attributed to the background characteristics of reviewers (Campion, 1993).

Some individuals within the research community have called for the examination of the methods and statistical analyses of submitted studies through audits of investigators' original records and data. In fact, certain journals now require authors to agree that their data will be kept available for consultation for a certain period of time, such as five years, after publication (St James-Roberts, 1976). The widespread adoption of such a requirement could prove informative, especially in light of the findings of an early investigation conducted by Wolins (1962). When Wolins requested original data from the authors of 37 published studies, he obtained data from only 7. When these data were reanalyzed, three included statistical errors large enough to influence the original research conclusions. Further, 21 of the researchers who were contacted claimed that their data had either been misplaced or destroyed. Assuming these results are not atypical, a review process that regularly includes experimental audits of submitted research could prove to be an effective deterrent against scientific misconduct.

Other issues in the publication of research

In addition to the problems inherent in the publication process relative to peer review which we considered in the previous section, researchers have certain ethical responsibilities when they attempt to publish their findings in scientific journals. Among the important issues in communicating knowledge to others that we have not yet considered are those dealing with the proper assignment of publication credit, protection of the anonymity of research participants, and the potential misuse of research results (see Box 9.6). Each of these issues is briefly addressed below.

Publication authorship assignments
Scholarly publications in the scientific disciplines matter, not only because their contents are assumed to add to scientific knowledge but also because the very act of publishing in reputable (e.g. peer-reviewed) journals is typically taken as a sign that a researcher is productive and is carrying out worthwhile work. In the disciplines of psychology, sociology, and political science, about 80 percent of the papers submitted for journal publication are rejected (in contrast to the physical sciences where acceptance rates are high) (Zuckerman and Merton, 1971). As a result, "getting published" generally is essential to the academic researcher's professional status and may be used as a criterion for continued employment and job tenure. Unfortunately, this often means that the *quantity* of publications may take on more significance than the *quality* of their content for individual researchers.

A professional emphasis on publication quantity can lead to ethically questionable conduct, such as attempts to publish the same research in multiple journals in order to "fatten" one's curriculum vitae (perhaps by making only minor modifications to the report or by changing its title so that it will not be obvious that the same research has been submitted to more than one journal). On this issue, most professional standards and journal publication policies are clear that the duplicate publication of original manuscripts or data, unless acknowledged as such, is unacceptable (e.g. see Box 9.2). The following statement taken from the *Personality and Social Psychology Bulletin* is typical of journal submission guidelines:

> Journal policy prohibits an author from submitting the same manuscript for concurrent consideration by another journal, and does not allow duplicate publication (i.e. publication of a manuscript that has been published in whole or in substantial part in another journal).

Multiple submissions are generally deplored by journal editors because republishing the same material unnecessarily overburdens the scientific literature and makes it more difficult for scientists to

---

| Box 9.6 |

Footnotes that somehow got left out of published manuscripts

As a means of elucidating some of the ethically questionable behavior associated with the communication of research, industrial-organizational psychologist Edwin Locke (1993), with tongue clearly in cheek, recently drew up a list of ten footnotes which have been "lost in the archives." The serious nature of such behavior notwithstanding, Locke's humorous footnotes are reproduced below.

Footnote
Number

1 The second author designed the study. The third author carried it out and wrote it up. The first author had the power.

2 Previous reviews by associate editors of three other journals, all of whom rejected the manuscript, totalled 27 single-spaced pages of comments, all of which we ignored. We don't thank any of them for their dumb comments.

3 Many of the references in this paper are totally unrelated to the topic of the study, but we added them to make the paper look scholarly.

4 The hypotheses were invented after-the-fact to explain the totally unpredicted and seemingly nonsensical results we obtained.

5 Sixty-seven subjects were discarded for non-compliance – with the hypotheses.

6 The original questionnaire included 100 predictor scales. This study reports the results for the five that worked.

7 We tried 37 different analytic techniques, some invented in ancient China. The one reported here (the Kawasaki Inverted-Listerine Analysis) was the only one that got significant results.

8 The full results of our 1/2-tailed tests are available, but see footnote number 9.

9 Our data are available for other scientists to look at. However, they are temporarily in Pakistan or Afghanistan, I am not sure which. Write us again in five years. ("Thank you for your belated inquiry about our data. Unfortunately they have been discarded, in Pakistan or Afghanistan, I am not sure which, because they are more than five years old." The Authors.)

10 We ran 12 pilot studies and finally got the design to work after threatening the subjects with bodily harm.

Source: Locke, E. A. (1993). Footnotes that somehow got left out of published manuscripts, *The Industrial–Organizational Psychologist*, 31, 2, p. 53. Reprinted by permission.

keep up, wastes the time of editors and reviewers, and delays the publication of original manuscripts.

More subtle issues also emerge in the publishing of scholarly research papers, with resolutions that are not always obvious. One of the more difficult areas involves the assignment of publication credit. Although at first glance this would appear to be a relatively straight-forward procedure, disputes concerning this issue are among the most common complaints to ethics committees initiated by academic researchers (Holaday and Yost, 1993; Keith-Spiegel and Koocher, 1985). A growing interest in the development of guidelines for authorship credit determination reflects an increase in multiple-authored journal articles (Garfield, 1978). A six-page research report appearing in a recent issue of the journal *Psychological Science*, for example, was attributed to no fewer than 13 authors (McCloskey et al., 1995).

In general, professional standards maintain that credits are to be assigned proportional to professional contribution, such that the main contributor to a work is listed first. Those who played a minor role in the research, such as persons who assisted in the coding of questionnaires or offered helpful ideas, are usually acknowledged in a footnote. In one survey, substantial agreement was obtained regarding authorship and footnote citation among psychologists who were asked to respond to a set of cases dealing with credit assignment (Spiegel and Keith-Spiegel, 1970). Although no consensus was reached, the respondents typically interpreted "principal contribution" in joint effort research consistent with the following rank-ordering: (1) the person who generated the hypothesis and design; (2) the person who wrote the material for publication; (3) the person who established the procedure and collected the data; and (4) the person who analyzed the data.

Publication credit tends to be an important concern among researchers for reasons that are tied to the vested interests of the individuals involved, such as the need to enhance one's career and to gain status and recognition among one's peers through publication output. "Senior" or first-named authors are often presumed to be the originators and principal contributors to the published work. In certain situations, ethical issues can emerge as a result of the power differential that often exists between collaborators on a research project (Fine and Kurdek, 1993). For example, researchers who have power and authority over their junior co-authors may take advant-age of their status to claim senior authorship despite having been only indirectly involved or minor contributors to the research.

Another possibility is that lower-status researchers who play a significant role in the research may not be listed by their senior collaborators as co-authors at all, but rather acknowledged in a footnote. Charges along these lines are occasionally voiced by graduate students who claim that their doctoral research advisors unfairly insist on being listed as co-authors on publications based on the dissertation. In some students' views, the advice they receive from their advisors constitutes a teaching obligation and thus should not be considered as a principal contribution (Keith-Spiegel and Koocher, 1985). Others have argued that faculty input into research might not be substantial enough to warrant authorship credit (Crespi, 1994).

In contrast to concerns about the possible exploitation of students in the determination of authorship credit, recent attention has focused on the issue of providing students with undue credit (Fine and Kurdek, 1993; Thompson, 1994). That is, some faculty researchers might be overly generous in assigning publication credit to promising students in hopes of boosting the students' formative careers. The ethical problem in such cases pertains to the potential risks involved, including false representation of the student's scholarly expertise, providing the student with an unfair professional advantage, and overly inflating expectations for the student's subsequent performance (Fine and Kurdek, 1993).

While recognizing the ethical implications of these issues, some researchers have advised that the assignment of authorship credit can be fair without requiring the formulation of rigid principles that introduce unnecessary precision, such as proposed point systems evaluating the magnitude of research contribution (see Fine and Kurdek, 1993). Thompson (1994) has suggested that the authorship order for research involving two authors, both of whom have made substantial contributions, can be determined in several mutually acceptable ways. For instance, first authorship could be randomly designated, a note could be added to indicate that the authors contributed equally, or it could be decided that one of the authors made a greater contribution. In other cases, such as faculty – student collaborations in which both parties contribute significantly to the work, Thompson recommends giving greater credit to the less established and less powerful contributor. In Thompson's view, while credit in scholarship may not always be fair or precise, the determination of authorship credit should be based on "mutually acceptable agreements that have a rational basis" (p. 1095). Disputes often can be avoided altogether when researchers decide on authorship credit at the outset of a collaboration.

Protecting the anonymity of research participants
The importance of maintaining the anonymity of those individuals who participate in research has been emphasized at various points in this book. Although a researcher may take great pains to protect participants' anonymity and guarantee the confidentiality of research results during the data collection stage of an investigation, it must be recognized that confidentiality can be lost in various ways when the research eventually is communicated to others. The American Psychological Association (1982) has identified several situations in which this may occur, including disclosure to third parties (such as parents or friends), to organizations of which the participant is a member (including employers, schools, and clinics), and to professional associates of the investigator. Confidentiality also can be lost as a result of a court order (e.g. when the investigator is legally required to provide information about research participants to the police and the courts) and through research publication.

In most of these cases, researchers often can avoid potential ethical dilemmas by obtaining the participant's permission prior to disclosure and by establishing a clear agreement with the participant and relevant third parties prior to the collection of data. Problems are likely to arise when associates of the participant (such as a teacher or therapist) claim a right to certain information obtained during an investigation, based on the grounds that the communication of data obtained under conditions of confidentiality can be used to the benefit of the participant. Although situations involving the loss of confidentiality on the basis of a court order may be more difficult to handle, it is advised that researchers investigating sensitive topics should explain the potential legal ramifications to the participant in advance and take care to store sensitive data in ways that make the identification of individual subjects impossible (American Psychological Association, 1982).

As mentioned, threats to participant anonymity and the confidentiality of data also can stem from the publication of research results. In behavioral science research in which data are likely to be presented in aggregate form (such as group averages), problems involving the identification of individual participants are rare. However, when case material on well-known persons or aggregate data on narrowly defined populations (such as organizations or small communities) is published, the identity of specific participants may be revealed, along with sensitive information about them. Although the risks of such revelations usually are not very great, these disclosures may be quite embarrassing to the individuals involved. Such was the case in a famous field study of the political and social life of a small town in upstate New York, identified by the fictitious name "Springdale" (Vidich and Bensman, 1958). Prior to carrying out their observations, the researchers assured the community's residents that no participants would be identified in published reports of the investigation and that their privacy and anonymity would be protected. In fact, a code of ethics was specifically devised for the purpose of safeguarding the integrity and welfare of the participants and for assuring that all data collected would remain confidential (Bell and Bronfenbrenner, 1959).

Despite the assurances given, publication of *Small Town in Mass Society*, Vidich and Bensman's (1958) book describing the Springdale project, was greeted by vehement objections from the townspeople. Although individuals mentioned in the book were given fictitious names, there were charges that many of the descriptions were embarrassing and failed to effectively disguise the identities of various members of the community. Residents claimed that the researchers' promise of anonymity had been betrayed by the published account of the study, that they had not been treated with respect and dignity in the account, and that the tone of the book was overly condescending and patronizing. Questions also were raised about characterizations of the attitudes and motives of various community members (Reynolds, 1979). The citizens of Springdale further reacted to the published report by publicly lampooning the researchers in a parade and by refusing future cooperation with any social scientists whatsoever, thereby destroying the possibility of replication and long-term follow up.

The Springdale case serves as a classic example of the potential consequences that can result when attempts to safeguard the anonymity of subjects in published reports are insufficient. When participants can recognize themselves in embarrassing or compromising descriptions in print, resulting negative effects may include a reduction in self-esteem and loss of confidence in the scientific process.

Preventing the misuse of research results
The final issue to be considered in regard to the communication of research findings has to do with the responsibilities of behavioral scientists to protect against the misuse of scientific knowledge derived from their studies. There is always the possibility that misunderstanding, incompetence, or vested interests will result in the improper use of research results by other researchers, business

enterprises, politicians, the media, or other individuals. Thus, special care must be taken when publicizing one's research to state conditions pertinent to its usefulness in both theoretical and applied contexts. Of course, simply stating the conditions of application and urging extreme caution in terms that even non-scientists can understand will not always guarantee that research results will be properly used.

Some illustrations of how the misuse of research results can arise have been provided by psychologist Stuart Cook (see Judd et al., 1991). One example involves misguided attempts to explain interracial group differences based on studies comparing the achievement test scores of blacks and whites. A beneficial outcome of such comparisons has been to stimulate a search for the causes and remedies of the documented inadequacies in the achievement of lower-class black children. However, some social scientists have warned of the inherent dangers of such research by suggesting that whites may inevitably interpret the results as indicative of inherited black inferiority, without considering numerous other explanations. Such an interpretation could then be used to support segregation and other racially discriminatory social policies. The heated debate that has ensued over recent publications pertaining to the decline of intelligence test scores, such as Herrnstein and Murray's *The Bell Curve* (1994) and Rushton's *Race, Evolution, and Behavior* (1994), has given rise to similar concerns about the consequences of research on group differences.

As another example of the potential misuse of research, Cook pointed to the research conducted on participative decision making in industrial and organizational settings. From this research various factors, such as participation in job-related decision making, have been linked to an increase in worker morale and job satisfaction. There are fears, however, that applications of this research in the development of new management approaches and leadership styles might result in the exploitation of workers. Given that improved morale is sometimes accompanied by an increase in worker productivity, employers could use the research in order to maximize company profits and to increase worker resistance to unionization.

As these two examples suggest, research results can be misinterpreted and used for purposes contrary to the investigator's intentions. Although it appears that many behavioral scientists now feel a greater ethical responsibility than in the past for the knowledge they produce, there is disagreement in the scientific community about the extent of that responsibility and the degree to which researchers should actively promote the utilization of their research findings. While some researchers believe the best they can do to prevent the possible misuse of their research is to anticipate likely misinterpretations and to counter them in their published reports, others choose to more actively oppose apparent misapplications of their work. Most behavioral scientists likely would agree that investigators should be held responsible for the adverse effects that result when they fail to inform decision makers of known negative consequences that can accompany application (Reynolds, 1979). The decision not to conduct research because of fears that the findings will be misused is likely to be an unsatisfactory one because of the possible loss of important research benefits (Haywood, 1976).

In recent years, researchers have found it necessary to improve their skills for interacting with journalists so that their research can more thoroughly and accurately be presented to the public through the mass media. Because of extensive public interest in topics and issues that often provide the focus of behavioral science investigations, researchers have increasingly been called on to communicate through mass media channels, including newspapers and magazines ranging from national papers to local weeklies, and radio and television shows that vary from science and public affairs programs to audience call-in shows (the latter in which the intent may be to have the scientist serve as little more than a scapegoat for public dissection). Some excellent advice is available to scientists interested in sharing their research and knowledge with the public via the mass media. For the interested reader, three publications are highly recommended: Gastel's *Presenting Science to the Public* (1983), Goldstein's *Handbook for Science Communication* (1986), and Miller's *The Scientist's Responsibility for Public Information* (1979).

Summary

Although the major focus of this chapter has been placed on the problems inherent in the scientific peer review system, this should not be interpreted as suggesting that the review process is so seriously flawed that it does not effectively provide protections against unethical conduct in the behavioral (and other) sciences. In fact, the general consensus among those who have carefully assessed the peer review system in science is that it appears to work quite well in most cases (cf. Altman, 1989; Fogg and Fiske, 1993). However, like most complex systems, peer review – whether it involves ethics committees, journal editors, research referees, or colleagues – is imperfect and, consequently, mistakes are made.

In recent years, apparent deficiencies in the approaches taken to protect the rights of research participants and the integrity of science have prompted attempts to ameliorate identifiable flaws within the peer review system. Steps have been taken by institutions to promote responsible research and to evaluate investigations of misconduct; academic departments and granting agencies have been encouraged to adopt new decision-making policies that emphasize the quality of a researcher's work, rather than quantity; and scientific journals have begun to institute policies intended to promote responsible author-ship practices ("Fraud in medical research," 1989).

These developments are typical of ongoing ethical progress in the behavioral sciences. Whenever ethical sensitivities have been raised as a result of an increased attention to moral issues within society or because of publicized research abuses, many positive changes within the scientific disciplines have been sure to follow. Yet one cannot expect too much from the further development of mechanisms for promoting and monitoring the ethicality of behavioral scientists, so long as there are researchers who deem it fit to place personal ambitions before scientific truth. Despite the progress that has been made to improve the process of peer review and the regulation of behavioral research, the burden of responsibility for ethical conduct continues to rest squarely on the shoulders of the researchers themselves.

Appendix 1
Research Principles for American Behavioral Scientists

Research Principles From the Ethical Principles of Psychologists and Code of Conduct *of the American Psychological Association*

6.06   *Planning Research*

(a)   Psychologists design, conduct, and report research in accordance with recognized standards of scientific competence and ethical research.
(b)   Psychologists plan their research so as to minimize the possibility that results will be misleading.
(c)   In planning research, psychologists consider its ethical acceptability under the Ethics Code. If an ethical issue is unclear, psychologists seek to resolve the issue through consultation with institutional review boards, animal care and use committees, peer consultations, or other proper mechanisms.
(d)   Psychologists take reasonable steps to implement appropriate protections for the rights and welfare of human participants, other persons affected by the research, and the welfare of animal subjects.

Source: Ethical Principles of Psychologists and Code of Conduct, *American Psychologist, 47,* 1597–1611.

6.07 *Responsibility*

(a) Psychologists conduct research competently and with due concern for the dignity and welfare of the participants.
(b) Psychologists are responsible for the ethical conduct of research conducted by them or by others under their supervision or control.
(c) Researchers and assistants are permitted to perform only those tasks for which they are appropriately trained and prepared.
(d) As part of the process of development and implementation of research projects, psychologists consult those with expertise concerning any special population under investigation or most likely to be affected.

6.08 *Compliance With Law and Standards*

Psychologists plan and conduct research in a manner consistent with federal and state law and regulations, as well as professional standards governing the conduct of research, and particularly those standards governing research with human participants and animal subjects.

6.09 *Institutional Approval*

Psychologists obtain from host institutions or organizations appropriate approval prior to conducting research, and they provide accurate information about their research proposals. They conduct the research in accordance with the approved research protocol.

6.10 *Research Responsibilities*

Prior to conducting research (except research involving only anonymous surveys, naturalistic observations, or similar research), psychologists enter into an agreement with participants that clarifies the nature of the research and the responsibilities of each party.

6.11 *Informed Consent to Research*

(a) Psychologists use language that is reasonably understandable to research participants in obtaining their appropriate informed consent (except as provided in Standard 6.12, Dispensing With

Informed Consent). Such informed consent is appropriately documented.

(b) Using language that is reasonably understandable to participants, psychologists inform participants of the nature of the research; they inform participants that they are free to participate or to decline to participate or to withdraw from the research; they explain the foreseeable consequences of declining or withdrawing; they inform participants of significant factors that may be expected to influence their willingness to participate (such as risks, discomfort, adverse effects, or limitations on confidentiality, except as provided in Standard 6.15, Deception in Research); and they explain other aspects about which the prospective participants inquire.

(c) When psychologists conduct research with individuals such as students or subordinates, psychologists take special care to protect the prospective participants from adverse consequences of declining or withdrawing from participation.

(d) When research participation is a course requirement or opportunity for extra credit, the prospective participant is given the choice of equitable alternative activities.

(e) For persons who are legally incapable of giving informed consent, psychologists nevertheless (1) provide an appropriate explanation, (2) obtain the participant's assent, and (3) obtain appropriate permission from a legally authorized person, if such substitute consent is permitted by law.

6.12 *Dispensing With Informed Consent*

Before determining that planned research (such as research involving only anonymous questionnaires, naturalistic observations, or certain kinds of archival research) does not require the informed consent of research participants, psychologists consider applicable regulations and institutional review board requirements, and they consult with colleagues as appropriate.

6.13 *Informed Consent in Research Filming or Recording*

Psychologists obtain informed consent from research participants prior to filming or recording them in any form, unless the research involves simply naturalistic observations in public places and it is not anticipated that the recording will be used in a manner that could cause personal identification or harm.

6.14   *Offering Inducements for Research Participation*

(a)   In offering professional services as an inducement to obtain research participants, psychologists make clear the nature of the services, as well as the risks, obligations, and limitations. (See also Standard 1.18, Barter [With Patients or Clients].)

(b)   Psychologists do not offer excessive or inappropriate financial or other inducements to obtain research participants, particularly when it might tend to coerce participation.

6.15   *Deception in Research*

(a)   Psychologists do not conduct a study involving deception unless they have determined that the use of deceptive techniques is justified by the study's prospective scientific, educational, or applied value and that equally effective alternative procedures that do not use deception are not feasible.

(b)   Psychologists never deceive research participants about significant aspects that would affect their willingness to participate, such as physical risks, discomfort, or unpleasant emotional experiences.

(c)   Any other deception that is an integral feature of the design and conduct of an experiment must be explained to participants as early as is feasible, preferably at the conclusion of their participation, but no later than at the conclusion of the research. (See also Standard 6.18, Providing Participants With Information About the Study.)

6.16   *Sharing and Utilizing Data*

Psychologists inform research participants of their anticipated sharing or further use of personally identifiable research data and of the possibility of unanticipated future uses.

6.17   *Minimizing Invasiveness*

In conducting research, psychologists interfere with the participants or milieu from which data are collected only in a manner that is warranted by an appropriate research design and that is consistent with psychologists' roles as scientific investigators.

6.18   *Providing Participants With Information About the Study*

(a)   Psychologists provide a prompt opportunity for participants to obtain appropriate information about the nature, results, and conclusions of the research, and psychologists attempt to correct any misconceptions that participants may have.
(b)   If scientific or humane values justify delaying or withholding this information, psychologists take reasonable measures to reduce the risk of harm.

6.19   *Honoring Commitments*

Psychologists take reasonable measures to honor all commitments they have made to research participants.

Research principles from the Code of Ethics *of the American Sociological Association (ASA, 1989)*

B.   *Disclosure and Respect for the Rights of Research Populations*

Disparities in wealth, power, and social status between the sociologist and respondents and clients may reflect and create problems of equity in research collaboration. Conflict of interest for the sociologist may occur in research and practice. Also to follow the precepts of the scientific method – such as those requiring full disclosure – may entail adverse consequences of personal risks for individuals and groups. Finally, irresponsible actions by a single researcher or research team can eliminate or reduce future access to a category of respondents by the entire profession and its allied fields.

1   Sociologists should not misuse their positions as professional social scientists for fraudulent purposes or as a pretext for gathering intelligence for any organization or government. Sociologists should not mislead respondents involved in a research project as to the purpose for which that research is being conducted.
2   Subjects of research are entitled to rights of biographical anonymity.

3 Information about subjects obtained from records that are opened to public scrutiny cannot be protected by guarantees of privacy or confidentiality.

4 The process of conducting sociological research must not expose respondents to substantial risk of personal harm. Informed consent must be obtained when the risks of research are greater than the risks of everyday life. Where modest risk or harm is anticipated, informed consent must be obtained.

5 Sociologists should take culturally appropriate steps to secure informed consent and to avoid invasions of privacy. Special actions may be necessary where the individuals studied are illiterate, have very low social status, or are unfamiliar with social research.

6 To the extent possible in a given study sociologists should anticipate potential threats to confidentiality. Such means as the removal of identifiers, the use of randomized responses and other statistical solutions to problems of privacy should be used where appropriate.

7 Confidential information provided by research participants must be treated as such by sociologists, even when this information enjoys no legal protection or privilege and legal force is applied. The obligation to respect confidentiality also applies to members of research organizations (interviewers, coders, clerical staff, etc.) who have access to the information. It is the responsibility of administrators and chief investigators to instruct staff members on this point and to make every effort to insure that access to confidential information is restricted.

8 While generally adhering to the norm of acknowledging the contributions of all collaborators, sociologists should be sensitive to harm that may arise from disclosure and respect a collaborator's wish or need for anonymity. Full disclosure may be made later if circumstances permit.

9 Study design and information gathering techniques should conform to regulations protecting the rights of human subject, irrespective of source of funding, as outlined by the American Association of University Professors (AAUP) in "Regulations Governing Research On Human Subjects: Academic Freedom and the Institutional Review Board," *Academe*, December 1981: 358-370.

10 Sociologists should comply with appropriate federal and institutional requirements pertaining to the conduct of research. These requirements might include but are not necessarily limited to

failure to obtain proper review and approval for research that involves human subjects and failure to follow recommendations made by responsible committees concerning research subjects, materials, and procedures.

Appendix 2: Psychology Codes of Ethics: An International Survey

Discussions of ethical issues in psychological research have largely been limited to conferences and publications in scientific journals within the United States and other English-speaking nations. Only within the past 15 years or so has an interest in these issues begun to emerge in other parts of the world. In 1977 German psychologist Heinz Schuler surveyed a number of the professional organizations for psychology around the world and obtained information about the existence of ethical codes in nine nations, including the United States, Canada, the Federal Republic of Germany, Great Britain, the Netherlands, Poland, Austria, Sweden, and France. According to Schuler, with the exception of the American ethical code, most of the others were labeled as preliminary and in preparation for revision. However, three basic principles were identified in all of the ethical codes for psychological research: (1) protection from physical harm; (2) protection from psychological harm; and (3) confidentiality of data (Schuler, 1982).

In light of recent ethical developments within the professions, the author conducted an informal follow-up survey of international ethics codes by obtaining the current codes of professional conduct from the following countries included in Schuler's review: Canada (Canadian Psychological Association, 1991), Great Britain (the British Psychological Society, 1995a), Germany (German Association of Professional Psychologists, 1986), France (French Psychological Society, 1976), the Netherlands (Netherlands Institute of Psychologists, 1988), and the United States (American Psychological Association, 1992). Additionally, codes were obtained from the Australian

Psychological Society (1986), the General Assembly for the Scandinavian Psychological Associations (1989), the Psychological Association of Slovenia (1982), the Spanish Psicólogo (Colegio Oficial de Psicólogos, 1987), and the Swiss Federation of Psychologists (Fédération Suisse des Psychologues, 1991). In most cases, these codes are generic in nature and provide the standards of conduct for psychologists who are engaged in a variety of professional activities in addition to research, such as consulting and therapy. A summary of the content of the codes is presented below, focusing especially on the research principles contained within them.

United States

The activities of the American Psychological Association in promulgating an ethics code for American psychologists were described at length in chapter 2. The interested reader should consult that chapter for a description of the current APA research principles and the process by which they were developed. Much attention is given to the APA principles in this text because they have served to a great extent as a model upon which other ethical codes are based.

It should be noted that another important American psychological association, the American Psychological Society (APS), has chosen not to draft a separate code of ethics for its members. This decision was largely based on considerations regarding an enforcement mechanism that would need to be developed as part of a code and the possibility of lawsuits stemming from judgments pertaining to ethical conduct (A. G. Kraut, personal communication, December 8, 1994). In lieu of a formal ethics code, the APS Board of Directors recently approved a brief, general statement of principle that is consistent with the society's interests in scientific and personal integrity but which does not put APS in the position of judging the conduct of its members. The statement reads as follows:

> The Board of Directors expects APS members to adhere to all relevant codes of ethical behavior and legal and regulatory requirements.

Australia

The current *Code of Professional Conduct* of the Australian Psychological Society was adopted in August 1986 and represents a revision of an original code dating back to 1968. Several amendments to the

1986 standards were approved in September 1990 and October 1991. The code sets forth three general principles intended to safeguard the integrity of the profession and the welfare of consumers of Australian psychological services. The principles pertain to (1) responsibility (involving the consequences of professional decisions), (2) competence (in the activities that individual psychologists choose to undertake), and (3) propriety (with regard to the treatment of clients, students, and research participants). In essence, the three Australian principles are akin to the 1992 APA *Principles*, in that they represent unenforceable ideals of the profession. The generic principles are followed by sections providing specific applications in the areas of assessment procedures, consulting relationships, teaching of psychology, supervision and training, research, public statements, and professional relationships.

Section E of the Australian code contains 11 statements for human subject and animal research. The similarity of these statements to previous versions of the APA research guidelines is striking. Responsibility is placed on the investigator for carefully evaluating the ethicality of a planned study; protection of research participants from psychological risks and disclosures of confidential data is emphasized; and the necessity to obtain participants' voluntary informed consent and to debrief them at appropriate stages of the investigation are included as essential elements of ethical conduct. Two statements specifically pertain to the need for researchers to treat laboratory animals humanely and with procedures that minimize discomfort and pain.

As a guide for assisting the investigator in ethical decision making, an appendix is provided with additional principles and general guidelines relating to psychological research with human participants. The principles point out the necessity for researchers to balance the costs to subjects against the welfare of others who might benefit from the results of a study and suggest that psychological investigations can be classified into three ethical categories.

Category 1: no ethical problems inherent in procedures used

This category includes benign research procedures that pose little, if any, physical or mental discomfort or risks to subjects. Experiments carried out to investigate basic psychological processes in the areas of perception, information processing, language acquisition, and learning represent examples of category 1 research. According to the accompanying discussion, more than 90 percent of Australian psychological research with human subjects is of this nature.

Category 2:   ethical questions raised by procedures used

The sorts of investigations classified into this category are those that involve deceptive procedures or that impose a certain degree of physical or mental stress or discomfort onto subjects (for example, as a result of conditions of sleeplessness, sensory deprivation, or physically painful stimuli).

Category 3:   complex and/or difficult ethical problems raised by procedures used

This category includes the small percentage of research investigations posing "unusual and/or difficult ethical problems." Listed as examples of category 3 research are simulated prison experiments, obedience studies in which subjects are ordered to behave in callous ways, long-term sensory deprivation studies, and investigations involving the administration of non-prescription drugs.

A prescribed action plan is provided for the investigator depending on how one's study is classified. If a research project falls into category 1, the researcher accepts full responsibility for the investigation and is expected to keep on file a careful description of the procedures used. For category 2 research, the investigator is required to seek the advice of colleagues on proper ethical safeguards prior to conducting the study. And if a planned investigation falls into category 3, the researcher, in addition to consulting with colleagues, must also request that the Committee on Ethical and Professional Standards of the Australian Psychological Society establish an advisory committee for overseeing the investigation.

Canada

The *Canadian Code of Ethics for Psychologists* was adopted by the Canadian Psychological Association (CPA) in 1986 and was revised for the first time in 1991. A *Companion Manual* consisting of a variety of materials to assist in the interpretation and application of the code first appeared in 1988. The development of the code began during the late 1970s in response to a growing interest among various Canadian groups in having an ethics code that reflected the realities of the nation's research culture (Sinclair et al., 1987). Prior to 1977, Canadian psychologists were expected to adhere to the APA standards for American psychologists, which were adopted with minor changes in wording. The Canadian code was intended, in large part, to further establish the identity of Canadian psychology as a profession and to provide a statement of moral principle for assisting the profession in meeting its responsibilities and for guiding the individual in resolving ethical dilemmas.

A novel approach was utilized to develop the Canadian code. The process began with a careful examination of the 1977 version (and later revisions) of the APA code by a standing CPA committee. Thirty-seven hypothetical ethical dilemmas representing the applied, teaching, and research functions of psychologists then were developed. The dilemmas were used not only to explore how a representative group of CPA members would have responded in the hypothetical situations, but also to gauge the reasoning and ethical principles used in their decision making. Although this approach contrasts with that utilized by the APA, in which psychologists were asked to identify ethical dilemmas they actually had encountered in their research activities, the hypothetical dilemmas were intended to reflect all of the APA ethical principles in place at that time. Of the 125 Canadian psychologists sampled, 59 respondents completed questionnaires that included between two to four of the hypothetical cases. The rationales that were provided for selected courses of action in resolving the dilemmas were content analyzed and subsequently categorized into groups of statements representing specific superordinate principles. On this basis, a resulting document emerged containing ethical standards organized around the four ethical principles that were used most consistently by the respondents to resolve the hypothetical dilemmas: (I) respect for the dignity of persons; (II) responsible caring; (III) integrity in relationships; and (IV) responsibility to society.

The principles are ordered according to the weight each should be given when they conflict, with the highest weight generally given to principle I, "Respect for the dignity of persons," because of its emphasis on individual moral rights, and the lowest weight to principle IV, "Responsibility to society." In other words, although it is assumed that psychologists have a responsibility to work for the benefit of society in ways that do not violate individual rights, when individual welfare conflicts with benefits to society the former is to bear greater ethical weight in decision making.

Each of the code's four principles is followed by a detailed values statement which serves to give definition to the principle. For example, the statement corresponding to principle I ("Respect for the dignity of persons") emphasizes the responsibility of psychologists to "Respect the dignity of all persons with whom they come in contact in their role as psychologists," especially those in vulnerable positions, such as research participants, clients, and students. A list of specific ethical standards follows each values statement. The standards, which were developed on the basis of an extensive review of ethical codes from other countries and the psychological literature pertaining to ethical guidelines, are intended to illustrate how the principles and values can be applied to the activities of psychologists.

Specific ethical guidelines pertinent to research conduct appear throughout the Canadian code as standards associated with each of the four superordinate ethical principles. Listed under the principle of respect for the dignity of persons are informed consent, freedom of consent, vulnerabilities, privacy, and confidentiality. Standards pertaining to risk/benefit analysis, the maximizing of benefits and minimizing of harm, correcting harm, and the care of animals are associated with the principle of responsible caring. The principle of integrity in relationships includes standards pertaining to accuracy/honesty, objectivity/lack of bias, and avoidance of deception. The principle of responsibility to society includes standards associated with the development of knowledge. Some examples of these research-related principles appear in Box A2.1.

The 284-page *Companion Manual* is a comprehensive guide that was developed to assist users of the CPA ethics code. The 1991 manual reproduces the code along with a running interpretive commentary, followed by a discussion of how the principles and standards might be applied to resolve ethical dilemmas. In the latter section it is suggested that ethical decision making can occur on three levels: (1) ethical behavior that is virtually automatic (that is, ethical decisions that may not appear as such to the decision maker, such as the routine preparation of consent forms prior to an impending research investigation); (2) choices that can be made relatively easily by reference to the code (in situations in which training and experience have not fully prepared the psychologist, as when a researcher needs to know the elements that constitute adequate informed consent); and (3) dilemmas in which ethical principles seem to conflict (the most difficult situations for which there is no specific standard to guide resolution of the dilemma). Suggestions as to how one might go about resolving difficult ethical dilemmas are provided in the context of a discussion of seven steps for ethical decision making, which appear in the preamble to the code (see Table A2.1).

The manual also includes an extensive bibliography on ethics, a set of vignettes of ethical dilemmas, and additional guidelines pertaining to the use of animals in research, instruction in psychology, and other areas of conduct. The CPA *Companion Manual* may well be the most

---

Box A2.1

The Canadian Code of Ethics for Psychologists: examples of research-related standards

PRINCIPLE I: RESPECT FOR THE DIGNITY OF PERSONS
In adhering to the Principle of Respect for the Dignity of Persons, psychologists would:

Informed Consent

I.14 Obtain informed consent for all research activities which involve obtrusive measures, invasion into the private lives of research participants, risks to the participant, or any attempt to change the behavior of research participants.

I.15 Establish and use signed consent forms which specify the dimensions of informed consent or which acknowledge that such dimensions have been explained and are understood, if such forms are required by law or if such forms are desired by the psychologist, the person(s) giving consent, or the organization for whom the psychologist works.

I.18 Assure, in the process of obtaining informed consent, that at least the following points are understood: purpose and nature of the activity; mutual responsibilities; likely benefits and risks; alternatives; the likely consequences of non-action; the option to refuse or withdraw at any time, without prejudice; over what period of time the consent applies; and, how to rescind consent if desired.

Privacy

I.32 Explore and collect only that information which is germane to the purpose(s) for which consent has been obtained.

Source: Canadian Psychological Association (1991). *Canadian Code of Ethics for Psychologists*. Old Chelsea, Quebec.

Table A2.1 *The Canadian Code of Ethics for Psychologists*: steps for ethical decision making

1 Identification of ethically relevant issues and practices
2 Development of alternative courses of action
3 Analysis of likely short-term, ongoing, and long-term risks and benefits of each course of action on the individual(s)/group(s) involved or likely to be affected (e.g. client, family or employees, employing institution, students, research participants, colleagues, the discipline, society, self)
4 Choice of course of action after conscientious application of existing principles, values, and standards
5 Action, with a commitment to assume responsibility for the consequences of the action
6 Evaluation of the results of the course of action
7 Assumption of responsibility for consequences of action, including correction of negative consequences, if any, or re-engaging in the decision-making process if the ethical issue is not resolved

Source: Canadian Psychological Association (1991). *Canadian Code of Ethics for Psychologists*, Old Chelsea, Quebec.

comprehensive guide for ethical conduct in psychology in existence today. It serves as a potentially effective tool for meeting the basic objectives of the CPA code: "to give more explicit guidelines for action when ethical principles are in conflict" and "to explicitly reflect the most useful decision rules (i.e. ethical principles) for ethical decision making."

France

A document entitled *Code de Deontologie* was adopted by the French Psychological Society (Société Française de Psychologie) in May 1961 (article 7 was added in May 1976) and is applicable to all members of the Society as a code of professional conduct. The code is organized into eight separate sections, with article 0 (field of application) describing to whom the code applies, and articles 1 through 7 detailing deontological principles in general domains of applied and research-related activities.

As stated in the preamble, a primary purpose of the code is to facilitate the study and resolution of theoretical and practical problems in the science and application of French psychology. The seven core articles of the code, with a brief description of the coverage of each, are as follows: (1) ethics (emphasizes human dignity, the recognition of society's ethical norms, and the importance of objectivity in the labeling of individuals as normal, adaptive, etc.); (2) professional secrets (stresses the importance of maintaining confidentiality of information obtained from others in practice or research); (3) respect for others (details the importance of protecting others from physical or psychological harm, undue manipulation, or restrictions of autonomy); (4) science (requires that the researcher must maintain an awareness of advances in the field, apply appropriate scientific methodologies, and communicate findings as completely and accurately as possible); (5) technical autonomy (emphasizes that the psychologist must be autonomous in the use of his or her techniques, refusing commitments that exceed the present state of knowledge or application, and maintaining responsibility for the choice of methods employed); (6) professional independence (maintains that psychologists should avoid work conditions in which the application of the ethical principles would not be possible or else threaten their professional independence); and (7) international ethics (condemns the use of notions such as "normal" or "pathological" when it has repressive goals in social or political contexts, whatever the country).

An examination of the French principles reveals that little attention is devoted to the conduct of research or specific research principles such as informed consent and debriefing. Instead, emphasis is placed on the fundamental rights of individuals with whom psychologists interact in their helping functions. Also recognized are the basic rights of the psychologist to engage in professional activities, while taking care not to exceed the professional boundaries as defined by current knowledge in the field. The current principles, while reflecting a greater interest among French psychologists in clinical practice rather than human subject research, are general enough to apply broadly both to practical and scientific endeavors and at least serve to sensitize psychologists to the importance of moral conduct as they engage in their professional activities.

The situation for French psychologists is somewhat complicated by the fact that their research is now covered by French law for medical research (e.g. the recently revised *Code de la Santé Publique*). As a result, researchers are more apt to consult current legal standards than the deontological code of their profession. It also should be noted that some French psychologists have begun to give attention to ethical issues in the conduct of psychological experimentation and the need for more specific ethical principles to guide research psychologists (e.g. Beauvois et al. 1990). In fact, initial steps

Appendix 2

have been undertaken towards developing a more comprenensive
and utilitarian ethical code than the one currently in place
(M. Carlier, personal communication, April 12, 1995). Intended for
1997, the new code is expected to include specific ethical principles
pertinent to seven areas of psychological activity, including
research.

Germany

The code of ethics for German psychologists came into force in 1986
following a three-year period of development by an ethics commission headed by Heinz Schuler. (A new edition with minor revisions
was published in 1989.) The previous code, *Professional Obligations
for Psychologists*, had been in place since 1967. Like its predecessors,
the new *Professional Code of Ethics for Psychologists* was modeled
after the APA ethical principles. In addition to including a generic set
of ethical principles, the German code of ethics also contains interpretations of a number of statutory requirements affecting German
psychologists, such as the Pharmaceutical Publicity Law (which
prohibits clinical psychologists from advertising), the "duty of
silence" (i.e. professional secrecy), and the adoption of professional
titles.

The German ethics code is organized into twelve main sections,
including a preamble (section I) which describes in general terms the
psychologist's obligation to the profession, responsibility to those
who place trust in his or her professional competencies, and the
necessity to maintain competence by keeping informed of developments in the field. The principles that most closely pertain to research
appear in the section on "research and training" (Section X) and are
grouped into three subsections: (1) planning the research project;
2) research procedure; and (3) publication of experimental results.
These research principles are broadly stated and are accompanied by
only a few illustrative examples.

The research principles underscore the importance of recognizing
during the planning of research the potential effects of an investigation on individuals, groups, or institutions involved in or directly
affected by the project. Offered as an example is the possibility that
scientific psychology could serve to influence public opinion and the
development of social values. Also relevant to the planning of
research is the necessity for the investigator to weigh scientific
objectives against extra-scientific outcomes (such as the welfare of

research participants). The principles also encourage the researcher to consider alternative scientific methods during the planning of research and to select the most suitable approach.

In terms of research procedure, the principles listed reflect familiar concerns: the researcher's responsibility to his or her subjects; the necessity to take special precautions when subjects are impaired (such as addicts) or else placed in a research situation that reduces their ability to act responsibly; the need to guarantee the safety and welfare of research participants and to eliminate unforeseeable risks to the greatest possible extent; and the requirement that participation in research is based on voluntary informed consent. It also is stated that experiments in which subjects show signs of unexpected stress reactions must be halted and attempts should be made to have any undesirable consequences of participation eliminated. A final principle emphasizes the ethical treatment of animals used for research purposes.

The third set of principles relevant to research procedure has to do with publication of research findings, focusing on the necessity of (1) communicating one's results to the professional public in a clear and complete manner; and (2) acknowledging all of the principal contributors to the research project. The code is undergoing another revision which is expected to be completed in the near future (H. Schuler, personal communication, March 22, 1993).

Great Britain

The British Psychological Society's (BPS) 1995 *Code of Conduct*, in contrast to the ethical codes considered thus far, is unique in the sense that it gives primary attention to research. The first section of the code consists of a set of general principles of conduct in professional practice, covering issues related to competence, the obtaining of consent, confidentiality, and personal conduct. Other sections of the code provide guidelines for the use of non-sexist language, the advertising of psychological services, and the use of professional titles.

The remainder of the code is comprised of standards for the ethical conduct of human and animal research. The Society's "Ethical principles for conducting research with human participants" and "Guidelines for the use of animals in research" appear as two separate sections. The 15 animal research guidelines were jointly proposed by the Scientific Affairs Board of the BPS and the Committee of the Experimental Psychology Society, having been based on guidelines previously developed by two British animal behavior societies. The animal guidelines, in place since 1985, are discussed in chapter 8 of this volume.

The current principles for human research are a result of the first major revision of the 1978 "Ethical principles." The revision was carried out by a BPS Standing Committee on Ethics in Research with Human Participants and was approved in February 1990. An introduction to the revised principles describes some of the issues that concerned the Committee during the revision process. Guiding the Committee during its revision was the recognition that psychologists owe a debt to the individuals who participate in research and that in return those participants should expect to be "treated with the highest standards of consideration and respect."

The issues of deception, debriefing, and risk were of primary concern during the revision process. While acknowledging that many persons view the use of research deception as inappropriate conduct, it was recognized that many psychological processes could not be studied if individuals were fully aware of the research hypothesis in advance. In its attempt to resolve this issue, the Committee distinguished between withholding some details of the research hypothesis and deliberately providing false information to research participants. It was concluded that researchers should attempt to provide as much information as possible to research participants and that if deception must be used, the true test of its appropriateness should be the reaction of participants once the deception was revealed to them. The importance of debriefing research participants also was emphasized, especially in cases in which information has been withheld or falsely provided.

In dealing with the issue of risk, it was recognized that participants must be protected from undue risk, but that to prohibit all research involving risk would serve to make much important research impossible. Thus, the Committee chose to define "undue risk" in terms of the risks that individuals are apt to encounter in their normal lives. This decision was intended to reduce the likelihood that individuals would be induced to take part in research involving greater risks than those normally encountered in their everyday lives.

These various considerations are reflected in the "Ethical principles" document, where principles are organized into the following sections: (1) introduction; (2) general; (3) consent; (4) deception; (5) debriefing; (6) withdrawal from the investigation; (7) confidentiality; (8) protection of participants; (9) observational research; (10)

giving advice; and (11) colleagues. Each section consists of one ("Colleagues") to nine ("Consent") statements clarifying appropriate conduct. For example, statement 4.2 pertains to proper conduct regarding the use of deliberate deception:

> Intentional deception of the participants over the purpose and general nature of the investigation should be avoided whenever possible. Participants should never be deliberately misled without extremely strong scientific or medical justification. Even then there should be strict controls and the disinterested approval of independent advisors.

As is apparent in this principle, investigators are urged to consult with others when they are unclear as to the appropriate course of conduct in their research endeavors.

The Netherlands

As stated in the Preamble to the 1988 *Professional Code for Psychologists* of the Netherlands Institute for Psychologists (Nederlands Instituut voor Psychologen, NIP), the present Dutch ethical code has a long history behind it, dating back to 1957 when work first began on a professional code. An initial code was accepted in 1960 and underwent minor adjustment in 1976. The 1988 document exists as a complete revision of the earlier versions and is intended to apply to all psychologists in the Netherlands, including one-third of the practicing psychologists who are not members of the NIP. (Professional sanctions are only possible in the case of psychologists who are members of the NIP.)

The Netherlands code is primarily oriented to ethical issues related to professional practice, although the guidelines are intended to apply to all professional relationships involving treatment, research, advice, or counseling. The code is comprised of an extended preamble, describing the functions of the principles (defined here as "rules") and assumptions underlying the development of the rules. In short, the rules are intended to serve in the best interests of those both inside and outside of the profession, by offering guidelines for proper conduct within the ranks of professional psychology and by protecting those who enter into a relationship with a psychologist from unethical conduct. At the heart of the code is the underlying principle of respect for human beings and the implied necessity for psychologists to exercise "the expected prudence" in all their professional relationships.

In addition to the preamble, the code consists of a set of "general rules," including statements pertinent to dignity, expertise, equality, confidentiality, accountability, acceptability, image, and relationship with colleagues. Additional rules are grouped under the following headings: rules on entering into a professional relationship and information to clients; freedom to take part; interdisciplinary cooperation, cooperation between colleagues and assistance; rules during the professional relationship; file; reporting in relationships involving advice, counseling or research; and supply of information to a third party. Rules pertaining to research are scattered throughout these sections and, as may be apparent from the section headings, largely focus on the issues of consent and confidentiality of data. These guidelines tend to be generally stated in order to apply to more than one professional activity. (In fact, the term "client" is used throughout the code to apply to any person who is the subject of research, advice, counseling, or treatment.) For example, rule 5.1–1, which deals in part with the issue of protection from harm, can apply to a research, clinical, or counseling professional relationship:

> During the professional relationship the psychologist shall not employ methods that are in any way detrimental to the client's dignity or that penetrate into the client's private life deeper than necessary for the objectives set.

With further regard to risk, the rules do not prohibit intentional exposure of individuals to negative experiences so long as several conditions have been fulfilled. Such conditions are spelled out in other rules, including those for informed consent and the right of the participant to withdraw from a psychological relationship at any time. Rule 2.4 reads as follows:

> The psychologist shall give information with regard to the nature and the objectives of the professional relationship to the client with whom he intends to enter into a professional relationship. The psychologist shall do anything within his power to ensure that the client is entirely free to decide in a responsible manner whether to enter into the professional relationship.

With principles as broadly stated as these, the Netherlands rules are no doubt somewhat difficult to apply in resolving specific research dilemmas. A statement from the preamble, recognizing the right of research psychologists to pursue knowledge in whatever ways they view as most appropriate, places the ethical responsibility for resolving specific research dilemmas firmly in the hands of the investigator.

One additional point about the Dutch code is that it includes a supplementary section specifying disciplinary measures that may be taken following a formal complaint against a member of the NIP. Consistent with most professional psychological societies, such measures ultimately can result in the psychologist's suspension or expulsion from membership in the professional organization.

Scandinavia

An ethics code for members of the psychological associations in the Nordic countries of Denmark, Norway, Finland, Sweden, and Iceland was approved by the General Assembly for the Scandinavian Psychological Associations in 1989. The code, *Ethical Principles for Scandinavian Psychologists*, is intended to regulate the professional activity of all Scandinavian psychologists, but is administered and interpreted separately by the psychological associations within each participating country.

The composition of the Scandinavian code is similar to the makeup of the ethics codes described above. The specific principles are preceded by a preamble, which explains that the principles were formulated to support psychologists in their attempts to resolve ethical questions, to protect clients and research participants from unethical treatment, and as a means for maintaining confidence in psychological research and practice. All facets of professional psychological activity are covered by the principles, including teaching, research, clinical work, consultation, and work in various applied settings. The principles are organized into eight sections, each of which covers a fundamental concern relating to important aspects of psychologists' professional activities: (1) responsibility; (2) competence; (3) obligations towards clients; (4) confidentiality; (5) psychological methods, investigations, and statements, (6) public statements; (7) professional relationships; and (8) research. Each specific principle within these sections is followed by a set of more detailed clarifications and examples. For example, the section on confidentiality includes one broadly stated principle emphasizing the importance of respecting the individual's right to confidentiality, followed by 11 clarifying points which stipulate how the principle should be applied in some specific circumstances. Section 8, which pertains to research, is comprised of five principles emphasizing psychologists' obligations to investigate important research questions in scientifically sound ways, to protect the well-being of research participants, to interpret the meaning of their research clearly, and to take precautions so that their animal subjects are not subjected to unnecessary suffering. The 11 clarifying points in Section 8 concern the importance of evaluating costs and benefits prior to carrying out a research project, the necessity to have research proposals reviewed by an ethics committee if there are doubts about their ethicality, and include requirements pertaining to anonymity, voluntary participation, and the obtaining of informed consent.

Slovenia

A *Code of Ethics for Psychologists* was adopted by the Psychological Association of Slovenia (Slovene Psychologists' Society) in 1982. The code consists of a list of 14 principles intended to serve as guidelines for psychologists in their theoretical and practical work, consistent with an overriding respect for humanity and the common goal of benefiting human welfare. An emphasis on human dignity, democratic principles, the integrity of the human personality, and the fostering of human relations is apparent throughout the code. The code is divided into sections that tend to emphasize ethical principles in the professional practice of psychologists, such as "psychological aid," "competence," "confidentiality," "professional independence," and "relations with the public." A unique section on "language" specifies the need for psychologists to take special care in their use of correct Slovene language in their written and verbal activities.

Principle 11, which comprises the section labeled "research," consists of four short paragraphs specifically dealing with research concerns. Voluntary informed consent is described as the basic ethical principle in the conduct of human subject research. Also emphasized is the requirement that psychologists must carry out their research under the auspices and authority of their working organizations. Complete anonymity must be assured when cases from practice are incorporated as part of a research presentation. Although the section on research is brief, it is recognized that other principles within the code (such as the one pertaining to the obligation to respect the confidentiality of information revealed in the course of psychological work) pertain to research as well as practice. Another relevant point is that the misuse of professional knowledge for the purpose of personal gain is viewed as a serious violation of the ethical principles.

Spain

The *Deontological Code of the Psychologist* (*Código Deontológico del Psicólogo*) was ratified by a governing body of the Colegio Oficial de Psicólogos in 1987 and received final approval in 1993, following further study and public comment. The Spanish code, which is unavailable in English, consists of 65 "articles" appearing within eight general areas and an introductory section. Section 4 ("De la investigacion y docencia") pertains specifically to research and teaching activities and is made up of six articles (numbers 33 to 38).

The first of the research articles states that an investigation should contribute to the progress of science and be carried out and communicated in scientifically sound ways. The remaining articles in section 4 describe, respectively, the researcher's responsibility (1) to protect research participants from irreversible or permanent harm; (2) to take care not to coerce persons into participating in studies involving temporary discomfort, by obtaining their informed consent and allowing them to discontinue participation at any time; (3) to debrief individuals who have participated in research involving deception; (4) to respect the dignity of subjects and their sense of decency or modesty, especially when the research involves the investigation of sensitive behaviors; and (5) to protect animal subjects from unnecessary suffering, harm, and pain. Standards pertaining to the confidentiality of information obtained from others and issues of professional secrets appear among the 11 articles in section 5 ("De la obtencion y uso de la informacion"). Section 1 ("Principios generales") articles specifically emphasize the need to protect individual rights and to respect the well-being of persons with whom Spanish psychologists interact in the performance of their professional activities. A final section of the code presents the official regulations of the Comisión Deontológica Estatal del Colegio Oficial de Psicólogos for enforcing the ethical principles and investigating charges of professional misconduct.

Switzerland

The *Deontological Code* (French version) of the Swiss Federation of Psychologists has been in effect since 1991. The basis objectives of the code, as described in a preamble, are to provide a means for protecting the public against abusive applications of psychology and to serve as a device for orienting Swiss psychologists in their professional conduct. The code's deontological principles are said to reflect the different aspects of responsibilities of professional psychology in research, teaching, and practice. The principles are divided into sections pertaining to the general responsibilities of Swiss psychologists, professional competence, professional secrets and the protection of data, the establishment of professional relationships, and the advertising of psychological services.

While the ethical principles are stated broadly enough to be applicable to various aspects of research procedure (such as the protection of confidential information obtained from research participants), there are no specific guidelines directly referring to the conduct of research *per se*, beyond a statement that one maintains honesty ("la véracité") in research, teaching, and publishing. Following the ethical principles is a separate section comprised of a set of detailed rules and regulations relative to the investigation of complaints by the Swiss Commission of the Professional Order.

Summary

Our brief survey of 11 ethics codes from around the world has revealed some similarities as well as differences (see Table A2.2). However, as Schuler (1982) cautioned in his earlier comparison, comparative surveys of ethical codes are limited by a number of considerations, including the differing structures of the codes, which make comparisons difficult; differences in the degree of specificity with which principles are formulated; and differences in the basic positions taken in the codes, with stronger emphasis placed either on research benefits or risks to research participants. Thus, care should be taken in drawing conclusions about the merits of a particular code without first referring to it directly.

One apparent commonality among the various codes described here is the fact that a number have been revised or updated since their initial development. In each case, the revisions tended to reflect newly emergent issues in the discipline or inadequacies noted in the earlier version(s) of the code. Separate research principles were included in most of the codes considered (USA, Australia, Canada, Great Britain, Germany, Scandinavia, Slovenia, Spain), but were merely implied by more broadly stated principles in other codes (France, The Netherlands, Switzerland). Of those codes that presented research principles apart from those dealing with other pro-

Table A2.2 Ethical codes in psychology: a comparative summary

	USA	Australia	Canada	France	UK	Germany	Netherlands	Scandinavia	Slovenia	Spain	Switzerland
First ethics code	1953	1968	1986	1961	1978	1967	1960	1989	1982	1993[c]	1991
Current revision	1992	1986[a]	1991	NA[b]	1991	1989	1988	NA	NA	NA	NA
Companion manual	Yes	No	Yes	No	No	No	No	No	No	No	No
Empirically based	Yes	No	Yes	No	No	No	No	No	No	No	No
Specific (S) or general (G) principles	S	S	S	G	S	S	S	S	G	G	G
Separate research principles	Yes	Yes	Yes	No	Yes	Yes	No	Yes	Yes	Yes	No
Steps or guidelines for ethical decision making	Yes	Yes	Yes	No	No	No	No	No	No	No	No

[a] Amended, 1990–1.
[b] NA = Non-applicable (the 1961 code was amended in 1976).
[c] Originally ratified in 1987; final approval in 1993.

fessional activities, in some cases the principles were specifically
stated with clarifications and examples (USA, Australia, Canada,
Great Britain, Scandinavia), while in others they were more generally
described (Slovenia, Spain). Only the American and Canadian codes
included an accompanying handbook with case examples, applications, and interpretations of the research principles. (At the time of
this writing, the American handbook was undergoing extensive
revision). Interestingly, these were the only two codes developed on
the basis of an empirical approach involving members of the profession who had been asked to describe their own approaches to
resolving ethical dilemmas. Thus, an empirical approach can be
recommended as a critical ingredient in the formulation of specific
ethical guidelines of practical use to research scientists for choosing
appropriate research procedures.

Also clear from our survey is that a single European perspective on
professional ethics does not yet exist, although for years it has been
intended to find a common set of regulations. One problem in
developing mutually acceptable ethical guidelines among nations has
to do with the confounding of professional and ethical topics (Heinz
Schuler, personal communication, February 3, 1993). Whereas
mutual ethical principles can be formulated rather easily, it is difficult
to adopt them on the professional level. This is because in the various
European countries, different systems have existed for scientific study
and different regulatory conditions have been in place for obtaining
a license for professional practice. However, with the emergence of
the European Community, participating nations have begun to establish common guidelines for the professions; thus, it is not unreason-
able to expect the creation of a practical European ethical code in the
not too distant future. One first step towards an international code is
the *Charter of Professional Ethics for Psychologists*, a one-page
document which briefly outlines general principles of conduct for
psychologists in all of their professional activities. The Charter
resulted from a series of meetings involving the participation of Italy,
France, Greece, Spain, Portugal, and Malta during the early 1990s.
The apparently successful implementation of an ethics code for
Scandinavian psychologists suggests that it is possible for several
countries to conform to a common set of ethical principles.

As a final point, each ethics code surveyed revealed an overriding
high regard for the well-being and dignity of research participants
and users of psychological services. This emphasis was reflected in
the attention given to such topics as informed consent (and the

---

Box A2.2

Obtaining copies of the ethical codes

1 *Australia*: copies of the 23-page *Australian Code of Professional Conduct* can be obtained by writing to The Australian Psychological Society Limited, National Science Centre, 191 Royal Parade, Parkville, Victoria, Australia, 3052.

2 *Canada*: revised versions of *A Canadian Code of Ethics for Psychologists* and the *Companion Manual* are available from The Canadian Psychological Association, 151 Slater St, Suite 205, Ottawa, Ontario K1P 5H3, Canada.

3 *France*: copies of the *Code de Deontologie* can be obtained by writing to the Société Française de Psychologie, 28–32 rue Serpente, 75006 Paris, France. (An English translation of the French code of ethics is not available.) Those readers who are interested in a French perspective on ethical dilemmas in human subject research should consult Beauvois et al. (1990). "Deontologie de la recherche en psychologie," pp 205–23 (also unavailable in English translation).

4 *Germany*: copies of the *Professional Code of Ethics for Psychologists* can be obtained from the German Association of Professional Psychologists, Berufsverband-Deutscher Psychologen, Heilsbachstrasse 22, D-5300 Bonn 1. (Available in English translation.)

5 *Great Britain*: copies of The British Psychological Society's 36-page *Code of Conduct*, which includes as a supplement the "Ethical principles for conducting research with human participants" can be obtained by writing The British Psychological Society, St Andrews House, 48 Princess Road East, Leicester, LE1 7DR, Great Britain. The "Ethical principles" also appear in the June 1990 issue of the British journal *The Psychologist* (Vol. 3, No. 6).

6 *The Netherlands*: copies of the English text of the code on ethics of the Netherlands Institute of Psychologists are available by writing to the Nederlands Instituut Van Psychologen, Postbus 9921, 1006 AP Amsterdam, The Netherlands.

7 *Scandinavia*: copies of the *Ethical Principles for Scandinavian Psychologists* (in English) are available by writing to the Dansk Psykolog Forening, Bjerregards Sidevej 4, 2500 Valby, Denmark.

8 *Slovenia*: copies of the *Code of Ethics for Psychologists* (in English) can be obtained by writing to the Psychological Association of Slovenia, Prusnikova 74, 61210 Ljubljana, Slovenia.

9 *Spain*: the *Codigo Deontologico del Psicólogo* is available by writing to the Colegio Official de Psicólogos, Junta de Governo Estatal, Junez de Balboa 58, 28001 Madrid, Spain. (The code currently is available only in Spanish.)

10 *Switzerland*: copies of the *Code Deontologique* are available by writing to the Fédération Suisse des Psychologues, BFSH 2, CH-1015 Lausanne, Switzerland. (The text of the code is available in French and German versions.)

11 *United States*: copies of the current American ethical code, *Ethical Principles of Psychologists and Code of Conduct*, are available by writing to the American Psychological Association, 750 First Street, NE, Washington, DC 20002, United States. The code also appears in the December 1992 issue of the *American Psychologist*, Vol 47, pp. 1597–1611. Interested readers also may wish to consult Canter et al.'s *Ethics for Psychologists: A Commentary on the APA Ethics Code* (1994). An accompanying handbook solely pertaining to the code's research principles is under preparation and should be available from the APA in the near future.

12 *Southern European countries*: copies of the 1994 *Charter of Professional Ethics for Psychologists*, adopted by Italy, Spain, France, Greece, Portugal, and Malta can be obained by writing to the Fédération Européenne des Associations de Psychologues, c/o Sveriges Psykologforbund, Box 3287, 103 65 Stockholm, Sweden. The charter is reproduced in English and the languages of the six countries involved.

related issue of deception), protection from harm, and privacy issues (including confidentiality) in all of the codes.

Details describing how to obtain copies of the ethical codes described in this appendix appear in Box A2.2.

Appendix 3:
Sample Research Application
and Informed Consent Forms

FITCHBURG STATE COLLEGE

HUMAN STUDIES COMMITTEE

APPLICATION FOR APPROVAL OF RESEARCH INVOLVING
HUMAN SUBJECTS

PRINCIPAL INVESTIGATOR: _____

CO-INVESTIGATORS: _____

_____

TITLE OF PROJECT: _____

_____

_____

1.   BRIEF DESCRIPTION OF PROJECT.

2. OUTLINE POTENTIAL BENEFIT OF THIS PROJECT TO THE INDIVIDUAL SUBJECT, GROUP OF SUBJECTS OR SOCIETY IN GENERAL.

3. OUTLINE POTENTIAL RISKS TO SUBJECT AND THE MEASURES THAT WILL BE TAKEN TO MINIMIZE SUCH.

4. BRIEFLY EXPLAIN WHEN AND IN WHAT MANNER YOU WILL OBTAIN WRITTEN INFORMED CONSENT OF YOUR SUBJECTS.

5. HOW ARE YOUR RESEARCH SUBJECTS SELECTED OR RECRUITED?

6. WHAT INCENTIVES OR REMUNERATIONS DO YOUR RESEARCH SUBJECTS RECEIVE?

7. DOES YOUR PROJECT INVOLVE THE INGESTION OF SUBSTANCES SUCH AS NUTRIENTS OR DRUGS?

8. GIVE APPROXIMATE NUMBER AND AGE OF SUBJECTS TO BE INVOLVED IN YOUR PROJECT.

9. WHAT MEASURES WILL BE TAKEN TO PROTECT THE RIGHTS AND PRIVACY OF THE SUBJECTS?

10. WHAT MEASURES WILL BE TAKEN TO ENABLE THE SUBJECT TO OMIT SPECIFIC PROCEDURES OR TO LEAVE THE STUDY?

SIGNATURE OF THE PRINCIPAL INVESTIGATOR

FITCHBURG STATE COLLEGE

INFORMED CONSENT FORM

SUBJECT'S NAME _____

PRINCIPAL INVESTIGATOR _____

CO-INVESTIGATOR _____

This portion is to be reviewed and signed by the subject:

I, THE ABOVE NAMED SUBJECT, UNDERSTAND THAT THE PURPOSE OF THIS PROJECT IS:

I, THE ABOVE-NAMED SUBJECT, UNDERSTAND THAT THE FOLLOWING IS A DESCRIPTION OF THE PROCEDURES I WILL BE UNDERGOING IN THIS PROJECT:

I, THE ABOVE-NAMED SUBJECT, UNDERSTAND THAT THE FOLLOWING IS AN EXPLANATION OF FORESEEABLE ATTENDANT DISCOMFORT AND/OR RISKS:

I, THE ABOVE-NAMED SUBJECT, UNDERSTAND THAT THE FOLLOWING IS AN EXPLANATION OF THE BENEFITS I MAY EXPECT OR THAT MAY BE EXPECTED FOR OTHERS AS A RESULT OF THIS PROJECT:

Informed Consent Form - Page 2

I have been informed of and understand the purpose of the above described project and its procedures. I also have been informed of and understand the forseeable discomfort, risks and benefits. I have further been advised that unforeseen effects may occur. I volunteer and agree to assume responsibility for any medical care required as a result of any injury sustained as a result of this study. Nevertheless, I wish to participate in this investigation.

I have been informed of and understand that the data collected in this study may be published or otherwise disseminated for scientific purposes. However, I understand that my name will not be published and that every effort will be made to protect my confidentiality.

I further understand that I may withdraw from the project at any time without prejudice to me.

Date _____   Subject _____

Date _____   Subject's _____
                         Representative

Date _____   Witness _____

Investigator's Declaration

I have explained to the above-named subject the nature and purpose of the procedures described above, and the foreseeable risks, discomforts, and benefits that may result. I have considered and rejected alternative procedures for obtaining this information. I have asked the subject if any questions have arisen regarding the procedures and have answered these questions to the best of my ability.

Date   Principal or Co-investigator

Sample informed consent form for student research in experimental psychology: subjects' rights

In order to ensure that your rights as a human subject are protected according to the established guidelines of the American Psychological Association, please consider the subjects' rights listed below, all of which you are entitled to expect. After you have finished, you may ask me to explain any issues that are unclear to you. When you are satisfied that you have read and fully understand your guaranteed rights, please sign your name below.

Subjects have the right:

1 to participate voluntarily, free from any coercion
2 to be informed of the general nature of the research
3 not to be deceived unnecessarily or in any way that might be harmful
4 to withdraw from the study at any time, without incurring any penalty
5 to be protected from physical and/or psychological discomfort, harm, and danger
6 to be informed (debriefed) at the conclusion of the study, regarding the intent of the research
7 to expect that any information divulged during the study will be considered confidential and private
8 to expect that reports of the experimental results will reflect group performances rather than individual performances, and that no participant will be individually identified

---

Subject Consent Form

Researcher (Please read and sign):

I, _____ (name of researcher) agree to abide by all of the guidelines and standards for conducting research with human participants as described by the American Psychological Association.

Date: _____

Subject (Please read and sign:)

I, _____ (name of subject) have been
informed about the general nature of this study and agree voluntarily
to participate. I have read and understand the Human Subjects'
Rights described by the American Psychological Association, and I
understand that all such rights will be guaranteed to me.

Date: _____

[*Source*: Reprinted with permission from the Behavioral Sciences Department, Fitchburg
State College, Fitchburg, Massachusetts, USA.]

Sample informed consent form for student research

This research is being conducted by _____,
Framingham State College, in partial fulfillment of the requirements
of course 42.450 Experimental Psychology.

In agreeing to participate in this research I understand the following:

The experiment will involve [A brief general description of the
research is given here. You need not give specific details of the
experiment, but you should give subjects a general overview of the
tasks involved.].

The entire experiment will take about _____ minutes.

There are no known expected discomforts or risks involved in the
experiment.[If discomfort or risk is expected, you must indicate
the nature of the discomfort and/or risk.]

There are no "disguised" or "trick" procedures involved in the
experiment.

All data from the experiment will remain anonymous. The data
from all subjects shall be compiled, analyzed, and submitted in a
laboratory report to the course instructor. No participant's data
shall be identified by name at any stage of the data analysis or in
the laboratory report.

At the conclusion of my participation in the experiment, I will be
given information concerning my performance and any questions
that I may have will be clearly and fully answered.

I may withdraw from this experiment at any time.

There are no special benefits to me for participating in this
experiment, and it will not affect my status at Framingham State
College.

_____   _____
(Experimenter)   (Experimental Participant)

                                         _____
                                         (Date)

[*Source*: Reprinted with permission from the Psychology Department, Framingham State
College, Framingham, Massachusetts, USA.]

References

Aaker, D. A. and Day, G. S. (1990). *Marketing Research* (4th ed.). New York: John Wiley and Sons.

Abramovitch, R., Freedman, J. L., Thoden, K., and Nikolich, C. (1991). Children's capacity to consent to participation in psychological research: Empirical findings. *Child Development, 62,* 1100–9.

Adair, J. G. (1973). *The Human Subject: The Social Psychology of the Psychological Experiment.* Boston: Little, Brown.

Adair, J. G., Dushenko, T. W., and Lindsay, R. C. L. (1985). Ethical regulation and their impact on research practice. *American Psychologist, 40,* 59–72.

Adair, J. G. and Lindsay, R. C. L. (1983). *Debriefing Subjects: A Comparison of Published Reports With a Self-Report Survey.* Unpublished manuscript, University of Manitoba, Winnipeg, Canada.

Adair, J. G., Lindsay, R. C. L., and Carlopio, J. (1983). Social artifact research and ethical regulations: Their impact on the teaching of experimental methods in psychology. *Teaching of Psychology, 10,* 159–62.

Aitkenhead, M. and Dordoy, J. (1983). Research on the ethics of research. *Bulletin of the British Psychological Society, 36,* 315–18.

Allen, V. (1966). Effect of knowledge of deception on conformity. *Journal of Social Psychology, 69,* 101–6.

Allport, G. W. and Postman, L. (1947). *The Psychology of Rumor.* New York: Holt.

Altemeyer, R. A. (1972). Subject pool pollution and the postexperimental interview. *Journal of Experimental Research in Personality, 5,* 79–84.

Altman, L. K. (1986, February 25). Peer review is challenged. *The New York Times,* p. C3.

Altman, L. K. (1989, June 6). Errors prompt proposals to improve "peer review" at science journals. *The New York Times,* p. C3.

American Anthropological Association (1983). *Professional Ethics: Statements and Procedures of the American Anthropological Association.* Washington, DC.

American Association of University Professors (1981, December). Regulations governing research on human subjects: academic freedom and the institutional review board. *Academe, 67,* 358–70.

American Marketing Association (1972). *Code of Ethics.* Chicago, IL.

American Psychological Association (1968). *Principles for the Care and Use of Animals.* Washington, DC.

American Psychological Association (1973). *Ethical Principles in the Conduct of Research With Human Participants.* Washington, DC.

American Psychological Association (1978). Report of the task force on the role of psychology in the criminal justice system. *American Psychologist, 33,* 1099–113.

American Psychological Association (1981). Ethical principles of psychologists. *American Psychologist, 36,* 633–8.

American Psychological Association (1982). *Ethical Principles in the Conduct of Research With Human Participants* (rev. ed.). Washington, DC.

American Psychological Association (1990). Ethical Principles of Psychologists (Amended June 2, 1989). *American Psychologist, 45,* 390–5.

American Psychological Association (1992). Ethical principles of psychologists and code of conduct. *American Psychologist, 47,* 1597–1611.

American Psychological Association (1993a). *Guidelines for Ethical Conduct in the Care and Use of Animals.* Washington, DC.

American Psychological Association (1993b). Ethical standards for the reporting and publishing of scientific information. *American Psychologist, 48,* 383.

American Psychological Association (1994). *Publication Manual of the American Psychological Association* (4th ed.). Washington, DC.

American Sociological Association (1971). *Code of Ethics.* Washington, DC.

American Sociological Association (1989). *Code of Ethics.* Washington, DC.

Animal research protection bill approved (1992, September). *APS Observer,* p. 4.

Animal testing for cosmetics banned by EC, beginning in 1998. (1992, November 4). *The Boston Globe,* p. 33.

Annas, G. J., Glantz, L. H., and Katz, B. F. (1977). *Informed Consent to Human Experimentation: The Subject's Dilemma.* Cambridge, MA: Ballinger.

Archer, J. (1986). Ethical issues in psychobiological research on animals. *Bulletin of the British Psychological Society, 39,* 361–4.

Arellano-Galdames, F. J. (1972). *Some Ethical Problems in Research on*

Human Subjects. Unpublished doctoral dissertation, University of New Mexico, Albuquerque.

Argyris, C. (1975). Dangers in applying results from experimental social psychology. *American Psychologist, 30*, 469–85.

Aronson, E. (1966). Avoidance of inter-subject communication. *Psychological Reports, 19*, 238.

Aronson, E. & Carlsmith, J. M. (1968). Experimentation in social psychology. In G. Lindzey and E. Aronson (eds.), *The Handbook of Social Psychology* (Vol. 2). Reading, MA: Addison-Wesley.

Arthur Conan Doyle is Piltdown suspect (1983, August 2). *The New York Times*, pp. C1, C6.

A scientific Watergate? (1991, March 26). *The New York Times*, p. A22.

Asch, S. (1955). Opinions and social pressure. *Scientific American, 193*, 31–5.

Asher, J. (1974, November). Can parapsychology weather the Levy affair? *APA Monitor, 5*, p. 4.

Atwell, J. (1981). Human rights in human subjects research. In A. J. Kimmel (ed.), *Ethics of Human Subject Research*. San Francisco: Jossey-Bass.

Australian Psychological Society (1986). *Code of Professional Conduct*. Parkville, Victoria.

Azar, B. (1994a, December). Animal research threatened by activism. *APA Monitor, 25*, p. 18.

Azar, B. (1994b, December). Research improves lives of animals. *APA Monitor, 25*, p. 19.

Azar, B. (1994c, December). Tactics of animal activists become more sophisticated. *APA Monitor, 25*, p. 20.

Azrin, N. H., Holz, W., Ulrich, R., and Goldiamond, I. (1961). The control of the content of conversation through reinforcement. *Journal of the Experimental Analysis of Behavior, 4*, 25–30.

Babbage, C. (1969). *Reflections on the Decline of Science in England, and on Some of Its Causes*. London: Gregg International.

Bacon, F. (1960). *Novum Organum*. New York: Bobbs-Merrill.

Baird, J. S., Jr. (1980). Current trends in college cheating. *Psychology in the Schools, 17*, 515–22.

Baldwin, E. (1993). The case for animal research in psychology. *Journal of Social Issues, 49*, 121–31.

Bales, J. (1992, November). New ethics code "practical and livable." *APA Monitor, 23*, pp. 6–7.

Baron, R. A and Byrne, D. (1987). *Social Psychology: Understanding Human Interaction* (5th ed.). Boston: Allyn and Bacon.

Baron, R. A. and Byrne, D. (1994). *Social Psychology: Understanding Human Interaction* (7th ed.). Boston: Allyn and Bacon.

Barrass, R. (1978). *Scientists Must Write*. London: Chapman and Hall.

Baumeister, R. F., Cooper, J., and Skib, B. A. (1979). Inferior performance as a selective response to expectancy: taking a dive to make a point. *Journal of Personality and Social Psychology, 37*, 424–32.

Baumrin, B. H. (1970). The immortality of irrelevance: the social role of science. In F. F. Korten, S. S. Cook, and J. I. Lacey (eds.), *Psychology and the Problems of Society*. Washington, DC: American Psychological Association.

Baumrind, D. (1964). Some thoughts on ethics of research: after reading Milgram's "Behavioral study of obedience." *American Psychologist, 19*, 421–3.

Baumrind, D. (1971). Principles of ethical conduct in the treatment of subjects: Reaction to the draft report of the Committee on Ethical Standards in Psychological Research. *American Psychologist, 26*, 887–96.

Baumrind, D. (1972). Reactions to the May 1972 draft report of the Ad Hoc Committee on Ethical Standards in Psychological Research. *American Psychologist, 27*, 1083–6.

Baumrind, D. (1975). Metaethical and normative considerations governing the treatment of human subjects in the behavioral sciences. In E. C. Kennedy (ed.), *Human Rights and Psychological Research: A Debate on Psychology and Ethics*. New York: Thomas Y. Crowell.

Baumrind, D. (1976). *Nature and definition of informed consent in research involving deception*. Paper prepared for the National Commission for the Protection of Human Subjects of Biomedical and Behavioral Research. Bethesda, MD: US Department of Health, Education, and Welfare.

Baumrind, D. (1977, April). *Snooping and duping: the application of the principle of informed consent to field research*. Paper presented at the meeting of the Society for Applied Anthropology, San Diego, CA.

Baumrind, D. (1981, August). The costs of deception. *Society for the Advancement of Social Psychology Newsletter, 7*, 1, 3, 6, 8–10.

Baumrind, D. (1985). Research using intentional deception: ethical issues revisited. *American Psychologist, 40*, 165–74.

Bayne, K. (1991). *National Institutes of Health Nonhuman Primate Management Plan*. Bethesda, MD: Office of Animal Care and Use, National Institutes of Health.

Beauvois, J.-L., Roulin, J.-L., and Tiberghien, G. (1990). *Manuel d'Etudes Pratiques de Psychologie I: Pratique de la Recherche*. Paris: Presses Universitaires de France.

Bechtel, H. K., Jr. and Pearson, W., Jr. (1985). Deviant scientists and scientific deviance. *Deviant Behavior, 6*, 237–52.

Beecher, H. K. (1959). *Experimentation in Man*. Springfield, IL: Thomas.

Beecher, H. K. (1966). Ethics and clinical research. *New England Journal of Medicine, 274*, 1354–60.

Bell, E. H. and Bronfenbrenner, U. (1959). Freedom and responsibility in research: comments. *Human Organization, 18*, 49–52.

Bell, R. (1995, April). Les cobayes humaines du plutonium [The human plutonium guinea-pigs]. *La Recherche, 26,* 384–93.

Bem, D. J. (1965). An experimental analysis of self-persuasion. *Journal of Experimental Social Psychology, 1,* 199–218.

Bentham, J. (1789). *An Introduction to the Principles of Morals and Legislation.* Edinburgh, William Tate.

Bergin, A. E. (1962). The effect of dissonant persuasive communications upon changes in a self-referring attitude. *Journal of Personality, 30,* 423–36.

Berk, R. A., Boruch, R. F., Chambers, D. L., Rossi, P. H., and Witte, A. D. (1987). Social policy experimentation: a position paper. In D. S. Cordray and M. W. Lipsey (eds.), *Evaluation Studies Review Annual* (Vol. 11). Newbury Park, CA: Sage.

Berkowitz, L. (1962). *Aggression: A Social Psychological Analysis.* New York: McGraw-Hill.

Berkowitz, L. (1969). The frustration-aggression hypothesis revisited. In L. Berkowitz (ed.), *Roots of Aggression.* New York: Atherton.

Berkun, M., Bialek, H. M., Kern, P. R., and Yagi, K. (1962). Experimental studies of psychological stress in man. *Psychological Monographs: General and Applied, 76,* 1–39.

Bermant, G., and others. (1974). The logic of simulation in jury research. *Criminal Justice and Behavior, 1,* 224–33.

Bernstein, D. A., Clarke-Stewart, A., Roy, E. J., Srull, T. K., and Wickens, C. D. (1994). *Psychology* (3rd ed.). Boston: Houghton-Mifflin.

Bickman, L. (1974). The social power of a uniform. *Journal of Applied Social Psychology, 4,* 47–61.

Bickman, L. (1981). Some distinctions between basic and applied approaches. In L. Bickman (ed.), *Applied Social Psychology Annual* (Vol. 2). Beverly Hills, CA: Sage.

Bickman, L. and Rosenbaum, D. P. (1977). Crime reporting as a function of bystander encouragement, surveillance, and credibility. *Journal of Personality and Social Psychology, 35,* 577–86.

Bickman, L. and Zarantonello, M. (1978). The effects of deception and level of obedience on subjects' ratings of the Milgram study. *Personality and Social Psychology Bulletin, 4,* 81–5.

Blanck, P. D., Bellack, A. S., Rosnow, R. L., Rotheram-Borus, M. J., and Schooler, N. R. (1992). Scientific rewards and conflicts of ethical choices in human subjects research. *American Psychologist, 47,* 959–65.

Blumberg, M. (1980). Job switching in autonomous work groups: An exploratory study in a Pennsylvania coal mine. *Academy of Management Journal, 23,* 287–306.

Blumberg, M. and Pringle, C. D. (1983). How control groups can cause loss of control in action research: the case of Rushton coal mine. *Journal of Applied Behavioral Science, 19,* 409–25.

Blumberg, M. S. and Wasserman, E. A. (1995). Animal mind and the argument from design. *American Psychologist, 50*, 133–44.

Boffey, P. M. (1981, October 27). Animals in the lab: protests accelerate, but use is dropping. *The New York Times*, pp. C1–C2.

Bok, S. (1978). *Lying: Moral Choice in Public and Private Life*. New York: Pantheon.

Boruch, R. F. and Cecil, J. S. (1979). *Assuring the Confidentiality of Social Research Data*. Philadelphia: University of Pennsylvania Press.

Boruch, R. F. and Cecil, J. S. (1982). Statistical strategies for preserving privacy in direct inquiry. In J. E. Sieber (ed.), *The Ethics of Social Research: Surveys and Experiments*. New York: Springer-Verlag.

Bowd, A. D. (1980). Ethical reservations about psychological research with animals. *Psychological Record, 30*, 201–10.

Bowd, A. D. and Shapiro, K. J. (1993). The case against laboratory animal research in psychology. *Journal of Social Issues, 49*, 133–42.

Bowen, D. D., Perloff, R., and Jacoby, J. (1972). Improving manuscript evaluation procedures. *American Psychologist, 27*, 221–5.

Bramel, D. (1962). A dissonance theory approach to defensive projection. *Journal of Abnormal and Social Psychology, 64*, 121–9.

Bramel, D. (1963). Selection of a target for defensive projection. *Journal of Abnormal and Social Psychology, 66*, 318–24.

Brehm, S. S. and Kassin, S. M. (1993). *Social Psychology* (2nd ed.). Boston: Houghton Mifflin.

Brill, S. (1976, July 19). A New York cabby story you won't believe. *New York*, pp. 8, 10.

British Psychological Society (1995a). *Code of Conduct, Ethical Principles and Guidelines*. Leicester, UK.

British Psychological Society (1995b). *Guidelines for the Use of Animals in Research*. Leicester, UK.

Britton, B. K. (1979). Ethical and educational aspects of participating as a subject in psychology experiments. *Teaching of Psychology, 6*, 95–8.

Broad, W. and Wade, N. (1982). *Betrayers of the Truth*. New York: Simon and Schuster.

Broad, W. J. (1983, December 13). New attack on Galileo asserts major discovery was stolen. *The New York Times*, pp. C1, C8.

Brock, T. C. and Becker, L. A. (1966). Debriefing and susceptibility to subsequent experimental manipulations. *Journal of Experimental Social Psychology, 2*, 314–23.

Bronowski, J. (1956). *Science and Human Values*. New York: Messner.

Broome, J. (1984). Selecting people randomly. *Ethics, 95*, 38–55.

Brown, R. (1962). Models of attitude change. In R. Brown, E. Galanter, E. Hess, and G. Mandler (eds.), *New Directions in Psychology: I*. New York: Holt, Rinehart and Winston.

Bryan, J. H. and Test, M. A. (1967). Models and helping: naturalistic studies in aiding behavior. *Journal of Personality and Social Psychology, 6,* 400–7.

Buckhout, R. (1965). Need for approval and attitude change. *Journal of Psychology, 60,* 123–8.

Burbach, D. J., Farha, J. G., and Thorpe, J. S. (1986). Assessing depression in community samples of children using self-report inventories: ethical considerations. *Journal of Abnormal Child Psychology, 14,* 579–89.

Burghardt, G. M. and Herzog, H. A., Jr. (1980). Beyond conspecifics: is "Brer Rabbit" our brother? *BioScience, 30,* 763–7.

Cameron, P. (1969). Frequency and kinds of words in various social settings, or what the hell's going on? *Pacific Sociological Review, 12,* 101–4.

Campbell, D., Sanderson, R. E., and Laverty, S. G. (1964). Characteristics of a conditioned response in human subjects during extinction trials following a single traumatic conditioning trial. *Journal of Abnormal and Social Psychology, 68,* 627–39.

Campbell, D. T. and Cecil, J. S. (1982). A proposed system of regulation for the protection of participants in low-risk areas of applied social research. In J. E. Sieber (ed.), *The Ethics of Social Research: Fieldwork, Regulation and Publication.* New York: Springer-Verlag.

Campbell, D. T., Boruch, R. F., Schwartz, R. D., and Steinberg, J. (1977). Confidentiality-preserving modes of access to files and to interfile exchange for useful statistical analysis. *Evaluation Quarterly, 1,* 269–300.

Campion, M. A. (1993). Are there differences between reviewers on the criteria they use to evaluate research articles? *The Industrial-Organizational Psychologist, 31*(2), 29–39.

Canadian Psychological Association (1988). *Canadian Code of Ethics for Psychologists: Companion Maunal.* Old Chelsea, Quebec.

Canadian Psychological Association (1991). *Canadian Code of Ethics for Psychologists* (rev. ed.). Old Chelsea, Quebec.

Canter, M. B., Bennett, B. E., Jones, S. E., and Nagy, T. F. (1994). *Ethics for Psychologists: A Commentary on the APA Ethics Code.* Washington, DC: American Psychological Association.

Carlson, J., Cook, S. W., and Stromberg, E. L. (1936). Sex differences in conversation. *Journal of Applied Psychology, 20,* 727–35.

Carlson, R. (1971). Where is the person in personality research? *Psychological Bulletin, 75,* 203–19.

Carroll, M. A., Schneider, H. G., and Wesley, G. R. (1985). *Ethics in the Practice of Psychology.* Englewood Cliffs, NJ: Prentice-Hall.

Ceci, S. J., Peters, D., and Plotkin, J. (1985). Human subjects review, personal values, and the regulation of social science research. *American Psychologist, 40,* 994–1002.

Chalk, R., Frankel, M. S., and Chafer, S. B. (1980). *AAAS Professional Ethics Activities in the Scientific and Engineering Societies.* Washington, DC: American Association for the Advancement of Science.

Christakis, N. A. (1988). Should IRBs monitor research more strictly? *IRB: A Review of Human Subjects Research*, *10*, 8–10.

Christensen, L. (1988). Deception in psychological research: When is its use justified? *Personality and Social Psychology Bulletin*, *14*, 664–75.

Clark, R. D. and Word, L. E. (1974). Where is the apathetic bystander? Situational characteristcs of the emergency. *Journal of Personality and Social Psychology*, *29*, 279–87.

Clingempeel, W. G., Mulvey, E., and Repucci, N. D. (1980). A national study of ethical dilemmas of psychologists in the criminal justice system. In J. Monahan (ed.), *Who is the Client?*. Washington, DC: American Psychological Association.

Cohen, I. B. (1957). Galileo. In I. B. Cohen (ed.), *Lives in Science*. New York: Simon and Schuster.

Coile, D. C. and Miller, N. E. (1984). How radical animal rights activists try to mislead humane people. *American Psychologist*, *39*, 700–1.

Cole, J. R. (1992). Animal rights and wrongs. In B. Slife and J. Rubinstein (eds.), *Taking Sides: Clashing Views on Controversial Psychological Issues* (7th ed.). Guilford, CT: Dushkin.

Colegio Oficial de Psicólogos. (1993). *Código Deontológico del Psicólogo* [*Deontological Code of the Psychologist*]. Madrid.

Collins, F. L., Jr., Kuhn, I. F., Jr., and King, G. D. (1979). Variables affecting subjects' ethical ratings of proposed experiments. *Psychological Reports*, *44*, 155–64.

Comroe, J. H., Jr. (1983). *Exploring the Heart: Discoveries in Heart Disease and High Blood Pressure*. New York: W. W. Norton.

Committee for the Use of Animals in School Science Behavior Projects (1972). Guidelines for the use of animals in school science behavior projects. *American Psychologist*, *27*, 337.

Cook, S. W. (1970). Motives in a conceptual analysis of attitude related behavior. In W. J. Arnold and D. Levine (eds.), *Nebraska Symposium on Motivation*. Lincoln: University of Nebraska Press.

Cook, T. D., Bean, R. B., Calder, B. J., Frey, R., Krovetz, M. L., and Reisman, S. R. (1970). Demand characteristics and three conceptions of the frequently deceived subject. *Journal of Personality and Social Psychology*, *14*, 185–94.

Cook, T. D. and Campbell, D. T. (1979). *Quasi-Experimentation: Design and Analysis Issues for Field Settings*. Boston: Houghton Mifflin.

Cooper, H. and Hedges, L. V. (eds.) (1994). *The Handbook of Research Synthesis*. New York: Russell Sage Foundation.

Cooper, J. (1976). Deception and role playing: on telling the good guys from the bad guys. *American Psychologist*, *31*, 605–10.

Coutu, W. (1951). Role-playing vs. role-taking: an appeal for clarification. *American Sociological Review*, *16*, 180–7.

Crespi, T. D. (1994). Student scholarship: in the best interests of the scholar. *American Psychologist*, *49*, 1094–5.

Crusco, A. H. and Wetzel, C. G. (1984). The Midas touch: The effects of interpersonal touch on restaurant tipping. *Personality and Social Psychology Bulletin, 10*, 512–17.

Culliton, B. J. (1974). The Sloan–Kettering affair: a story without a hero. *Science, 184*, 644–50.

Cummings, L. L. and Frost, P. J. (eds.) (1985). *Publishing in the Organizational Sciences.* Homewood, IL: Irwin.

Curie-Cohen, M., Luttrell, L., and Shapiro, S. (1979). Current practice of artificial insemination by donor in the US. *New England Journal of Medicine, 11*, 585–90.

Curran, W. J. (1969). Governmental regulation of the use of human subjects in medical research: the approach of two federal agencies. *Daedalus, 98*, 542–94.

Curran, W. J. and Beecher, H. K. (1969). Experimentation in children. *Journal of the American Medical Association, 10*, 77–83.

Darley, J. M. and Latané, B. (1968). Bystander intervention in emergencies: Diffusion of responsibility. *Journal of Personality and Social Psychology, 8*, 377–83.

Davis, F. (1961). Comment on "Initial interaction of newcomers in alcoholics anonymous." *Social Problems, 8*, 364–5.

Davis, J. R. and Fernald, P. S. (1975). Laboratory experience versus subject pool. *American Psychologist, 30*, 523–4.

Davis, S. F., Grover, C. A., Becker, A. H., and McGregor, L. N. (1992). Academic dishonesty: prevalence, determinants, techniques, and punishments. *Teaching of Psychology, 19*, 16–20.

Day, R. L. (1975). A comment on ethics in marketing research. *Journal of Marketing Research, 12*, 232–3.

Devenport, L. D. (1989). Sampling behavior and contextual change. *Learning and Motivation, 20*, 97–114.

Devenport, L. D. and Devenport, J. A. (1990). The laboratory animal dilemma: a solution in our backyards. *Psychological Science, 1*, 215–16.

Diamond, S. S. and Morton, D. R. (1978). Empirical landmarks in social psychology. *Personality and Social Psychology Bulletin, 4*, 217–21.

Dickersin, K., Min, Y., and Meinert, C. L. (1992). Factors influencing publication of research results. *Journal of the American Medical Association, 267*, 374–8.

Dickson, J. P., Casey, M., Wyckoff, D., and Wind, W. (1977). Invisible coding of survey questionnaires. *Public Opinion Quarterly, 41*, 100–6.

Diener, E. and Crandall, R. (1978). *Ethics in Social and Behavioral Research.* Chicago: The University of Chicago Press.

Diener, E., Matthews, R., and Smith, R. (1972). Leakage of experimental information to potential future subjects by debriefed subjects. *Journal of Experimental Research in Personality, 6*, 264–7.

DiFonzo, N., Bordia, P., and Rosnow, R. L. (1994). Reining in rumors. *Organizational Dynamics*, 23, 47–62.

Domjan, M. and Purdy, J. E. (1995). Animal research in psychology: more than mets the eye of the general psychology student. *American Psychologist*, 50, 496–503.

Drewett, R. and Kani, W. (1981). Animal experimentation in the behavioral sciences. In D. Sperlinger (ed.), *Animals in Research*. New York: Wiley.

Eaton, W. O. (1983). The reliability of ethical reviews: some initial empirical findings. *Canadian Psychologist*, 24, 14–18.

Eckholm, E. (1985, May 7). Fight over animal experiments gains intensity on many fronts. *The New York Times*, pp. C1, C3.

Edsall, G. A. (1969). A positive approach to the problem of human experimentation. *Daedalus*, 98, 463–78.

Efran, M. G. (1974). The effect of physical appearance on the judgment of guilt, interpersonal attraction, and severity of recommended punishment in a simulated jury task. *Journal of Research in Personality*, 8, 45–54.

Eibl-Eibesfeldt, I. (1975). *Ethology: The Biology of Behavior*. New York: Holt, Rinehart and Winston.

Ellsworth, P. C. (1977). From abstract ideas to concrete instances. *American Psychologist*, 32, 604–15.

Epstein, Y. M., Suedfeld, P., and Silverstein, S. J. (1973). The experimental contract: subjects' expectations of and reactions to some behaviors of experimenters. *American Psychologist*, 28, 212–21.

Erikson, K. T. (1967). A comment on disguised observation in sociology. *Social Problems*, 14, 366–73.

Errera, P. (1972). Statement based on interviews with forty "worst cases" in the Milgram obedience experiments. In J. Katz (ed.), *Experimentation With Human Beings*. New York: Russell Sage Foundation.

Ethics Resource Center (1979). *Codes of Ethics in Corporations and Trade Associations and the Teaching of Ethics in Graduate Business Schools*. Princeton, NJ: Opinion Research Corporation.

Evans, P. (1976, December). The Burt affair: sleuthing in science. *APA Monitor*, 7, p. 1, 4.

Eysenck, H. J. (1981). On the fudging of scientific data. *American Psychologist*, 36, 692.

Faden, R. R. and Beauchamp, T. L. (1986). *A History and Theory of Informed Consent*. New York: Oxford University Press.

Farrow, J. M., Lohss, W. E., Farrow, B. J., and Taub, S. I. (1975). Intersubject communication as a contaminating factor in verbal conditioning. *Perceptual and Motor Skills*, 40, 975–82.

Federation Suisse des Psychologues (1991). *Code Déontologique*. Lausanne, Switzerland.

Feeney, D. (1987). Human rights and animal welfare. *American Psychologist, 42,* 593–9.

Festinger, L. and Carlsmith, J. M. (1959). Cognitive consequences of forced compliance. *Journal of Abnormal and Social Psychology, 58,* 203–10.

Fidler, D. S. and Kleinknecht, R. E. (1977). Randomized response versus direct questioning: two data-collection methods for sensitive information. *Psychological Bulletin, 84,* 1045–9.

Field, P. B. (1989, October). Animal images in college psychology texts. *The Animals' Agenda,* p. 14.

Field, P. B. (1990) The most invasive animal experiments in psychology textbooks. *PsyETA Bulletin, 9,* 6–8.

Fillenbaum, R. S. (1966). Prior deception and subsequent experimental performance: the faithful subject. *Journal of Personality and Social Psychology, 4,* 532–7.

Fine, M. A. and Kurdek, L. A. (1993). Reflections on determining authorship credit and authorship order on faculty-student collaborations. *American Psychologist, 48,* 1141–7.

Fisher, K. (1982, November). The spreading stain of fraud. *APA Monitor, 13,* pp. 7–8.

Fletcher, R. (1991). *Science, Ideology, and the Media: The Cyril Burt Scandal.* New Brunswick, NJ: Transaction.

Fo, W. S. and O'Donnell, C. R. (1975). The buddy system: effects of community intervention on delinquent offenses. *Behavior Therapy, 6,* 522–4.

Fogg, L. and Fiske, D. W. (1993). Foretelling the judgments of reviewers and editors. *American Psychologist, 48,* 293–4.

Forsyth, D. R. (1980). A taxonomy of ethical ideologies. *Journal of Personality and Social Psychology, 39,* 175–84.

Forsyth, D. R. (1987). *Social Psychology.* Pacific Grove, CA: Brooks/Cole.

Forsyth, D. R. and Strong, S. R. (1986). The scientific study of counseling and psychotherapy: a unificationist view. *American Psychologist, 41,* 113–19.

Forward, J., Canter, R., and Kirsch, N. (1976). Role enactment and deception methodologies: alternative paradigms. *American Psychologist, 31,* 595–604.

Fossey, D. (1981). Imperiled giants of the forest. *National Geographic, 159,* 501–23.

Fossey, D. (1983). *Gorillas in the Mist.* Boston: Houghton-Mifflin.

Fowler, F. J., Jr. (1993). *Survey research methods* (rev. ed.). Newbury Park, CA: Sage.

Fox, J. A. and Tracy, P. E. (1980). The randomized response approach. *Evaluation Review, 4,* 601–22.

Fox, J. A. and Tracy, P. E. (1984). Measuring associations with randomized response. *Social Science Research, 13,* 188–97.

Frankel, M. S. (1975). The development of policy guidelines governing human experimentation in the United States. *Ethics in Science and Medicine, 2,* 43–59.

Frankena, W. K. (1973). *Ethics* (2nd ed.). Englewood Cliffs, NJ: Prentice-Hall.

Franklin, A. D. (1981). Milikan's published and unpublished data on oil drops. *Historical Studies in the Physical Sciences, 11,* 185–201.

Fraud in medical research tied to lax rules. (1989, February 14). *The New York Times,* p. C2.

Freedman, J. (1969). Role-playing: psychology by consensus. *Journal of Personality and Social Psychology, 13,* 107–14.

Freeman, L. C. and Ataov, T. (1960). Invalidity of indirect and direct measures of attitude toward cheating. *Journal of Personality, 28,* 443–7.

French Psychological Society (1976). *Code de Deontologie.* Paris.

Friendly, J. (1983, June 12). Does a "public figure" have a right to privacy? *The New York Times,* p. 8E.

Gaertner, S. L. and Bickman, L. (1971). Effects of race on the elicitation of helping behavior: the wrong number technique. *Journal of Personality and Social Psychology, 20,* 218–22.

Gallistel, C. R. (1981). Bell, Magendie, and the proposals to restrict the use of animals in neurobehavioral research. *American Psychologist, 36,* 357–60.

Gallo, P. S., Smith, S., and Mumford, S. (1973). Effects of deceiving subjects upon experimental results. *Journal of Social Psychology, 89,* 99–107.

Gallup, G. G. (1979). Self-awareness in primates. *American Scientist, 67,* 417–21.

Gallup, G. G. and Suarez, S. (1980). On the use of animals in psychological research. *Psychological Record, 30,* 211–18.

Galvin, S. and Herzog, H. A., Jr. (1992). Ethical ideology, animal rights activism and attitudes toward the treatment of animals. *Ethics and Behavior, 2,* 141–9.

Garfield, E. (1978). The ethics of scientific publication. *Current Contents, 40,* 5–12.

Gastel, B. (1983). *Presenting Science to the Public.* Philadelphia: ISI Press.

Gaston, J. (1971). Secretiveness and competition for priority of discovery in physics. *Minerva, 9,* 472–92.

Gaylin, W. and Blatte, H. (1975). Behavior modification in prisons. *American Criminal Law Review, 13,* 11–35.

Gazzaniga, M. S. and LeDoux, J. E. (1978). *The Integrated Mind.* New York: Plenum Press.

Geen, R. G. (1968). Effects of frustration, attack, and prior training in aggressiveness upon aggressive behavior. *Journal of Personality and Social Psychology, 9,* 316–21.

Geller, D. M. (1978). Involvement in role-playing simulations: a demonstration with studies on obedience. *Journal of Personality and Social Psychology, 36,* 219–235.

Geller, D. M. (1982). Alternatives to deception: why, what, and how? In J. E. Sieber (ed.), *The Ethics of Social Research: Surveys and Experiments.* New York: Springer-Verlag.

General Assembly for the Scandinavian Psychological Associations. (1989, March). Ethical principles for Scandinavian psychologists. *News From EFPPA, 3*(1).

Georgoudi, M. and Rosnow, R. L. (1985). Notes towards a contextualist understanding of social psychology. *Personality and Social Psychology Bulletin, 11,* 5–22.

Gerdes, E. P. (1979). College students' reactions to social psychological experiments involving deception. *Journal of Social Psychology, 107,* 99–110.

Gergen, K. J. (1973a). The codification of research ethics: Views of a doubting Thomas. *American Psychologist, 28,* 907–12.

Gergen, K. J. (1973b). Social psychology as history. *Journal of Personality and Social Psychology, 26,* 309–20.

German Association of Professional Psychologists. (1986). *Professional Code of Ethics for Psychologists.* Bonn.

Gersten, J. C., Langner, T. S., and Simcha-Fagan, O. (1979). Developmental patterns of types of behavioral disturbance and secondary prevention. *International Journal of Mental Health, 7,* 132–49.

Glass, G., Tiao, G. C., and Maguire, T. O. (1971). Analysis of data on the 1900 revision of German divorce laws as a time-series quasi-experiment. *Law and Society Review, 4,* 539–62.

Glinski, R. J., Glinski, B. C., and Slatin, P. T. (1970). Nonnaivety contamination in conformity experiments: Sources, effects, and implications for control. *Journal of Personality and Social Psychology, 16,* 478–85.

Golding, S. L. and Lichtenstein, E. (1970). Confession of awareness and prior knowledge of deceptions as a function of interview set and approval motivation. *Journal of Personality and Social Psychology, 14,* 213–23.

Goldman, J. and Katz, M. D. (1982). Inconsistency and institutional review boards. *Journal of the American Medical Association, 248,* 197–202.

Goldstein, J. H. (ed.) (1986). *Reporting Science: The Case of Aggression.* Hillsdale, NJ: Erlbaum.

Goldstein, R. (1981). On deceptive rejoinders about deceptive research: a reply to Baron. *IRB: A Review of Human Subjects Research, 3,* 5–6.

Gray, B. H., Cooke, R. A., and Tannenbaum, A. S. (1978). Research involving human subjects. *Science, 201,* 1094–1101.

Gray, B. and Cooke, R. A. (1980). The impact of institutional review boards on research. *Hastings Center Report, 10,* 36–41.

Green, B. F. (1992). Exposé or smear? The Burt affair. *Psychological Science*, 3, 328–31.

Greenberg, M. (1967). Role-playing: an alternative to deception? *Journal of Personality and Social Psychology*, 7, 152–7.

Griffin, J. H. (1961). *Black Like Me*. Boston: Houghton Mifflin.

Grodin, M. A., Zaharoff, B. E., and Kaminow, P. V. (1986). A 12-year audit of IRB decisions. *Quality Review Bulletin*, 12, 82–6.

Gross, A. E. & Fleming, I. (1982). Twenty years of deception in social psychology. *Personality and Social Psychology Bulletin*, 8, 402–8.

Haines, V. J., Diekhoff, G. M., LaBeff, E. E., and Clark, R. E. (1986). College cheating: immaturity, lack of commitment, and the neutralizing attitude. *Research in Higher Education*, 25, 342–54.

Hamsher, J. H. and Reznikoff, M. (1967). Ethical standards in psychological research and graduate training: A study of attitudes within the profession. *Proceedings of the 75th Annual Convention of the American Psychological Association*, 2, 203–4.

Haney, C., Banks, W. C., and Zimbardo, P. G. (1973). Interpersonal dynamics in a simulated prison. *International Journal of Criminology and Penology*, 1, 69–97.

Hatfield, F. (1977). Prison research: the view from inside. *Hastings Center Report*, 7, 11–12.

Haveman, R. H. and Watts, H. W. (1976). Social experimentation as policy research: a review of negative income tax experiments. In G. Glass (ed.), *Evaluation Studies Review Annual* (Vol. 1). Beverly Hills, CA: Sage.

Haviland, J. B. (1977). Gossip as competition in Zinacantan. *Journal of Communication*, 27, 186–91.

Haywood, H. C. (1976). The ethics of doing research . . . and of not doing it. *American Journal of Mental Deficiency*, 81, 311–17.

Hearnshaw, L. (1979). *Cyril Burt, Psychologist*. Ithaca, NY: Cornell University Press.

Henle, M. & Hubble, M. B. (1938). "Egocentricity" in adult conversation. *Journal of Social Psychology*, 9, 227–34.

Herbert, W. (1978, February). HEW tightens regulations on prisoner research. *APA Monitor*, 9, 12–15.

Herman, R. (1993, January 14). US weighs in on ethics as self-regulation fails. *International Herald Tribune*, p. 7.

Herrnstein, R. J. and Murray, C. (1994). *The Bell Curve: Intelligence and Class Structure in American Life*. New York: The Free Press.

Herzog, H. A., Jr. (1988). The moral status of mice. *American Psychologist*, 43, 473–4.

Herzog, H. A., Jr. (1990). Discussing animal rights and animal research in the classroom. *Teaching of Psychology*, 17, 90–4.

Herzog, H. A., Jr. (1993). "The movement is my life": the psychology of animal rights activism. *Journal of Social Issues*, 49, 103–19.

Higbee, K. L. and Wells, M. G. (1972). Some research trends in social psychology during the 1960s. *American Psychologist, 27,* 963–6.

Hilts, P. J. (1991, March 21). Crucial research data in report biologist signed are held false. *The New York Times,* pp. A1, B10.

Hilts, P. J. (1994a, April 30). Tobacco firm halted nicotine study. *International Herald Tribune,* p. 3.

Hilts, P. J. (1994b, October 13). Human radiation testing wider than US revealed. *International Herald Tribune,* pp. 1, 4.

Hite, S. (1987). *Women and Love: A Cultural Revolution in Progress.* New York: Knopf.

Hochschild, A. R. (1987, November 15). Why can't a man be more like a woman? *The New York Times Book Review,* pp. 3, 34.

Hogan, P. M. and Kimmel, A. J. (1992). Ethical teaching of psychology: One department's attempts at self-regulation. *Teaching of Psychology, 19,* 205–10.

Holaday, M and Yost, T. E. (1993). Publication ethics. *Journal of Social Behavior and Personality, 8,* 557–66.

Holden, C. (1982). New focus on replacing animals in the lab. *Science, 215,* 35–8.

Holden, C. (1986, July). NIH transfers disputed monkeys to regional primate center. *Science, 233,* p. 154.

Holmes, D. S. (1976). Debriefing after psychological experiments. *American Psychologist, 31,* 858–75.

Holmes, D. S. and Bennett, D. H. (1974). Experiments to answer questions raised by the use of deception in psychological research. *Journal of Personality and Social Psychology, 29,* 358–67.

Holton, G. (1978). Subelectrons, presuppositions, and the Millikan–Ehrenhaft dispute. *Historical Studies in the Physical Sciences, 9,* 161–224.

Horowitz, I. A. (1969). Effects of volunteering, fear arousal, and number of communications on attitude change. *Journal of Personality and Social Psychology, 11,* 34–7.

Horowitz, I. L. (1967). *The Rise and Fall of Project Camelot.* Cambridge, MA: MIT Press.

Horowitz, I. L. and Rothschild, B. H. (1970). Conformity as a function of deception and role playing. *Journal of Personality and Social Psychology, 14,* 224–6.

Humphreys, L. (1970). *Tearoom Trade.* Chicago: Aldine.

Hunt, S. D., Chonko, L. B., and Wilcox, J. B. (1984). Ethical problems of marketing researchers. *Journal of Marketing Research, 21,* 309–24.

Institute for Social Research. (1976). *Research Involving Human Subjects.* Ann Arbor: University of Michigan.

Isambert, F.-A. (1987). L'expérimentation sur l'homme comme pratique et comme représentation [Experiments on man as practice and as representation]. *Actes de la Recherche en Sciences Sociales, 60,* 15–30.

Jacoby, J. and Aranoff, D. (1971). Political polling and the lost-letter technique. *Journal of Social Psychology, 83,* 209–12.

Jamison, W. and Lunch, W. (1992). The rights of animals, science policy, and political activism. *Science, Technology, and Human Values, 17,* 438–58.

Jason, L. A. and Bogat, G. A. (1983). Preventive behavioral interventions. In R. D. Felner, L. A. Jason, J. N. Moritsugu, and S. S. Farber (eds.), *Preventive Psychology: Theory, Research and Practice.* Elmsford, NY: Pergamon.

Jasper, J. M. and Nelkin, D. (1992). *The Animal Rights Crusade: The Growth of a Moral Protest.* New York: Free Press.

Jendreck, M. P. (1989). Faculty reactions to academic dishonesty. *Journal of College Student Development, 30,* 401–6.

Jensen, A. R. (1974). Kinship correlations reported by Sir Cyril Burt. *Behavior Genetics, 4,* 1–28.

Jensen, A. R. (1978). Sir Cyril Burt in perspective. *American Psychologist, 33,* 499–503.

Johnson, D. (1990). Animal rights and human lives: time for scientists to right the balance. *Psychological Science, 1,* 213–14.

Jones, J. H. (1993). *Bad Blood: The Tuskegee Syphilis Experiment* (rev. ed.). New York: Free Press.

Jonsen, A. R., Parker, M. L., Carlson, R. J., and Emmott, C. B. (1977). Biomedical experimentation on prisoners. *Ethics in Science and Medicine, 4,* 1–28.

Jorgenson, J. (1971). On ethics and anthropology. *Current Anthropology, 12,* 347–50.

Jourard, S. M. (1967). Experimenter-subject dialogue: a paradigm for a humanistic science of psychology. In J. F. T. Bugental (ed.), *Challenges of Humanistic Psychology.* New York: McGraw-Hill.

Jourard, S. M. (1968). *Disclosing Man to Himself.* Princeton, NJ: Van Nostrand.

Joynson, R. B. (1989). *The Burt Affair.* London: Routledge.

Judd, C. M., Smith, E. R., and Kidder, L. H. (1991). *Research Methods in Social Relations* (6th ed.). Fort Worth, TX: Holt, Rinehart and Winston.

Jung, J. (1969). Current practices and problems in the use of college students for psychological research. *Canadian Psychologist, 10,* 280–90.

Jung, J. (1975). Snoopology. *Human Behavior, 4,* 56–69.

Kallgren, C. A. and Kenrick, D. T. (1990, March). Ethical judgments and nonhuman research subjects: the effects of phylogenetic closeness and affective valence, Paper presented at the Eastern Psychological Association meeting, Philadelphia, PA.

Kamin, L. J. (1974). *The science and politics of IQ.* New York: Wiley.

Kant, I. (1965). *The Metaphysical Elements of Justice* (J. Ladd, trans.). Indianapolis, IN: Bobbs-Merrill.

Katz, J. (1970). The education of the physician-investigator. In P. Freund (ed.), *Experimentation With Human Subjects*. London: Allen and Unwin.

Katz, J. (1972). *Experimentation With Human Beings*. New York: Russell Sage Foundation.

Keen, W. W. (1914). *Animal Experimentation and Medical Progress*. Boston: Houghton Mifflin.

Keith-Spiegel, P. and Koocher, G. P. (1985). *Ethics in Psychology: Professional Standards and Cases*. New York: Random House.

Kelman, H. C. (1967). Human use of human subjects: The problem of deception in social psychological experiments. *Psychological Bulletin*, 67, 1–11.

Kelman, H. C. (1968). *A Time to Speak: On Human Values and Social Research*. San Francisco: Jossey-Bass.

Kelman, H. C. (1972). The rights of the subject in social research: An analysis in terms of relative power and legitimacy. *American Psychologist*, 27, 989–1016.

Kershaw, D. N. and Small, J. C. (1972). Data confidentiality and privacy: lessons from the New Jersey Income Tax Experiment. *Public Policy*, 20, 258–80.

Kershaw, D. N. and Fair, J. (1976). *The New Jersey Income Maintenance Experiment*. New York: Academic Press.

Kiesler, C. A., Pallak, M. S., and Kanouse, D. E. (1968). Interactive effects of commitment and dissonance. *Journal of Personality and Social Psychology*, 8, 331–8.

Kimble, G. (1976). The role of risk/benefit analysis in the conduct of psychological research, Paper prepared for the National Commission for the Protection of Human Subjects of Biomedical and Behavioral Research. Bethesda, MD: US Department of Health, Education, and Welfare.

Kimble, G. (1987). The scientific value of undergraduate research participation. *American Psychologist*, 42, 267–8.

Kimmel, A. J. (1979). Ethics and human subjects research: a delicate balance. *American Psychologist*, 34, 633–5.

Kimmel, A. J. (1988a). *Ethics and Values in Applied Social Research*. Newbury Park, CA: Sage.

Kimmel, A. J. (1988b). Herbert Kelman and the ethics of social-psychological research. *Contemporary Social Psychology*, 12, 152–8.

Kimmel, A. J. (1991). Predictable biases in the ethical decision making of American psychologists. *American Psychologist*, 46, 786–8.

Kimmel, A. J. and Keefer, R. (1991). Psychological correlates of the transmission and acceptance of rumors about AIDS. *Journal of Applied Social Psychology*, 21, 1608–28.

King, D. J. (1970). The subject pool. *American Psychologist*, 25, 1179–81.

King, D. W. and King, L. A. (1991). Validity issues in research on Vietnam veteran adjustment. *Psychological Bulletin, 109*, 107–24.

King, F. A. and Yarbrough, C. J. (1985). Medical and behavioral benefits from primate research. *The Physiologist, 28*, 75–87.

Kinnear, T. C. and Taylor, J. R. (1991). *Marketing Research: An Applied Approach* (4th ed.). New York: McGraw-Hill.

Koenig, F. (1985). *Rumor in the Marketplace: The Social Psychology of Commercial Hearsay.* Dover, MA: Auburn House.

Koestler, A. (1972). *The Case of the Midwife Toad.* New York: Random House.

Kong, D. (1994, February 20). 1800 tested in radiation experiments. *The Boston Globe*, pp. 1, 22.

Koocher, G. P. (1977). Bathroom behavior and human dignity. *Journal of Personality and Social Psychology, 35*, 120–1.

Korn, J. H. (1984). Coverage of research ethics in introductory and social psychology textbooks. *Teaching of Psychology, 11*, 146–9.

Korn, J. H. (1988). Students' roles, rights, and responsibilities as research participants. *Teaching of Psychology, 15*, 74–8.

Korn, J. H. and Bram, D. R. (1987). *What is missing from methods sections of APA journal articles?* Unpublished manuscript, Saint Louis University, Saint Louis, MO.

Korn, J. H. and Hogan, K. (1992). Effect of incentives and aversiveness of treatment on willingness to participate in research. *Teaching of Psychology, 19*, 21–4.

Lamberth, J. and Kimmel, A. J. (1981). Ethical issues and responsibilities in applying scientific behavioral knowledge. In A. J. Kimmel (ed.), *Ethics of Human Subject Research.* San Francisco: Jossey-Bass.

Landis, C. and Hunt, W. A. (1939). *The Startle Pattern.* New York: Farrar and Rinehart.

Landis, M. H. and Burtt, H. E. (1924). A study of conversations. *Journal of Comparative Psychology, 4*, 81–9.

Larson, C. C. (1982, January). Animal research: striking a balance. *APA Monitor, 13*, pp. 1, 12–13.

Latané, B. and Darley, J. M. (1970). *The Unresponsive Bystander: Why Doesn't He Help?* New York: Appleton-Century-Crofts.

Lea, S. E. G. (1979). Alternatives to the use of painful stimuli in physiological psychology and the study of animal behavior. *ATLA Abstracts, 7*, 20–1.

Leak, G. K. (1981). Student perception of coercion and value from participation in psychological research. *Teaching of Psychology, 8*, 147–9.

Levin, J. (1981). Ethical problems in sociological research. In A. J. Kimmel (ed.), *Ethics of Human Subject Research.* San Francisco: Jossey-Bass.

Levin, J. and Arluke, A. (1985). An exploratory analysis of sex differences in gossip. *Sex Roles, 12*, 281–6.

Levine, R. J. (1975). The nature and definition of informed consent in various research settings, Paper prepared for the National Commission for the Protection of Human Subjects of Biomedical and Behavioral Research. Bethesda, MD: US Department of Health, Education, and Welfare.

Levy, C. M. and Brackbill, Y. (1979, March). Informed consent: getting the message across to kids. *APA Monitor*, 10, p. 3.

Levy, L. H. (1967). Awareness, learning, and the beneficient subject as expert witness. *Journal of Personality and Social Psychology*, 6, 365–70.

Lewin, K. (1947). Group decision and social change. In T. M. Newcomb and E. L. Hartley (eds.), *Readings in Social Psychology*. New York: Holt.

Lewin, K. (1951). Problems of research in social psychology. In D. Cartwright (ed.), *Field Theory in Social Science*. New York: Harper and Row.

Lichtenstein, E. (1970). Please don't talk to anyone about this experiment: disclosure of deception by debriefed subjects. *Psychological Reports*, 26, 485–6.

Lifton, R. J. (1986). *The Nazi Doctors: Medical Killing and the Psychology of Genocide*. New York: Basic Books.

Linden, L. E. and Weiss, D. J. (1994). An empirical assessment of the random response method of sensitive data collection. *Journal of Social Behavior and Personality*, 9, 823–36.

Linder, D. E., Cooper, J., and Jones, E. E. (1967). Decision freedom as a determinant of the role of incentive magnitude in attitude change. *Journal of Personality and Social Psychology*, 6, 245–54.

Lindsay, R. C. L. and Holden, R. R. (1987). The introductory psychology subject pool in Canada. *Canadian Psychology*, 28, 45–52.

Locke, E. A. (1993). Found in the SIOP archives: Footnotes that somehow got left out of published manuscripts. *The Industrial-Organizational Psychologist*, 31(2), 53.

Loo, C. M. (1982). Vulnerable populations: case study in crowding research. In J. E. Sieber (ed.), *The Ethics of Social Research: Surveys and Experiments*. New York: Springer-Verlag.

Lorion, R. P. (1983). Evaluating preventive interventions: Guidelines for the serious social change agent. In R. D. Felner, L. A. Jason, J. N. Moritsugu, and S. S. Farber (eds.), *Preventive Psychology: Theory, Research and Practice*. Elmsford, NY: Pergamon.

Lorion, R. P. (1984). Research issues in the design and evaluation of preventive interventions. In J. P. Bowker (ed.), *Education for Primary Prevention in Social Work*. New York: Council on Social Work Education.

Lowman, R. L. (1993). An ethics code for I/O psychology: for what purpose and at what cost? *The Industrial-Organizational Psychologist*, 31(1), 90–2.

Lowman, R. L. (ed.) (1985). *Casebook on Ethics and Standards for the*

Practice of Psychology in Organizations. College Park, MD: Society for Industrial and Organizational Psychology.

Lowman, R. P. and Soule, L. M. (1981). Professional ethics and the use of humans in research. In A. J. Kimmel (ed.), *Ethics of Human Subject Research.* San Francisco: Jossey-Bass.

Lueptow, L., Mueller, S. A., Hammes, R. R., and Master, L. S. (1977). The impact of informed consent regulations on response rate and response bias. *Sociological Methods and Research, 6,* 183–204.

MacCoun, R. J. and Kerr, N. L. (1987). Suspicion in the psychology laboratory: Kelman's prophesy revisited. *American Psychologist, 42,* 199.

McAskie, M. (1978). Carelessness or fraud in Sir Cyril Burt's kinship data? A critique of Jensen's analysis. *American Psychologist, 33,* 496–8.

McCabe, K. (1990, February). Beyond cruelty. *Washingtonian,* pp. 112–18, 153–7.

McCarthy, C. R. (1981). The development of federal guidelines for social research. In A. J. Kimmel (ed.), *Ethics of Human Subject Research.* San Francisco: Jossey-Bass.

McCloskey, M., Rapp, B., Yantis, S., Rubin, G., Bacon, W. F., Dagnelie, G., Gordon, B., Aliminosa, D., Boatman, D. F., Badeker, W., Johnson, D. N., Tusa, R. J., and Palmer, E. (1995). A developmental deficit in localizing objects from vision. *Psychological Science, 6,* 112–17.

McCord, D. M. (1991). Ethics-sensitive management of the university human subject pool. *American Psychologist, 46,* 151.

McCord, J. (1978). A thirty-year follow-up of treatment effects. *American Psychologist, 33,* 284–9.

McCormick, R. H. (1976). Experimentation on children: Sharing in sociality. *Hastings Center Report, 6,* 41–6.

McDonald, P. J. and Eilenfield, V. C. (1980). Physical attractiveness and the approach/avoidance of self-awareness. *Personality and Social Psychology Bulletin, 6,* 391–5.

McGuire, W. J. (1969). Suspiciousness of experimenter's intent. In R. Rosenthal and R. L. Rosnow (eds.), *Artifact in Behavioral Research.* New York: Academic Press.

McNamara, J. R. and Woods, K. M. (1977). Ethical considerations in psychological research: A comparative review. *Behavior Therapy, 8,* 703–8.

McNemar, Q. (1946). Opinion-attitude methodology. *Psychological Bulletin, 43,* 289–374.

McNemar, Q. (1960). At random: sense and nonsense. *American Psychologist, 15,* 295–300.

Mahoney, M. J. (1976). *Scientist As Subject: The Psychological Imperative.* Cambridge, MA: Ballinger.

Marshall, G. D. and Zimbardo, P. G. (1979). Affective consequences of inadequately explained physiological arousal. *Journal of Personality and Social Psychology, 37,* 970–88.

Masling, J. (1966). Role-behavior of the subject and psychologist and its effects upon psychological data. *Nebraska Symposium on Motivation, 14,* 67–103.

Medawar, P. B. (1976, May 22). Science and the patchwork mouse. *Science News, 109,* 335.

Meeus, W. H. J. and Raaijmakers, Q. A. W. (1986). Administrative obedience: carrying out orders to use psychological-administrative violence. *European Journal of Social Psychology, 16,* 311–24.

Meeus, W. H. J and Raaijmakers, Q. A. W. (1987). Administrative obedience as a social phenomenon. In W. Doise and S. Moscovici (eds.), *Current Issues in European Social Psychology* (Vol. 2). Cambridge: Cambridge University Press.

Mehren, E. (1987, November 16). Controversies surrounding self-proclaimed cultural historian Shere Hite. *The Boston Globe,* p. 2.

Meier, P. (1992, May 12). Ethical and scientific issues in drug trials with human participants. Presentation to the APA Committee on Standards in Research, New York.

Menges, R. J. (1973). Openness and honesty versus coercion and deception in psychological research. *American Psychologist, 28,* 1030–4.

Merritt, C. B. and Fowler, R. G. (1948). The pecuniary honesty of the public at large. *Journal of Abnormal and Social Psychology, 43,* 90–3.

Middlemist, D., Knowles, E. S., and Matter, C. F. (1976). Personal space invasions in the laboratory: Suggestive evidence for arousal. *Journal of Personality and Social Psychology, 33,* 541–6.

Middlemist, D., Knowles, E. S., and Matter, C. F. (1977). What to do and what to report: a reply to Koocher. *Journal of Personality and Social Psychology, 35,* 122–4.

Milgram, S. (1963). Behavioral study of obedience. *Journal of Abnormal and Social Psychology, 67,* 371–8.

Milgram, S. (1964). Issues in the study of obedience: a reply to Baumrind. *American Psychologist, 19,* 848–52.

Milgram, S. (1969, June). The lost letter technique. *Psychology Today,* pp. 3, 30, ff.

Milgram, S. (1974). *Obedience to Authority.* New York: Harper and Row.

Milgram, S., Mann, L., and Harter, S. (1965). The lost letter technique: a tool of social research. *Public Opinion Quarterly, 29,* 437–8.

Mill, J. S. (1957). *Utilitarianism.* New York: Bobbs-Merrill.

Miller, A. G. (1972). Role playing: an alternative to deception; a review of the evidence. *American Psychologist, 27,* 623–36.

Miller, N. E. (1979). *The Scientist's Responsibility for Public Information: A Guide to Effective Communication With the Media.* Washington, DC: Society for Neuroscience.

Miller, N. E. (1983). Understanding the use of animals in behavioral research: Some critical issues. In J. A. Sechzer (ed.), *The Role of Animals in Biomedical Research*. New York: New York Academy of Sciences.

Miller, N. E. (1985). The value of behavioral research on animals. *American Psychologist, 40,* 423–40.

Mills, J. (1976). A procedure for explaining experiments involving deception. *Personality and Social Psychology Bulletin, 2,* 3–13.

Mirvis, P. H. and Seashore, S. E. (1982). Creating ethical relationships in organizational research. In J. E. Sieber (ed.), *The Ethics of Social Research: Surveys and Experiments*. New York: Springer-Verlag.

Mishkin, B. (1985). On parallel tracks: protecting human subjects and animals. *Hastings Center Report, 15,* 36–7.

Mitchell, R. G., Jr. (1990). An unprincipled ethic? The missing morality of the ASA code. *The American Sociologist, 21,* 271–4.

Mitchell, S. C. and Steingrub, J. (1988). The changing clinical trials scene: the role of the IRB. *IRB: A Review of Human Subjects Research, 10,* 1–5.

Mitroff, I. I. (1974). Norms and counter-norms in a select group of the Apollo moon scientists: a case study of the ambivalence of scientists. *American Sociological Review, 39,* 579–95.

Mitscherlich, A. and Mielke, F. (eds.) (1960). *Medizin Ohne Menschlichkeit*. Frankfurt, Germany: Fischer.

Mixon, D. (1971). Behavior analysis treating subjects as actors rather than organisms. *Journal for the Theory of Social Behavior, 1,* 19–31.

Mixon, D. (1972). Instead of deception. *Journal for the Theory of Social Behavior, 2,* 145–77.

Mixon, D. (1974). If you won't deceive, what can you do? In N. Armistead (ed.), *Reconstructing Social Psychology*. Baltimore, MD: Penguin.

Montanye, T., Mulberry, R. F., and Hardy, K. R. (1971). Assessing prejudice toward Negroes at three universities using the lost-letter technique. *Psychological Reports, 29,* 531–7.

Mook, D. G. (1983). In defense of external validity. *American Psychologist, 38,* 379–87.

Moore, H. T. (1922). Further data concerning sex differences. *Journal of Abnormal and Social Psychology, 17,* 210–14.

Moriarty, T. (1975). Crime, commitment, and the unresponsive bystander: two field experiments. *Journal of Personality and Social Psychology, 31,* 370–6.

Mott, F. W. and Sherrington, C. S. (1895). Experiments on the influence of sensory nerves upon movement and nutrition of the limbs. *Proceedings of the Royal Society, 57,* 481–8.

Mulvey, E. P. and Phelps, P. (1988). Ethical balances in juvenile justice research and practice. *American Psychologist, 43,* 65–9.

Muñoz, R. F. (1983, May). *Prevention intervention research: a sample of ethical dilemmas*, Paper presented at the NIMH State of the Art

Workshop on Ethics and Primary Prevention, California State University, Northridge.

Muñoz, R. F., Glish, M., Soo-Hoo, T., and Robertson, J. (1982). The San Francisco mood survey project: Preliminary work toward the prevention of depression. *American Journal of Community Psychology*, *10*, 317–29.

National Academy of Sciences (1979). *Report of the Panel on Privacy and Confidentiality as Factors in Survey Response*. Washington, DC.

National Central Bureau of Statistics. (1977). *The National Central Bureau of Statistics and the General Public: Findings of an Interview Survey Taken in Sweden During the Spring of 1976*. Stockholm: SCB.

National Commission for the Protection of Human Subjects of Biomedical and Behavioral Research (1979). *The Belmont Report: Ethical Principles and Guidelines for the Protection of Human Subjects of Research*. Washington, DC: US Government Printing Office.

National Science Board (1991). *Science and Engineering Indicators*. Washington, DC: US Government Printing Office.

Netherlands Institute of Psychologists (1988). *Professional Code for Psychologists*. Amsterdam.

Neuliep, J. W. and Crandall, R. (1993). Reviewer bias against replication research. In J. W. Neuliep (ed.), *Replication Research in the Social Sciences* [Special Issue]. *Journal of Social Behavior and Personality*, *8*, 21–9.

Newberry, B. H. (1973). Truth-telling in subjects with information about experiments: who is being deceived? *Journal of Personality and Social Psychology*, *25*, 369–74.

Newton, R. (1977). *The Crime of Claudius Ptolemy*. Baltimore, MD: Johns Hopkins University Press.

Novak, E., Seckman, C. E., and Stewart, R. D. (1977). Motivations for volunteering as research subjects. *Journal of Clinical Pharmacology*, *17*, 365–71.

Novak, M. and Petto, A. J. (eds.). (1991). *Through the Looking Glass: Issues of Psychological Well-Being in Captive Nonhuman Primates*. Washington, DC: American Psychological Association.

Office for Protection From Research Risks (1986). *Public Health Service Policy on Humane Care and Use of Laboratory Animals*. Bethesda, MD: National Institutes of Health.

O'Leary, C., Willis, F., and Tomich, E. (1970). Conformity under deceptive and non-deceptive techniques. *Sociological Quarterly*, *11*, 87–93.

Olson, T. and Christiansen, G. (1966). *The Grindstone Experiment: Thirty-One Hours*. Toronto: Canadian Friends Service Committee.

Orem, J. M. (1990, March 14). Demands that research be useful threaten to undermine basic science in this country. *The Chronicle of Higher Education*, pp. B2-B3.

Orne, M. T. (1959). The nature of hypnosis: artifact and essence. *Journal of Abnormal and Social Psychology, 58*, 277–99.

Orne, M. T. (1962). On the social psychology of the psychological experiment: With particular reference to demand characteristics and their implications. *American Psychologist, 17*, 776–83.

Orne, M. T. (1969). Demand characteristics and the concept of quasi-controls. In R. Rosenthal and R. L. Rosnow (eds.), *Artifact in Behavioral Research*. New York: Academic Press.

Orne, M. T. and Holland, C. H. (1968). On the ecological validity of laboratory deceptions. *International Journal of Psychiatry, 6*, 282–93.

Page, M. M. (1973). On detecting demand awareness by postexperimental questionnaire. *Journal of Social Psychology, 91*, 305–23.

Patten, S. C. (1977). Milgram's shocking experiments. *Philosophy, 52*, 425–40.

Pechman, J. A. and Timpane, P. M. (eds.) (1975). *Work Incentives and Income Guarantees: The New Jersey Income Tax Experiment*. Washington, DC: The Brookings Institution.

Pepitone, A. (1981). Lessons from the history of social psychology. *American Psychologist, 36*, 972–85.

Percival, T. (1803). *Medical Ethics*. Manchester, UK: S. Russell.

Peters, D. P. and Ceci, S. J. (1982). Peer review practices of psychological journals: the fate of published articles, submitted again. *The Behavioral and Brain Sciences, 5*, 187–95.

Petrof, J. V., Sayegh, E. E., and Vlahopoulos, P. I. (1982). The influence of the school of business on the values of its students. *Journal of the Academy of Marketing Sciences, 10*, 500–13.

Phillips, D. P. (1977). Motor vehicle fatalities increase just after publicized suicide stories. *Science, 196*, 1464–5.

Phillips, M. T. and Sechzer, J. A. (1989). *Animal Research and Ethical Conflict*. New York: Springer-Verlag.

Pihl, R. O., Zacchia, C., and Zeichner, A. (1981). Follow-up analysis of the use of deception and aversive contingencies in psychological experiments. *Psychological Reports, 48*, 927–30.

Piliavin, I. M., Rodin, J., and Piliavin, J. A. (1969). Good Samaritanism: an underground phenomenon? *Journal of Personality and Social Psychology, 13*, 289–99.

Piliavin, J. A. and Piliavin, I. M. (1972). Effects of blood on reactions to a victim. *Journal of Personality and Social Psychology, 23*, 353–61.

Plous, S. (1991). An attitude survey of animal rights activists. *Psychological Science, 2*, 194–6.

Polich, J. M., Ellickson, P. L., Reuter, P., and Kahn, J. P. (1984). *Strategies for Controlling Adolescent Drug Use*. Santa Monica, CA: Rand.

Poll finds support for animal tests (1985, October 29). *The New York Times*, p. C4.

Powers, E. and Witmer, H. (1951). *An Experiment in the Prevention of*

Delinquency: The Cambridge-Somerville Youth Study. New York: Columbia University Press.

Pratt, D. (1976). *Painful Experiments on Animals*. New York: Argus Archives.

Prentice, E. D. and Antonson, D. L. (1987). A protocol review guide to reduce IRB inconsistency. *IRB: A Review of Human Subjects Research*, 9, 9–11.

President's Commission for the Study of Ethical Problems in Medicine and Biomedical and Behavioral Research (1982). *Compensating Research Injuries*. Washington, DC: US Government Printing Office.

Pruzan, A. (1976). Effects of age, rearing and mating experiences on frequency dependent sexual selection in *Drosophila Pseudoobscura*. *Evolution*, 30, 130–45.

Pruzan, A., Applewhite, P. B., and Bucci, M. J. (1977). Protein synthesis inhibition alters *Drosophila* mating behavior. *Pharmacology, Biochemistry and Behavior*, 6, 355–7.

Psychological Association of Slovenia (1982). *Code of Ethics for Psychologists*. Ljubljana, Slovenia.

Ramsey, P. (1976). The enforcement of morals: nontherapeutic research on children. *Hastings Center Report*, 6, 21–9.

Randall, W. C. (1983). Is medical research in jeopardy? *The Physiologist*, 26, 73–7.

Raupp, C. D. and Cohen, D. C. (1992). "A thousand points of light" illuminate the psychology curriculum: volunteering as a learning experience. *Teaching of Psychology*, 19, 25–30.

Ravitch, N., Veatch, R. M., Kohut, T. A., and Lifton, R. J. (1987). Symposium: The Nazi doctors: medical killing and the psychology of genocide. *The Psychohistory Review*, 16, 3–66.

Reese, H. W. and Fremouw, W. J. (1984). Normal and normative ethics in behavioral sciences. *American Psychologist*, 39, 863–76.

Regan, T. (1983). *The Case for Animal Rights*. Berkeley: University of California Press.

Reiss, A. J. (1976). Selected issues in informed consent and confidentiality with special reference to behavioral/social science research/inquiry, Paper prepared for the National Commission for the Protection of Human Subjects of Biomedical and Behavioral Research. Bethesda, MD: US Department of Health, Education, and Welfare.

Researcher admits he faked journal data (1977, March 5). *Science News*, 111, 150–1.

Resnick, J. H. and Schwartz, T. (1973). Ethical standards as an independent variable in psychological research. *American Psychologist*, 28, 134–9.

Reynolds, P. D. (1979). *Ethical Dilemmas and Social Science Research*. San Francisco: Jossey-Bass.

Riecken, H. W. (1975). Social experimentation. *Society*, 12, 34–41.

Riegel, K. F. (1979). *Foundations of Dialectical Psychology*. New York: Academic Press.

Ring, K. (1967). Experimental social psychology: some sober questions about some frivolous values. *Journal of Experimental and Social Psychology*, 3, 113–23.

Ring, K., Wallston, K., and Corey, M. (1970). Mode of debriefing as a factor affecting subjective reaction to a Milgram-type obedience experiment: an ethical inquiry. *Representative Research in Social Psychology*, 1, 67–88.

Rissman, E. F. (1995). An alternative animal model for the study of female sexual behavior. *Current Directions in Psychological Science*, 4, 6–10.

Robb, J. W. (1988). Can animal use be ethically justified? In H. N. Guttman, J. A. Mench, and R. C. Simmonds (eds.), *Proceedings From a Conference on Science and Animals: Addressing Contemporary Issues*. Bethesda, MD: Scientists' Center for Animal Welfare.

Robson, C. (1993). *Real World Research: A Resource for Social Scientists and Practitioner-Researchers*. Oxford, UK: Blackwell.

Rollin, B. (1981). *Animal Rights and Human Morality*. Buffalo, NY: Prometheus.

Rollin, B. E. (1985). The moral status of research animals in psychology. *American Psychologist*, 40, 920–6.

Rosenhan, D. L. (1973). On being sane in insane places. *Science*, 179, 250–8.

Rosenthal, R. (1991). Teacher expectancy effects: A brief update 25 years after the Pygmalion experiment. *Journal of Research in Education*, 1, 3–12.

Rosenthal, R. (1994). Interpersonal expectancy effects: a 30-year perspective. *Current Directions in Psychological Science*, 3, 176–9.

Rosenthal, R. and Jacobson, L. (1968). *Pygmalion in the Classroom*. New York: Holt.

Rosenthal, R. and Rosnow, R. L. (eds.) (1969). *Artifact in Behavioral Research*. New York: Academic Press.

Rosenthal, R. and Rosnow, R. L. (1975). *The Volunteer Subject*. New York: Wiley.

Rosenthal, R. and Rosnow, R. L. (1984). Applying Hamlet's question to the ethical conduct of research: a conceptual addendum. *American Psychologist*, 39, 561–3.

Rosenthal, R. and Rosnow, R. L. (1991). *Essentials of Behavioral Research: Methods and Data Analysis* (2nd ed.). New York: McGraw-Hill.

Rosnow, R. L. (1978). The prophetic vision of Giambattista Vico: Implications for the state of social psychological theory. *Journal of Personality and Social Psychology*, 36, 1322–31.

Rosnow, R. L. (1981). *Paradigms in Transition: The Methodology of Social Inquiry*. New York: Oxford University Press.

Rosnow, R. L. (1991). Inside rumor: a personal journey. *American Psychologist, 46*, 484–96.

Rosnow, R. L., Esposito, J. L., and Gibney, L. (1988). Factors influencing rumor spreading: replication and extension. *Language and Communication, 7*, 1–14.

Rosnow, R. L., Goodstadt, B. E., Suls, J. M., and Gitter, A. G. (1973). More on the social psychology of the experiment: When compliance turns to self-defense. *Journal of Personality and Social Psychology, 27*, 337–43.

Rosnow, R. L. and Rosenthal, R. (1993). *Beginning Behavioral Research: A Conceptual Primer.* New York: Macmillan.

Rosnow, R. L., Rotheram-Borus, M. J., Ceci, S. J., Blanck, P. D., and Koocher, G. P. (1993). The institutional review board as a mirror of scientific and ethical standards. *American Psychologist, 48*, 821–6.

Rosnow, R. L. and Suls, J. M. (1970). Reactive effects of pretesting in attitude research. *Journal of Personality and Social Psychology, 15*, 338–43.

Ross, H. L. (1973). Law, science and accidents: the British Road Safety Act of 1967. *Journal of Legal Studies, 2*, 1–75.

Ross, H. L., Campbell, D. T., and Glass, G. V. (1970). Determining the social effects of legal reform: the British "breathalyser" crackdown of 1967. *American Behavioral Scientist, 13*, 493–509.

Ross, L., Lepper, M. R., and Hubbard, M. (1975). Perseverence in self-perception and social perception: Biased attributional processes in the debriefing paradigm. *Journal of Personality and Social Psychology, 32*, 880–92.

Rothenberg, R. (1990, October 5). Surveys proliferate, but answers dwindle. *The New York Times*, pp. A1, D4.

Rotherham-Borus, M. (1991). *HIV Interventions for Adolescents.* Washington, DC: Surgeon General's Panel on HIV.

Rubenstein, C. (1982). Psychology's fruit flies. *Psychology Today, 16*, 83–4.

Ruebhausen, O. M. and Brim, O. G., Jr. (1966). Privacy and behavioral research. *American Psychologist, 21*, 423–37.

Rugg, E. A. (1975). *Ethical Judgments of Social Research Involving Experimental Deception.* Doctoral dissertation, George Peabody College for Teachers, Nashville, TN.

Ruling on animals in research is struck down (1994, September). *APS Observer*, p. 12.

Rushton, J. P. (1994). *Race, Evolution, and Behavior: A Life History Perspective.* New Brunswick, NJ: Transaction.

Ryder, R. (1975). *Victims of Science: The Use of Animals in Research.* London: Davis-Poynter.

Saks, M. J. and Blanck, P. D. (1992). Justice improved: the unrecognized benefits of aggregation and sampling in the trial of mass torts. *Stanford Law Review*, 44, 815–51.

Sarason, S. B. (1981). *Psychology Misdirected*. New York: Free Press.

Savin, H. B. (1973). Professors and psychological researchers: Conflicting values in conflicting roles. *Cognition*, 2, 147–9.

Schachter, S. (1959). *The Psychology of Affiliation*. Stanford, CA: Stanford University Press.

Schachter, S. and Singer, J. (1962). Cognitive, social, and physiological determinants of the emotional state. *Psychological Review*, 69, 379–99.

Schlenker, B. R. and Forsyth, D. R. (1977). On the ethics of psychological research. *Journal of Experimental Social Psychology*, 13, 369–96.

Schuler, H. (1982). *Ethical Problems in Psychological Research*. New York: Academic Press.

Schultz, D. P. (1969). The human subject in psychological research. *Psychological Bulletin*, 72, 214–28.

Sears, D. O. (1986). College sophomores in the laboratory: Influences of a narrow data base on psychology's view of human nature. *Journal of Personality and Social Psychology*, 51, 515–30.

Seberhagen, L. W. (1993). An ethics code for statisticians – what next? *The Industrial-Organizational Psychologist*, 30, 71–4.

Seeman, J. (1969). Deception in psychological research. *American Psychologist*, 24, 1025–28.

Seiler, L. H. and Murtha, J. M. (1980). Federal regulation of social research using "human subjects": a critical assessment. *The American Sociologist*, 15, 146–57.

Shanab, M. E. and Yahya, K. A. (1977). A behavioral study of obedience in children. *Journal of Personality and Social Psychology*, 35, 530–6.

Shapiro, K. (1983). Psychology and its animal subjects. *International Journal for the Study of Animal Problems*, 4, 188–91.

Shapiro, K. (1984). Response to APA's "Why animals?" *Psychology in Maine*, 1, 1–2.

Shapiro, K. J. (1989). The Silver Springs monkeys and APA. *PsyETA Bulletin*, 8, 1–6.

Shapiro, K. J. (1991, July). Use morality as basis for animal treatment. *APA Monitor*, 22, p. 5.

Sharpe, D., Adair, J. G., and Roese, N. J. (1992). Twenty years of deception research: a decline in subjects' trust? *Personality and Social Psychology Bulletin*, 18, 585–90.

Shaughnessy, J. J. and Zechmeister, E. B. (1990). *Research Methods in Psychology* (2nd ed.). New York: McGraw-Hill.

Sheets, T., Radlinski, A., Kohne, J., and Brunner, G. A. (1974). Deceived respondents: once bitten, twice shy. *Public Opinion Quaterly*, 36, 261–2.

Sherif, M., Harvey, O. J., White, B. J., Hood, W. E., and Sherif, C. W.

(1961). *Intergroup Conflict and Cooperation: The Robbers Cave Experiment*. Norman, OK: Institute of Group Relations.

Sherrington, C. S. (1906). *The Integrative Action of the Nervous System*. New Haven, CT: Yale University Press.

Shubin, S. (1981, January). Research behind bars: Prisoners as experimental subjects. *The Sciences*, 21, pp. 10–13, 29.

Sieber, J. E. (1982a). Deception in social research I: kinds of deception and the wrongs they may involve. *IRB: A Review of Human Subjects Research*, 4, 1–6.

Sieber, J. E. (1982b). Deception in social research III: the nature and limits of debriefing. *IRB: A Review of Human Subjects Research*, 6, 1–4.

Sieber, J. E. (1982c). Ethical dilemmas in social research. In J. E. Sieber (ed.), *The Ethics of Social Research: Surveys and Experiments*. New York: Springer-Verlag.

Sieber, J. E. and Saks, M. J. (1989). A census of subject pool characteristics. *American Psychologist*, 44, 1053–61.

Sieber, J. E. and Sorensen, J. L. (1991). Ethical issues in community-based research and intervention. In J. Edwards, R. S. Tindale, L. Heath, and E. J. Posavac (eds.), *Social Psychological Applications to Social Issues*. Vol. 2: *Methodological Issues in Applied Social Psychology*. New York: Plenum Press.

Sigall, H., Aronson, E., and Van Hoose, T. (1970). The cooperative subject: myth or reality? *Journal of Experimental Social Psychology*, 6, 1–10.

Sigma Xi, The Scientific Research Society (1984). *Honor in Science*. New Haven, CT.

Silverman, I. (1977). *The Human Subject in the Psychological Experiment*. New York: Pergamon.

Silverman, I., Shulman, A. D., and Wiesenthal, D. L. (1970). Effects of deceiving and debriefing psychological subjects on performance in later experiments. *Journal of Personality and Social Psychology*, 14, 203–12.

Silverstein, A. J. (1974). Compensating those injured through experimentation. *Federal Bar Journal*, 33, 322–30.

Sinclair, C., Poizner, S., Gilmour-Barrett, K., and Randall, D. (1987). The development of a code of ethics for Canadian psychologists. *Canadian Psychology*, 28, 1–8.

Singer, E. (1978). Informed consent: consequences for response rate and response quality in social surveys. *American Sociological Review*, 43, 144–62.

Singer, E. and Frankel, M. R. (1982). Informed consent procedures in telephone interviews. *American Sociological Review*, 47, 416–27.

Singer, P. (1975). *Animal Liberation*. New York: Avon.

Sjoberg, G. (1967). Project Camelot: selected reactions and personal reflections. In G. Sjoberg (ed.), *Ethics, Politics, and Social Research*. Cambridge, MA: Schenkman.

Skelly, F. (1977). Comment. *Public Opinion Quarterly, 41*, 110–11.

Slife, B. and Rubinstein, J. (1992). *Taking Sides: Clashing Views on Controversial Psychological Issues* (7th ed.). Guilford, CT: Dushkin.

Smith, C. P. (1981). How (un)acceptable is research involving deception? *IRB: A Review of Human Subjects Research, 3*, 1–4.

Smith, D. H. (1978). Scientific knowledge and forbidden truths - are there things we should not know? *Hastings Center Report, 8*, 30–5.

Smith, M. B. (1976). Some perspectives on ethical/political issues in social research. *Personality and Social Psychology Bulletin, 2*, 445–53.

Smith, N. L. (1985). Some characteristics of moral problems in evaluation practice. *Evaluation and Program Planning, 8*, 5–11.

Smith, S. S. and Richardson, D. (1983). Amelioration of deception and harm in psychological research: the important role of debriefing. *Journal of Personality and Social Psychology, 44*, 1075–82.

Snow, C. P. (1959). *The Search* (rev. ed.). New York: Charles Scribner's Sons.

Sommer, B. and Sommer, R. (1991). *A Practical Guide to Behavioral Research: Tools and Techniques* (3rd ed.). New York: Oxford University Press.

Sommer, R. and Sommer, B. (1989). Social facilitation effects in coffeehouses. *Environment and Behavior, 21*, 651–66.

Sperling, S. (1988). *Animal Liberators: Research and Morality*. Berkeley: University of California Press.

Sperry, R. W. (1968). Hemisphere deconnection and unity in conscious awareness. *American Psychologist, 23*, 723–33.

Spiegel, D. and Keith-Spiegel, P. (1970). Assignment of publication credits: Ethics and practices of psychologists. *American Psychologist, 25*, 738–47.

Spielberger, C. D. (1962). The role of awareness in verbal conditioning. *Journal of Personality, 30*, 73–101.

Spivak, G., Platt, J. J., and Shure, M. B. (1976). *The Problem-Solving Approach to Adjustment*. San Francisco: Jossey-Bass.

Spivak, G. and Shure, M. B. (1974). *Social Adjustment of Young Children: A Cognitive Approach to Solving Real-Life Problems*. San Francisco: Jossey-Bass.

Sprague, R. L. (1993). Whistleblowing: a very unpleasant avocation. *Ethics and Behavior, 3*, 103–33.

Stang, D. J. (1976). Ineffective deception in conformity research: Some causes and consequences. *European Journal of Social Psychology, 6*, 353–67.

Stanley, B. and Sieber, J. (eds.). (1992). *Social Research on Children and Adolescents: Ethical Issues*. Newbury Park, CA: Sage.

Stanton, A. L., Burker, E. J., and Kershaw, D. (1991). Effects of researcher follow-up of distressed subjects: Tradeoff between validity and ethical responsibility? *Ethics and Behavior, 1*, 105–12.

Stanton, A. L. and New, M. J. (1988). Ethical responsibilities to depressed research participants. *Professional Psychology: Research and Practice*, *19*, 279–85.

Steininger, M., Newell, J. D., and Garcia, L. T. (1984). *Ethical Issues in Psychology*. Homewood, IL: Dorsey.

St James-Roberts, I. (1976). Cheating in science. *New Scientist*, *60*, 466–9.

Stoke, S. M. and West, E. D. (1931). Sex differences in conversational interests. *Journal of Social Psychology*, *2*, 120–26.

Stricker, L. J. (1967). The true deceiver. *Psychological Bulletin*, *68*, 13–20.

Stricker, L. J., Messick, S., and Jackson, D. N. (1967). Suspicion of deception: implications for conformity research. *Journal of Personality and Social Psychology*, *5*, 379–89.

Strohmetz, D. B., Alterman, A. I., and Walter, D. (1990). Subject selection bias in alcoholics volunteering for a treatment study. *Alcoholism: Clinical and Experimental Research*, *14*, 736–8.

Strohmetz, D. B. and Rosnow, R. L. (1995). A mediational model of research artifacts. In J. Brzezinski (ed.), *Probability in Theory Building: Experimental and Nonexperimental Models of Scientific Research in Behavioral Sciences*. Amsterdam: Editions Rodopi.

Strutt, R. J. (1924). *John William Strutt, Third Baron Rayleigh*. London: Arnold.

Sullivan, D. S. and Deiker, T. A. (1973). Subject-experimenter perceptions of ethical issues in human research. *American Psychologist*, *28*, 587–91.

Suls, J. M. and Rosnow, R. L. (1981). The delicate balance between ethics and artifacts in behavioral research. In A. J. Kimmel (ed.), *Ethics of Human Subject Research*. San Francisco: Jossey-Bass.

Susman, G. I. (1976). *Autonomy at Work: A Sociotechnical Analysis of Participative Management*. New York: Praeger.

Swazey, J. P., Anderson, M. S., and Lewis, K. S. (1993). Ethical problems in academic research. *American Scientist*, *81*, 542–53.

Tanke, E. D. and Tanke, T. J. (1982). Regulation and education: The role of the institutional review board in social science research. In J. E. Sieber (ed.), *The Ethics of Social Research: Fieldwork, Regulation and Publication*. New York: Springer-Verlag.

de Tarde, G. (1901). *L'Opinion et la Foule*. Paris: Felix Alcan.

Taub, E., Bacon, R., and Berman, A. J. (1965). The acquisition of a trace-conditioned avoidance response after deafferentation of the responding limb. *Journal of Comparative and Physiological Psychology*, *58*, 275–9.

Tedeschi, J. T. and Rosenfeld, P. (1981). The experimental research controversy at SUNY: a case study. In A. J. Kimmel (ed.), *Ethics of Human Subject Research*. San Francisco: Jossey-Bass.

Tesch, F. (1977). Debriefing research participants: though this be method there is madness to it. *Journal of Personality and Social Psychology, 35,* 217–24.

Thomas, S. B. and Quinn, S. C. (1991). The Tuskegee syphilis study, 1932 to 1972: implications for HIV education and AIDS risk education programs in the Black community. *American Journal of Public Health, 81,* 1498–1505.

Thompson, B. (1994). The big picture(s) in deciding authorship order. *American Psychologist, 49,* 1095–96.

Thompson, R. A. (1990). Vulnerability in research: a developmental perspective on research risk. *Child Development, 61,* 1–16.

Tracy, P. E. and Fox, J. A. (1981). The validity of randomized response for sensitive measurements. *American Sociological Review, 46,* 187–200.

Triplett, N. (1897). The dynamogenic factors in pacemaking and competition. *American Journal of Psychology, 9,* 507–33.

Trist, E. L., Higgin, G. W., Murray, H., and Pollack, A. B. (1963). *Organizational Choice: Capabilities of Groups at the Coal Face Under Changing Technologies.* London: Tavistock.

Trochim, W. (1982). *Research Design for Program Intervention: The Regression-Discontinuity Approach.* Beverly Hills, CA: Sage.

Twedt, D. T. (1963). Why a marketing research code of ethics? *Journal of Marketing, 27,* 48–50.

Tybout, A. M. and Zaltman, G. (1974). Ethics in marketing research: their practical relevance. *Journal of Marketing Research, 11,* 357–68.

Tybout, A. M. and Zaltman, G. (1975). A reply to comments on "Ethics in marketing research: their practical relevance." *Journal of Marketing Research, 12,* 234–7.

US Department of Agriculture (1989, August 31). Animal welfare; final rules. *Federal Register,* pp. 36112–63.

US Department of Agriculture (1990, July 16). Animal welfare; guinea pigs, hamsters, and rabbits. *Federal Register,* pp. 28879–84.

US Department of Agriculture (1991, February 15). Animal welfare; standards; final rule. *Federal Register,* pp. 6426–505.

Veatch, R. M. (1982). Problems with institutional review board inconsistency. *Journal of the American Medical Association, 248,* 179–80.

Verplanck, W. S. (1955). The control of the content of conversation: reinforcement of statements of opinion. *Journal of Abnormal and Social Psychology, 55,* 668–76.

Vidich, A. J. and Bensman, J. (1958). *Small Town in Mass Society: Class, Power, and Religion in a Rural Community.* Princeton, NJ: Princeton University Press.

Vinacke, W. E. (1954). Deceiving experimental subjects. *American Psychologist, 9,* 155.

Wade, N. (1976). Animal rights: NIH cat sex study brings grief to New York museum. *Science, 194,* 162–67.

Wahl, J. M. (1972). *Role Playing vs. Deception: Differences in Experimental*

Realism as Measured by Subject's Level of Involvement and Level of Suspicion. Doctoral dissertation, University of Oregon, Eugene, OR.

Walsh, W. B. and Stillman, S. M. (1974). Disclosure of deception by debriefed subjects. *Journal of Counseling Psychology, 21,* 315–19.

Walster, E., Berscheid, E., Abrahams, D., and Aronson, V. (1967). Effectiveness of debriefing following deception experiments. *Journal of Personality and Social Psychology, 6,* 371–80.

Walton, R. E. (1978). Ethical issues in the practice of organization development. In G. Bermant, H. C. Kelman, and D. P. Warwick (eds.), *The Ethics of Social Intervention.* Washington, DC: Hemisphere.

Warner, S. L. (1965). Randomized response: a survey technique for eliminating evasive answer bias. *Journal of the American Statistical Association, 60,* 63–9.

Warwick, D. P. (1975, February). Social scientists ought to stop lying. *Psychology Today, 8,* pp. 38, 40, 105–6.

Waterman, A. S. (1974). The civil liberties of the participants in psychological research. *American Psychologist, 29,* 470–1.

Watson, J. D. (1969). *The Double Helix.* New York: New American Library.

Webb, E. J., Campbell, D. T., Schwartz, R. D., Sechrest, L., and Grove, J. B. (1981). *Nonreactive Measures in the Social Sciences* (2nd ed.). Boston: Houghton Mifflin.

Weber, S. J. and Cook, T. D. (1972). Subject effects in laboratory research: an examination of subject roles, demand characteristics, and valid inferences. *Psychological Bulletin, 77,* 273–95.

Weiner, J. S. (1955). *The Piltdown Forgery.* London: Oxford University Press.

Weinstein, D. (1979). Fraud in science. *Social Science Quarterly, 59,* 639–52.

Weinstein, D. (1981). *Scientific Fraud and Scientific Ethics* (CSEP Occasional Papers: No. 4). Chicago: Center for the Study of Ethics in the Professions.

Welsome, E. (1993). The plutonium experiment [special reprint]. *The Albuquerque Tribune,* 47 pp.

Welt, L. G. (1961). Reflections on the problems of human experimentation. *Connecticut Medicine, 25,* 75–9.

West, S. G., Gunn, S. P., and Chernicky, P. (1975). Ubiquitous Watergate: an attributional analysis. *Journal of Personality and Social Psychology, 32,* 55–65.

West, S. G. and Gunn, S. P. (1978). Some issues of ethics and social psychology. *American Psychologist, 33,* 30–8.

Westin, A. F. (1968). *Privacy and Freedom.* New York: Atheneum.

White, L. A. (1979). Erotica and aggression: the influence of sexual arousal, positive affect, and negative affect on aggressive behavior. *Journal of Personality and Social Psychology, 37,* 591–601.

Willis, R. H. and Willis, Y. A. (1970). Role playing versus deception: an experimental comparison. *Journal of Personality and Social Psychology,* *16*, 472–7.

Wilson, D. W. and Donnerstein, E. (1976). Legal and ethical aspects of nonreactive social psychological research: an excursion into the public mind. *American Psychologist, 31*, 765–73.

Winslow, J. H. and Meyer, A. (1983, September). The perpetrator at Piltdown. *Science 83, 4*, 32–43.

Wolins, L. (1962). Responsibility for raw data. *American Psychologist, 17*, 657–8.

Wuebben, P. L. (1967). Honesty of subjects and birth order. *Journal of Personality and Social Psychology, 5*, 350–2.

Yin, R. K. (1989). *Case Study Research: Design and Methods* (rev. ed.). Newbury Park, CA: Sage.

Young, S. C., Herrenkohl, L. R., Bibace, R., and Lin, A. W. (1995, March). Shifting from "experimenter subject" to "researcher participant" relationships: conceptual and methodological implications, Paper presented at the Eastern Psychological Association meeting, Boston.

Zak, S. (1989, March). Ethics and animals. *The Atlantic, 273*, pp. 68–74.

Zicklin, E. (1994, March). I. V. league. *Spy*, pp. 18–19.

Zimbardo, P. (1969). The human choice: individuation, reason and order versus deindividuation, impulse and chaos. In W. J. Arnold and D. Levine (eds.), *Nebraska Symposium on Motivation*, Vol. 17. Lincoln: University of Nebraska Press.

Zimbardo, P. (1973a). On the ethics of intervention in human psychological research: with special reference to the Stanford prison experiment. *Cognition, 2*, 243–56.

Zimbardo, P. (1973b). The psychological power and pathology of imprisonment. In E. Aronson and R. Helmreich (eds.), *Social Psychology*. New York: Van Nostrand.

Zimbardo, P., Anderson, S. M., and Kabat, L. G. (1981). Induced hearing deficit generates experimental paranoia. *Science, 212*, 1529–31.

Zimbardo, P., Haney, C., Banks, W., and Jaffe, D. (1973, April 8). The mind is a formidable jailer: a Pirandellian prison. *The New York Times Magazine*, pp. 38–60.

Zola, J. C., Sechzer, J. A., Sieber, J. E., and Griffin, A. (1984). Animal experimentation: issues for the 1980s. *Science, Technology, and Human Values, 9*, 40–50.

Zuckerman, H. and Merton, R. K. (1971). Patterns of evaluation in science: institutionalisation, structure, and functions of the referee system. *Minerva, 9*, 66–100.

Subject Index

action research, 160
American Anthropological Association
(AAA), 51–2
American Psychological Association
(APA)
  ethical principles of, 26–7, 31–41,
  47–9, 318–22, 326
  evolution of professional guidelines,
  31–41, 47–9
  guidelines for animal research, 37,
  244, 247, 263–6
  qualified nature of principles, 40, 42
  standards for publishing, 287, 288–9
American Psychological Society (APS),
326
American Sociological Association
(ASA), 50–1
  ethical principles of, 50–1, 230,
  322–4
  see also codes of ethics
animal research, 236–72
  alternatives to, 260, 267–71
  animal pain and suffering, 252–3,
  259–60
  arguments for and against, 249–57
  benefits of, 20, 237, 240, 256–9
  criticisms of, 20–1, 237, 240, 246,
  252, 253–4
  deafferented monkey studies, 19–21,
  243, 253, 266
  factors influencing judgments of,
  272

guidelines and regulations for, 242,
  247, 260, 261–7
prevalence of, 236, 239, 244, 260
see also animal rights movement
animal rights movement, 19, 21
  activists in, 245–6, 248
  emergence of, 242–3
  versus animal welfare movement,
  248
  see also animal research
Animal Welfare Act, 242, 261
anonymity, 141–2, 190–3
  in the communication of research
  findings, 313–14
  and privacy, 131–2
anti-vivisection movement
  see animal rights movement
applied psychology
  role in development of ethics code,
  31–2
  see also applied research
applied research
  ethical responsibilities of applied
  researchers, 164, 165, 181
  examples of, 165–9, 171–2,
  181–2
  goals of, 159, 161, 198
  unanticipated consequences of, 162,
  165, 171–4
  versus basic research, 159, 163, 164,
  165, 184
archival records, 141, 142, 144

Name Index